'Remarkable' *Eastern Eye*

'Compelling' *Guardian*

'Brilliant . . . A fascinating account' *New World*

'A sad and poignant story' *Focus*

'Miller is an excellent historian' *New York Times Book Review*

'Excellent . . . This is an engrossing and enlightening read, and not just for physicists'
Fortean Times

'A lively and exciting narrative . . . so beautifully written that I read it in one sitting . . . A 'must read' for anyone interested in astrophysics of the history of science'
Nature

Also by Arthur I. Miller

*Einstein, Picasso: Space, Time and the Beauty
That Causes Havoc*

*Insights of Genius: Imagery and Creativity
in Science and Art*

*Imagery in Scientific Thought: Creating
20th-Century Physics*

Arthur I. Miller lectures, writes and broadcasts on creativity in art and science. He is the author of _____ us: *Imagery and C_____ ., Imagery in Science, Thought: Creating 20th-Century Physics*; and *Einstein, Picasso: Space, Time and the Beauty That Causes Havoc*, which was nominated for the Pulitzer prize. He is Professor of the History and Philosophy of Science at University College London and lives in London with his wife, the author Lesley Downer.

'*Empire of the Stars* dramatically succeeds in conveying the clash of scientific ideas and the personal conflicts underlying Chandrasekhar's remarkable anticipation of the existence of black holes in our universe. This is a story that needed to be told'
Roger Penrose

'Arthur I. Miller, for so long the doyen of historians of modern science, has surpassed himself with this brilliant, elegantly written book. It is a profound story of friendship, disappointment and hope filled with truly remarkable characters; the narrative is also enormously broad in its scope, crossing continents and exploring questions at the heart of our understanding of the universe: what are black holes? Where do they come from? What do they mean?'
David Bodanis, author of *Electric Universe* and *E=mc2*

'A wonderful read. Empire of the Stars provides insight into the personalities of some of the great scientific minds of the last century together with a fascinating historical presentation of the development of astrophysics, specifically the collapse of matter into black holes. This is a must read for anyone with even a passing interest in modern astronomy and astrophysics'
Dr. Martin C. Weisskopf, Fellow of the American Physical Society and a 2004 winner of the Rossi Prize awarded by the American Astronomical Society's High Energy Astrophysics Division

'A fascinating book . . . A well researched chronicle of how powerful intellects confronted some of the most fascinating challenges in the world of science – and how they confronted each other as well'
Martin Rees, *Sunday Times*

EMPIRE
OF THE
STARS

Friendship, Obsession and Betrayal
in the Quest for Black Holes

ARTHUR I. MILLER

To Lesley

ABACUS

First published in Great Britain in 2005 by Little, Brown
This edition published in 2006 by Abacus
Reprinted 2007

ISBN 978-0-349-11627-3

Papers used by Abacus are natural, recyclable products made from
wood grown in sustainable forests and certified in accordance with
the rules of the Forest Stewardship Council.

Typeset in Goudy by M Rules
Printed and bound in Great Britain by Clays Ltd, St Ives plc

In my entire scientific life, extending over forty-five years, the most shattering experience has been the realisation that [Kerr's] exact solution of Einstein's equations of general relativity provides the *absolutely exact representation* of untold numbers of massive black holes that populate the universe. This 'shuddering before the beautiful', this incredible fact that a discovery motivated by a search after the beautiful in mathematics should find its exact replica in Nature, persuades me to say that beauty is that to which the human mind responds at its deepest and most profound.

SUBRAHMANYAN CHANDRASEKHAR

The conditions in the star are very extreme; but the ultimate things to be dealt with – electrons, atomic nuclei, X-rays – are the same in the star as in the laboratory, and we can apply our laboratory knowledge of them. Calculate 'according to the laws of physics as known to terrestrial experiment'; and then turn to the man with the telescope and ask, 'Is that anything like the stars *you* come across in the sky?'

ARTHUR STANLEY EDDINGTON

The thought that the laws of the macrocosmos in the small reflect the terrestrial world obviously exercises a great magic on mankind's mind; indeed its form is rooted in the superstition (which is as old as the history of thought) that the destiny of men could be read from the stars.

MAX BORN

CONTENTS

PART III

ACKNOWLEDGEMENTS

Researching the dramatic story of Chandra and Eddington, of Chandra's mathematical verification of black holes and his four-decade wait for the scientific community to accept it, involved getting to grips with esoteric theoretical physics and state-of-the-art technology as well as meeting as many as possible of the major players in the story.

Above all, I am deeply grateful to Chandra's widow, Lalitha, who generously spent many hours sharing with me her memories of her late husband. I am also much indebted to James Cronin, Valeria Ferrari, Robert Geroch, Roger Hildebrand, Norman Lebovitz, T. D. Lee, Eugene Parker, Edwin Salpeter, Noel Swirdlow, Saul Teukolsky, Kip Thorne, Michael Turner, Peter Vandervoort, Robert Wald and Don York, all very distinguished long-term colleagues of Chandra's who contributed personal reminiscences and details about his life and career.

Many thanks to Chandra's cousin, V. Radhakrishnan, of the Raman Institute in Bangalore, India, who filled in details of Chandra's life, provided superb and companionable hospitality, and showed me around his latest project, his catamaran. Dr U. Ponnambalam, chairman of the physics department at Presidency College, Chennai, very kindly gave me an informative tour of the college, where Chandra was an undergraduate.

At Santa Cruz, Stan Woosley patiently took me through the present state of research on supernovae. Don Osterbrock kindly filled me in on day-to-day life during Chandra's years in residence at Yerkes Observatory and shared his historical knowledge and acumen as an astrophysicist. He also read large chunks of the manuscript and made invaluable comments.

I enjoyed an unforgettable weekend at Los Alamos in the company of Stirling Colgate, who gave me a fascinating account of the relationship between nuclear weapons and the birth of modern research into supernovae and black holes, among much else. His colleague back in the 1960s, Richard White, kindly provided more details of the flavour of the times. Sergey Blinnikov generously filled me in on his work with Yakov Zel'dovich on supernovae in the Soviet Union. I am grateful to Edward Frieman, Richard Garwin, Marvin Goldberger and Conrad Longmuire, who told me much about Chandra's association with Los Alamos.

My work would have been quite impossible without access to the University of Chicago's magnificent Joseph Regenstein Library, where Chandra's correspondence and manuscripts are deposited. I was very fortunate to have Jay Satterfield's assistance there. I am also most grateful to Takeshi Oka for his splendid hospitality at the University of Chicago and generous assistance with archival matters.

Peter Hingley, Librarian at the Royal Astronomical Society, Burlington House, London, generously devoted much time to dealing with my frequent requests for information.

As in all my other historical research, it was invaluable to have access to interviews carried out over more than four decades by the American Institute of Physics' Center for the History of Physics. Every historian of physics owes an enormous debt to the Center and to its director, Spencer Weart.

I would also like to offer my thanks to the following archivists and librarians who provided valuable assistance: Judith Goodstein (California Institute of Technology), Shaun J. Hardy (Carnegie Institution of Washington), Mark Hurn (Institute of Astronomy, Cambridge University), Dan Lewis (Huntington Library, San Marino, California), Kathi Murphy (McGill University), Jonathan Smith (Trinity College, Cambridge University), Tony Hogarth Smith (Christ's Hospital, Horsham, England) and G. Madhaven (executive secretary, Indian Academy of Sciences, Bangalore, India).

K.C. Wali's biography of Chandra, written with the benefit of extensive interviews with Chandra, was extremely helpful.

Marianne Douglas and Patrick Douglas, relatives of Eddington's biographer Allie Vibert Douglas, kindly provided further infor-

ACKNOWLEDGEMENTS xiii

Eddington's daily life at the Cambridge Observatory.

Many thanks as always to my close friend, Mike Brady, who
read an early version of this book in its entirety and offered much
sagacious advice as well as his usual support. One of the unexpected rewards of this book was that I renewed my old friendship
with Gary Steigman, whose 'speed of light' email correspondence
helped clarify many issues. And warm thanks to my pal Dave
Scott for filling me in on his excursion to the Moon on Apollo 15
as well as for good cheer and many companionable dinners.

Many colleagues generously took time off from their busy
schedules to respond to my queries and to read portions of the
manuscript. I am very grateful to Mitch Berger, Stirling Colgate,
Andy Fabian, Chris Fryer, Roy Garstang, Jeremy Gray, Chryssa
Kouveliotou, Keith Mason, Leon Mestel, Martin Rees, Virginia
Trimble, Steve Weinberg, Martin Weisskopf and Stan Woosley.

My appreciation to Ken Brecher, Steve Brush, David Cassidy,
Emma Goddard, Andrew Gregory, Simon Singh, Steve Miller,
Roger Stuewer and Andrew Warwick for helpful and enlightening
conversations.

I am hugely indebted to my agent, Peter Tallack of Conville &
Walsh Literary Agency, for his encouragement, enthusiasm and
faith in this project. The Agency has been immensely helpful in
smoothing out the route from first to final draft and all the travail
that goes on in between. Particular thanks too to Patrick Walsh.
I owe an enormous debt to my editors: Tim Whiting at Little,
Brown, who provided much support and helpful criticisms; and
Amanda Cook at Houghton Mifflin, whose insightful comments
and suggestions were invaluable in shaping the final version. This
book would not be what it is without their help.

When I decided to shift from physics research to the history of
scientific ideas, it was in order to explore the way in which scientists think as they seek out deep truths about the cosmos, both on
the large scale and the small, and to examine what happens when
scientist meets nature in a one-on-one confrontation. Looking
into the story I tell in this book, I could not avoid becoming aware
of the dauntingly hard and very solitary work involved in research
at the cutting edge of science. At this level, scientists often find

themselves in thrall to their intellectual endeavours with a degree of intensity and competitiveness that can exact a high psychological toll. Playing for such high stakes, they sometimes act and react in ways which may seem inexplicable to lesser human beings. This is how it was for the major players in my story. Working out why has involved spinning a scenario based on what can be inferred from the information available. For this I bear full responsibility, as well as for any scientific errors.

A little way into the research for this book I happened to go to an art gallery opening in London where I met a beautiful and intriguing woman. Lesley became my inspiration, my muse, my wife and my soulmate. She has greatly improved this book, and my life too.

Arthur I. Miller
London, 2004

NOTE ON CHANDRA'S ENGLISH

Writing in English, Chandra often – sometimes deliberately – used idiosyncratic spellings and forms of expression. These I have retained so that we may better hear his voice and experience his feelings.

PROLOGUE

Or, if there were a sympathy in choice,
War, death, or sickness did lay siege to it,
Making it momentary as a sound,
Swift as a shadow, short as any dream,
Brief as the lightning in the collied night,
That, in a spleen, unfolds both heaven and earth,
And ere a man hath power to say, 'Behold!'
The jaws of darkness do devour it up:
So quick bright things come to confusion.

WILLIAM SHAKESPEARE, *A MIDSUMMER NIGHT'S
DREAM* (ACT I, SCENE 1)

Ever since the evocative term 'black hole' was coined in 1967, these mysterious voids in the universe have assumed an almost mystical appeal. The attraction of a vast emptiness that imprisons not only matter but also light is quite literally inescapable.

Imagine that you are an astronaut, seduced by the grandeur of a black hole into straying too close. Trapped by its immense gravitational field and the tornado-like swirling of space around it, you sweep feet first over the horizon. As you fall what you see is truly awesome. Just before you slip over the edge, the whole universe of stars and galaxies appears to rush together into one bright spot. The intense gravity of the black hole funnels the light from distant objects into a tighter and tighter cone, like tunnel vision. All the while you are entranced by the firework display of atoms snared by the black hole's immense gravity. They dance in a cosmic traffic jam, bumping into each together, becoming hotter and hotter until they too blaze with the brightness of a million suns as they stray too close and plunge with you into nowhere. Then you start to feel the irresistible attraction of the collapsed star deep inside, at once unimaginably small and infinitely dense. As the collapsed star sucks you deeper and deeper into the black hole, the gravitational pull grows stronger and stronger. You stretch like a piece of toffee, longer and longer and thinner and thinner, until you are torn apart. The potent gravitational force around the black hole means that light takes longer and longer to reach distant observers. They see you poised on the edge of the black hole, forever frozen in space and time.

A black hole is a well in space, the final resting place of a collapsed star. For decades, scientists resisted the very idea as a theoretical freak that couldn't actually exist, an ugly solution to the most beautiful theory ever created, Albert Einstein's general theory of relativity. But astronomers now know that the universe is littered with these monsters and that a giant black hole sits at the centre of our own Galaxy. What's more, we can actually observe black holes by detecting the X-rays emitted by particles as they spiral towards the event horizon before plummeting in.

Having taken their place in the fabric of nature, black holes have opened our minds to staggering and sometimes frightening speculation: to questions such as whether they are spawning baby universes, of which ours may be one; or whether they might open a short cut to a distant part of the universe, or even be portals for time travel; and to just how we might devise an experiment to create a black hole in a laboratory here on Earth.

Many scientists now believe that black holes hold the key to understanding how our universe has evolved and how nature behaves at its most extreme level. At the very edge of space and time, black holes are the engines that power the brightest objects in the universe: quasars, brighter than a trillion suns. Black holes have pushed our knowledge of the cosmos to its limits.

Black holes may ultimately reveal the microstructure of matter and the fate of the universe itself. For it is in these gravitational plugholes that atoms are crushed into their basic building blocks, even possibly right out of existence; that the laws of physics as we know them may break down completely; and that the two great scientific theories of the twentieth century – general relativity (which describes the world of the very large) and quantum mechanics (which describes the very small) – meet head on.

In recounting the epic story of the discovery of black holes, *Empire of the Stars* is a window through which we can glimpse humankind's remarkable quest to understand how stars are born, live and die, and the way in which this knowledge has profoundly altered our scientific and cultural views of the world. It is the story of one man's fight with the scientific establishment for recognition of his idea – an episode which sheds light on what science is, how it works and where it can go wrong.

Fascinated by the stars from time immemorial, we have come to believe that their fate is ultimately ours. What is so extraordinary about the early research into the life history of stars is that it consisted almost purely of theoretical speculation, of the imaginings of scientists with the confidence to make grandiose assumptions about some of the largest objects ever known. With the experimental data on stellar evolution so meagre, scientists were forced to play God. The biggest assumption they made was that the science discovered by humans – ourselves the accidental products of the stars, and a mere fraction of the age of the universe – could be used to explore the evolution of stars thousands of trillions of miles distant: from their birth some billions of years ago, to their deaths untold billions of years in the future.

What breathtaking chutzpah that was! Tracing the story of those pioneering astrophysicists is an intellectual adventure, with twists and turns that will take us into the deepest realms of theoretical science, including designs for manufacturing hydrogen bombs and the effect of the arms race on astrophysical research. It also sheds light on several epic conflicts: between concepts of classical and modern science embodied in the vastly different outlooks of physicists and astrophysicists, and between cultures at the twilight of the British Empire in the 1930s. All this led to a thirty-year delay before the notion of black holes was finally accepted, by which time just about everyone had forgotten about the role played by the man who had originally provided clear-cut evidence for their very existence.

Subrahmanyan Chandrasekhar's flash of inspiration came when he was an unknown nineteen-year-old, in the hot summer of 1930. In ten minutes, sitting in a deckchair overlooking the Arabian Sea, Chandra (as he was universally known) carried out some calculations that augured a disturbing fate for small, dense stars known as white dwarfs. At the time, scientists assumed that these were the final resting states of dead stars. Those which had been found were more or less the mass of the Sun but no bigger than the Earth. Chandra's sums showed that there was an upper limit to the mass of these white dwarfs. But what would happen to a star that ended up more massive than that at the end of its life, after it had burnt up all its fuel? Unable to end its life as an inert rock, it might begin an endless process of collapse, crunched by its own gravity into a singularity – a minuscule point of infinite density and zero volume, many trillions of times smaller than the full stop at the end of this sentence and many trillions of times denser than the Earth.

Only one person understood the full implications of Chandra's discovery: Sir Arthur Stanley Eddington, the greatest astrophysicist in the world at that time. Eddington himself had flirted with the idea that a dead star might collapse indefinitely in this manner, so he should have been delighted with Chandra's mathematical verification. Instead, without any warning, he used a meeting of the Royal Astronomical Society, on 11 January 1935, to savage Chandra's result cynically and unmercifully. The encounter cast a shadow over both their lives and hindered progress in astrophysics for nearly half a century.

As a teenager I read Eddington and was sufficiently taken by his writings to want to become a scientist. Although I didn't understand much of what he described, it all sounded exciting. The sheer sweep of the subject matter – from atoms to the life and death of stars – was breathtaking, and the language vivid and gripping. Chandra's writings were inspiring in another way. They exemplified how a superbly gifted scientist could use mathematics to study the nature of the stars.

Yet the more I discovered about Chandra's story, the more intriguing it became. For all his brilliance, his life was tinged with tragedy. As a young man growing up in India he had been a prodigy, recognised by many as a genius. He spent his youth absorbed in abstruse problems of mathematics and physics. Then he received a scholarship to Trinity College, Cambridge, where the great scientific minds of the day held court. It was on the way to England that he made his discovery about the fate of white dwarf stars. But to his shock, Eddington refused even to take it seriously and subjected him to public ridicule in a persecution that went on for years. Chandra never really regained his confidence. Despite a long and incredibly productive scientific career, no amount of recognition could ever satisfy him. I wondered what other great discoveries he might have made, had his early life not been blighted by disappointment. By all accounts he was a reserved, deeply private and highly serious man. Who was the real man behind that stern façade?

And what of Eddington? Why did the greatest astrophysicist in the world choose to demolish the youthful Indian in such a vicious way? Eddington was famous for his sharp tongue. He was harsh and cynical with other scientists, too. But with Chandra, for reasons that seemed at first 'mysterious' (as a colleague of Chandra's put it), Eddington's criticisms took on a different and darker tinge.

By the time I met Chandra, he was eighty-three. The occasion was a conference on creativity at the Chicago Academy of Sciences, where Chandra was to give the keynote lecture. In the ballroom-sized auditorium, it was standing room only. There was a buzz of excitement as a distinguished Indian man walked through the massive double doors. Here was a Nobel laureate, one of the most important scientists of the age, who had come to address the audience on matters of high scientific creativity. Despite his slight figure he was a man of enormous presence. Elegantly dressed and no more than five foot six in height, he carried himself with great dignity, though his shoulders were beginning to stoop. His sparse white hair was carefully combed across the top of his still handsome dark face, with its prominent forehead, piercing eyes and full mouth clamped tightly in an expression of iron determination.

As he spoke, he looked up from his notes from time to time to reminisce with obvious delight about great scientists of the past whom he had known. The audience was spellbound. Afterwards I was able to exchange a few words with him and shake his hand. It was a thrilling moment. This was the man who had transformed our understanding of the heavens and been a great inspiration to me throughout my life. He spoke of the book he was writing on his intellectual hero, Isaac Newton, and of the recent exciting discoveries that had been made researching black holes. Thoughtlessly, I mentioned the Eddington episode – and his face clouded. He graciously shook my hand, and we agreed to speak again.

Later reports of the altercation with Eddington make out that Chandra rapidly put it behind him and that the two men were actually firm friends. But even a cursory glance at Eddington's scientific articles and the correspondence between the two reveals a very different story. Again and again, Chandra expressed deep anger, frustration and resentment, while Eddington clung stubbornly to his own view of the universe, ridiculing Chandra's discovery as 'stellar buffoonery'.[1] Throughout his life, Chandra never missed an opportunity to recount the events of that fateful day at the Royal Astronomical Society, repeating that he had been right and Eddington wrong, even though Eddington refused ever to admit it. He was always careful to speak in glowing terms about Eddington the man. But in interviews he revealed how profoundly he had suffered.

After that memorable meeting in 1993, I continued to puzzle over Chandra and his complex, tragic story. A couple of years ago I decided to explore it more deeply. Sadly, by this time Chandra was dead. But I was fortunate enough to meet Lalitha, his devoted wife of over fifty years. I also interviewed many of his colleagues, some of whom had studied under him. In Bangalore I met his cousin, the eminent astrophysicist V. Radhakrishnan, son of Chandra's uncle, C. V. Raman, India's first physics Nobel laureate. From their recollections and from letters and documents that Chandra and, later, Lalitha deposited at the University of Chicago's Joseph Regenstein Library, little by little the real Chandra – the man behind the persona he built up around himself – began to emerge.

An important step in getting to know Chandra was to

experience the tropical heat and dust of south India, where he was born and grew up and lived until he was nineteen, when he finished his undergraduate schooling at Presidency College, Madras (now Chennai). Eager to do his doctoral degree at Trinity College, Cambridge, he travelled by sea to enter a remarkably different culture – a frenetic and highly competitive world, startlingly different from the comfortable, rather lethargic atmosphere of Madras. It was on that voyage that he made his great discovery.

I walked across the burning sands of the Marina, the long beach that edges Chennai. In Chandra's time it was elegant and fashionable. He would cycle there in the evenings with his brothers to escape the stifling heat, and gaze in wonder at the brilliant stars. As he remembered in a letter to his brother Balakrishnan, he sometimes came alone to the silent beach, to fling himself on the sand and pray to his God that he might be another Einstein.

Empire of the Stars begins with Chandra's traumatic confrontation with Eddington on 11 January 1935. From there we move back to Chandra's earliest days in India – the only place where he ever felt at home – and the ground-breaking discovery he made on his way to England. There he was plunged into the thrilling atmosphere of European science in the 1930s, when dinner among the elite academics at high table at Trinity was dominated not by gossip or matters of personal advancement but by feverish scientific discussion. This was the time when Eddington and his colleagues and rival astrophysicists, James Jeans and Edward Arthur Milne, were laying the foundations of astrophysics, not through the study of real stars but, astonishingly, by using fiendishly complicated mathematical calculations to create theoretical models of how stars ought to be.

At Niels Bohr's Institute in Copenhagen, Chandra found the cut and thrust of scientific debate even more ferocious. It was there that ideas such as Heisenberg's uncertainty principle, quantum mechanics and nuclear physics were hammered out. The phrase the assembled scientists most dreaded from the great Bohr was 'very interesting'. 'Very interesting' meant 'just not interesting enough'. Consumed by their magnificent obsession with understanding the workings of the universe, great and small, these men spent their days grappling with mighty problems of the essence of being.

While many of these problems have been explored ever since, Chandra's work lay dormant for three decades, only to be looked at anew when an entirely different development in world affairs – the race to design the hydrogen bomb – sparked renewed interest in the possibility of black holes. The explosive power that the scientists were looking for turned out to be the very same power that had created these massive holes in space, and the universe itself. Chandra, now living in the United States, and his discovery returned to their proper place at the forefront of scientific endeavour. Towards the end of his life, he received a Nobel prize for his achievement. But the price was terrible; his great discovery had been overlooked for almost forty years.

This book is the biography of an idea, rather than of a man. Nevertheless, science is a human endeavour, driven by hopes, dreams and aspirations, especially at the highest level, where scientists are playing for the highest stakes. These people may be brilliant, some may even be considered geniuses. But as human beings they may also be seriously flawed. Such were Chandra and Eddington.

PART I

1

Fatal Collision

It had been a momentous meeting of the Royal Astronomical Society at Burlington House, just off Piccadilly, that Friday 11 January.[1] In a mere forty-five minutes, two men's lives were changed forever and astrophysics was set back more than thirty years. Yet, as the hundred distinguished members filed out of the hall, chatting excitedly, no one could put their finger on exactly what had happened. What is certain is that at 6.15 p.m., a shy, boyish twenty-four-year-old astrophysicist from Cambridge, Dr Subrahmanyan Chandrasekhar, rose to present a dramatic discovery that had been ignored for nearly five years.

What Chandra had to say was entirely new and went against all the accepted tenets of science. He was aware that it might be met with frowns and criticisms, even downright opposition. A lot was at stake, both scientifically and personally. But he was not worried. He was sure he could hold his own in any scientific give-and-take. Young though he was, he had a maturity and a command of astrophysics far beyond his years. He had given his first lectures when he was just eighteen, at Presidency College in Madras. There he regularly held forth at physics colloquia normally dominated by mature research scientists. In India, his contemporaries spoke of him as a genius. Chandra had a full thirty minutes to argue the case for his ground-breaking discovery (most were only given half that time). Sir Arthur Eddington would then deliver the next paper.

The two men could not have been more different. Chandra – a diminutive young Indian, fresh-faced and immaculately dressed – was cherubically handsome, so dark he was practically black. With a broad nose, full mouth and glistening black hair which swept over his forehead, he had the aristocratic bearing of a Brahmin. He radiated youthful innocence, in stark contrast to Eddington's air of worldly superciliousness.

Eddington – then fifty-two – was the holder of the prestigious Plumian Professorship of Astronomy and Experimental Philosophy and the recipient of practically every award science could bestow. A tall, etiolated figure with a lugubrious pale face, thin-lipped and sardonic, he kept a pince-nez carefully balanced on his narrow nose and stood ramrod straight. In his rumpled three-piece tweed suit with a fob watch in his breast pocket and an air of carefully cultivated dishevelment, he was the archetypal Oxbridge don. He carried himself like a representative of the British Empire in its grandest days. Eddington was renowned for his ruthlessness and rapier wit. Many of those present had come simply to see him perform. He never disappointed.

Chandra's great discovery concerned nothing less than the ultimate fate of the universe. What became of stars at the end of their lives, when they had finally burnt up all their fuel? Astrophysicists assumed that they shrank and shrank until they became white dwarfs, so small and dense that a star with the same mass as the Sun might shrink to be no bigger than the Earth. But what happened then? Chandra had already published two papers on white dwarfs a few years earlier, in 1931 and 1932. The problem they examined was one that Eddington had raised in his influential 1926 book, *The Internal Constitution of the Stars*. Nevertheless, the young Indian's work was totally ignored.

Chandra was just twenty when his first brief paper on white dwarfs was published, in 1931. He had written it just as he was embarking on his postgraduate career at Cambridge, with no time for corrections or improvements. In his next paper he laid out his arguments in a more polished form. But still the scientific establishment refused to pay any attention. There was no encouragement from anyone – not from Eddington, or from his thesis supervisor

Ralph Fowler, or from his new friend Edward Arthur Milne, Rouse Ball Professor of Mathematics at Oxford. Were they trying to protect a young innocent from foolishly publishing an outrageous idea? Was it professional jealousy – were they trying to safeguard their own positions in the scientific cosmos? Or was there some other, more sinister factor, to do with Chandra's colour and racial origin? Such a thought would be taboo today, but in those days the Raj still ruled India, and Englishmen still basked in the belief in their innate superiority. Could it be that these stalwarts of Empire found it unacceptable to be overtaken by a young man from one of the colonies and refused to accept that he might have anything to teach them?

The train ride from Cambridge to London seemed interminable that freezing January Friday. It was not that Chandra was nervous about delivering a paper before such an august body as the RAS; he had already done so several times. All those presentations had been well received and duly published in the society's proceedings, the *Monthly Notices of the Royal Astronomical Society*. But his new paper was different. In his previous work he had simply elaborated on other scientists' research, mostly filling in details; no mean task, the sort of work that novices undertake to prove their technical worth. But this latest paper contained his own dramatic, ground-breaking discovery. Here was his chance to prove himself, the day when everyone would sit up and take notice. He was full of excitement and anticipation.

Nevertheless, something was not quite right. The previous day, the assistant secretary of the RAS had told Chandra in confidence that Eddington would speak after him. From the title of Eddington's paper, Chandra knew that it concerned his discovery. He wondered uneasily whether Eddington would be critical. When the two had met over dinner at high table the previous evening, Eddington had been evasive, refusing to reveal anything about his own talk. He had arranged, he said, for Chandra to have extra time to speak because he understood that the young man had so much to say. Chandra later recalled how he was 'really annoyed because here was Eddington coming to see me, week after week, about my work while he was writing a paper himself and he never told me about it'.[2] But perhaps some criticism at the well-attended RAS meeting from someone of Eddington's stature would be taken as a mark of the work's importance. Chandra had not the slightest inkling of what would actually happen.

Huddled in his dark overcoat, collar turned up against the chill January winds, Chandra took a taxi from Liverpool Street Station to South Kensington, as he did every Friday before an RAS meeting, to have lunch with his friend William McCrea. Six years older than Chandra, McCrea was a cheery young man with dark hair and a trim moustache. A Reader in mathematics at Imperial College, he too had studied under Fowler. Having completed his Ph.D. in 1929 on the constitution of the Sun's outer layers, he was a little more experienced than Chandra in the ways of academia, so Chandra naturally confided in him. McCrea tried to ease his friend's anxiety. He gently reminded him that the RAS meeting was a wonderful opportunity for him to air his important and intriguing result, and that he would undoubtedly learn a great deal from the ensuing discussion.

After lunch they walked briskly over to Burlington House and arrived at four, in time for tea. Chandra was eager to discover what research others were doing and to pick up the gossip. Suddenly, there was a commotion. Eddington appeared, moving through the crowded anteroom like Moses parting the Red Sea, giving an imperious nod here and there. McCrea knew Eddington slightly from his Cambridge days. Boldly, he stepped into his path and asked him about his lecture. Without breaking his stride and without so much as a 'How are you?' to Chandra, Eddington replied, archly, 'That's a surprise for you!'[3] Eddington's brusqueness shook Chandra. It reminded him of something that had happened the first time he had met the great man, back in 1931. Eddington was discussing a comment by his chief rival, Sir James Jeans, on his correlation between the mass of a star and its brightness. He suddenly sat bolt upright and barked that Jeans was 'a damned liar'.[4]

The white-walled meeting room was intimidatingly large and high-ceilinged. At 4.25 p.m. the audience began to filter in, crowding into the rows of seats in a sea of starched white shirts, ties and dark jackets. The seats were arranged in tiers, as in a theatre, with a narrow aisle down one side. On the stage at the front was a podium and a long table with chairs where the president and other leading figures sat, looking out over the audience. The walls were hung with portraits of past presidents of the society and great astronomers. There was a portrait of Newton over the president's chair, and one of the English astronomer Sir William Herschel, the discoverer of Uranus, on the wall to the right. There was not a single window.[5] As the sessions progressed, the room became increasingly stuffy. The first row was reserved for eminent figures such as Eddington, Fowler and Jeans. Milne, on whose theory of stellar structure Chandra had written several papers, was in the second row. Junior figures such as Chandra and McCrea were relegated to the seats high up at the back. At a lecture on whether there was air on Venus, someone was heard to murmur, 'Never mind Venus, is there air in here?'

The room was packed. Just about every major player in British astronomy and astrophysics was there. Not only were these monthly meetings of great academic interest, they were enjoyable social occasions too. The ferocious debates were a popular spectacle, making it well worth the trip. On the surface all seemed playful. The protagonists walked the boards like thespians making grand gestures and turning out exquisite literary flourishes, while ensuring that English understatement and indirectness prevailed. Each used his razor-sharp wit to snare his opponent or score points off them. In public, speakers such as Eddington gave no quarter, but in private everyone was assumed to be on excellent terms. This was not always the case, however. Some were particularly hurt by Eddington's acid wit. A great deal of animosity seethed beneath the veneer of high manners. A reviewer of G. H. Hardy's wonderful little book A Mathematician's Apology wrote that it 'is a reminder of a way of life where the participants did their best to hurt each other by day and dined together at night'.[6]

At 4.30 the President, F. J. M. Stratton, called the meeting to order. Stratton, holder of the chair of astrophysics at Cambridge and director of the solar physics laboratory there, was an imposing figure both as a scientist and in person. He had an exemplary war record and in public life was known as 'Colonel Stratton'. But to friends, this short, solid man with his rumpled suit buttoned to the neck, and happiest in the company of soldiers, was known simply as 'Chubby'. Many of his former students held prestigious positions in astronomy.[7]

Chandra remembered the first time he met Stratton. It was at a tea party in May 1933 in Eddington's garden. Eddington had invited him specially to meet his close friend, the famous American astrophysicist Henry Norris Russell, who was visiting Cambridge from Princeton University. To Chandra's relief, Russell was complimentary about his recent calculations on the shape of stars, work that drew on the research of their mutual friend Milne. Chandra wrote proudly to his father that he had 'had a very pleasant time discussing with them "big guns"'.[8] At the time, Chandra felt he had Eddington's support, for why else would the great man have gone so far out of his way to be kind to him? Eddington and Russell were the leading figures in British and American astronomy. They controlled appointments; they could make or break a man's career. After the party, Chandra had begun to feel more at home.

One of the first formalities at the meeting was to award Milne the society's gold medal for his research achievements. Chandra was delighted. Milne had been one of those most deeply offended by Eddington's cutting wit, so much so that it had affected his career and personal life. Chandra recalled Milne telling him that he had no confidence in his own mathematical ability, because his university training had been interrupted by the Great War.[9] Eddington had sniffed out this weakness. In every showdown they had, at the RAS or elsewhere, he homed in unerringly on the mathematical elements in Milne's ideas. Just recently, during one of Chandra's visits to Milne at Oxford, Milne had blurted out that his run-ins with Eddington had 'sapped his energy'.[10]

Six fifteen-minute papers preceded Chandra's, followed by comments from the floor. Finally, at 6.15, Stratton called Chandra to the podium. By then the air in the room was so dank and heavy it was almost tropical, like that day in late July, almost five years earlier, when Chandra had set off from Bombay to make the long ocean journey from India to England.

Nineteenth-century science held that stars form when the inward crush of gravity is balanced by the outward pressure of the star's own gas particles and the radiation pressure they emit. But as a star grows older and burns up its fuel, it becomes dimmer, and gravity begins to take over, crushing the star into a dense ball.

Could it be that stars eventually collapsed entirely? It was a tantalising mystery, and one that Eddington made every effort to play down. After all, it seemed to suggest that at least some stars did not end up as inert rocks. And this was a troublesome notion, indeed. Surely it was totally impossible for something as big as a star to disappear into nothingness. What on earth would become of it if that were the case?

In 1926, a colleague of Eddington's, Ralph Fowler, had proposed using quantum physics to resolve the situation. Fowler was the first theoretical physicist in England to grasp the deeper meaning of the new atomic physics, and he was unique in that he walked a thin line between astrophysics and physics. At that time, physicists looked upon astrophysics as a backwater. While physicists were concocting mind-boggling theories on the unpredictable world of quanta, astrophysicists remained resolutely classical. For physicists the most interesting question about stars was what made them shine – in other words, what were they burning? Where did they get their fuel from? Everyone assumed that stars generated their energy in the nuclei of their atoms, but no one knew exactly how. Nuclear physics was a subject that would remain murky until the discovery of the neutron in 1932. Astrophysicists, on the other hand, set this problem aside. Instead, they attempted to explore the structure of stars by creating models that avoided looking too deeply into the problem of what exactly stars are made of. Applying one of the remarkable new laws of quantum theory allowed Fowler to show that a star's core can stem the gravitational contraction. It ceases to shrink and can die in peace, in keeping with the vision we have had of the universe since the beginning of time.

But Chandra realised that Fowler's solution was incomplete. What he'd left out – remarkably – was Einstein's special theory of relativity, which says that the speed of light (not space or time) is constant. When Chandra applied relativity to Fowler's results on the voyage from Madras, he came up with something extraordinary, something no one had done before. He calculated that there was an upper limit to the mass of a white dwarf. It took him just ten minutes to deduce this, but he spent the rest of the journey puzzling over its implications. He knew he was on to something, but 'didn't understand what this limit meant, and . . . didn't know how it would end'.[11] It was beginning to dawn on him that there was only one inescapable though almost unimaginable conclusion: that white dwarf stars with a mass greater than the maximum mass he had just discovered simply could not exist. Their own gravity would crush them until they disappeared into nothingness.

Apart from his calculation of the limiting mass, Chandra's discovery confirmed the scenario that Eddington had suggested in one of his books but dismissed as absurd: that white dwarf stars might not end up as inert rocks but could collapse completely. He had solved Eddington's problem.

Standing at the podium, dwarfed by the white domed ceiling, the dark-skinned young Indian gazed around at the sea of starched shirts, dark jackets and European faces. At last he had the audience he wanted. Everything rode on this moment. He now fully understood the magnitude of what he had achieved as a young man of nineteen on that sunny summer's day, at sea in transit between two worlds, two civilisations, two cultures that could not have been more different.

Chandra must have felt rather as Einstein had in 1905, the year the young patent clerk discovered special relativity while working at the Swiss Federal Patent Office in the intellectual backwater of Berne. Like Einstein, he was convinced that the great scientists of the day were 'theorising out of their depth'.[12] Only he could see the truth. There he was, Subrahmanyan Chandrasekhar, a nonentity just as Einstein had been in that magic year. Like Einstein, he had lifted a corner of a great veil. In so doing he had revealed a majestic yet terrifying picture of the fate of the stars and of humanity.

The young man loosened his collar a little and brushed a drop of sweat from his forehead. The windowless room was hot and still, sealed against the January winds which whistled outside the venerable building. Glancing at his watch, he turned to the last page of his paper and read out his conclusion in confident tones: 'The life of a star of small mass must be essentially different from that of a star of large mass . . . For a star of small mass the natural white dwarf stage is an initial step towards complete extinction. A star of large mass cannot pass into the white dwarf stage and one is left speculating on other possibilities.'[13] He had finished. He gathered his papers together and returned to his seat near the back of the hall. The walk must have seemed an eternity. Perhaps he day-dreamed that this might be the moment that would launch him on a glorious career. He had given his all. He had done the best he could.

The discussion was thrown open to the floor. Milne compared his own research programme on white dwarfs with Chandra's, insisting that Chandra's was but a small part of his own. Chandra listened with half an ear. He was waiting to hear what Eddington had to say.

Ever theatrical, Eddington remained seated for a measured frac-tion of a minute, building up the dramatic tension. Then he rose from the front row. In one step he mounted the podium, turned, cocked his head and began with an arrogant flourish: 'I do not know whether I shall escape from this meeting alive!'[14] For the point of his paper was that the very basis of Chandra's theory was downright absurd. There was no such thing as an upper limit to the mass of a white dwarf. There were gasps. Chandra was shocked. Could he have heard Eddington correctly? If so, he was claiming that Chandra was utterly wrong and all his work was for nothing.

In essence Eddington was reverting to Fowler's theory, in which there are no restrictions on the white dwarf's mass – they always die a peaceful death rather than disappearing into nothingness. This was more to Eddington's taste. Eddington applauded Fowler for getting astrophysics out of a mess, and derided Chandra for reopening a can of worms. This was no English game-playing: it was a direct slap in the face, and Chandra felt it in just that way.

Nevertheless, Chandra had essentially solved the deeply impor-tant problem Eddington had proposed. Eddington would have had nothing to lose by saying, 'Yes, I was only joking in 1926 when I proposed that stars might collapse to nothingness. But now this brilliant young man has shown that this actually exists in Nature. He and I are going to explore this unexpected result.' Great sci-entists such as Einstein and Bohr would surely have done so. Eddington's reputation ranked alongside theirs. Had Eddington not been of great help to Chandra the previous year? To ease the burden of his extensive numerical calculations, he had pulled strings to procure him an advanced mechanical calculator. So what had happened? Why had Eddington revealed nothing of his disagreements – disagreements so strongly felt that he had presented them with an acidity and cynicism that went far beyond even the limits of the RAS's monthly skirmishes?

Chandra recalled the hours upon hours of conversation he had had with Eddington in Chandra's own rooms: 'I was telling him, "How can a star evolve? Massive stars must behave differently," and so on – all this was being talked about.'[15] Eddington had always appeared interested and thoughtful. Now he knew that Eddington's sphinx-like face hid a duplicity of shocking proportions.

Eddington's true agenda was revealed when he said, 'I think there should be a law of Nature to prevent a star from behaving in this absurd way!'[16] In other words, 'The hell with physics!' With an awful feeling in the pit of his stomach, Chandra realised that 'despite this man's incredible physical insight, he has always operated with preconceived ideas'.[17] While physicists were at that very moment happily taking on board quantum theory, with all its weirdness, contradictions, ambiguities and apparent impossibilities, astrophysics was, it seemed, bogged down in the past. Astrophysicists refused to follow the mathematics of their theories as they found themselves led inexorably into ever more seemingly implausible domains of physical reality. Instead, they insisted on ignoring results on the fate of the stars that upset their preconceived view of the universe as benign. There is an unwritten rule in science that when anything potentially observable is predicted to become infinite, it is a sure sign that the theory is breaking down. As Albert Einstein was fond of saying, 'Only two things are infinite, the universe and human stupidity, and I'm not sure about the former.' Astrophysicists such as Eddington simply refused to believe that something as big as a star could ever become infinitely tiny.

Chandra must have been deeply frustrated. Looking around at the nodding heads and mesmerised expressions, it was clear to him that everyone in the audience believed Eddington, even his own friend McCrea. How could this be? Just a few hours earlier, McCrea had been agreeing with him. But now McCrea was whispering to Chandra that he 'thought Eddington sounded plausible'.[18]

What was happening? Why didn't anybody speak up and tell Eddington that he was plain wrong? Chandra stood up to reply in his defence. He was dumbfounded when President 'Chubby' Stratton refused him any chance to respond. Instead, he brought down the curtain on Act I of the Chandra–Eddington confrontation with the curt words, 'The arguments of this paper will need to be very carefully weighed before we can discuss it.' And without further ado, he introduced the next presentation. Eddington's 'authority was so great, that people accepted him,' Chandra recalled ruefully. 'He made jokes about [my theory] and made me look like a fool.'[19] 'At the end of it, everybody came by and said, "Too bad." "Too bad."'[20]

Although minutes were taken of that fateful RAS meeting, and each man's presentation was duly published, there was no consensus on why Eddington had launched into this ferocious attack without any warning. The basis of his argument was not at all clear, yet no one in the audience questioned it. According to McCrea, Eddington was guided by instinct. If anyone other than Eddington had offered an argument based on this premise, McCrea and others would have had real doubts about its validity. But despite the flimsiness of Eddington's arguments, the academic community chose to support him rather than an outsider. Such was the force of Eddington's reputation and personality. In later years, Chandra always said bitterly that there was only one case in which Eddington was wrong and never admitted it: his own.[21] He never forgot the laughter and guffaws that rose on cue during Eddington's carefully orchestrated lecture.

It was all the more incredible to Chandra because Eddington understood his discovery better than anyone else, Chandra included. Eddington's insight and intuition were enormous, but he utterly refused to accept the physics. He stood by his preconceived conceptions of nature. Soon after 11 January, at Trinity, Eddington commented to Chandra, 'You look at it from the point of view of the star. I look at it from the point of view of nature.' Bewildered,

Chandra asked, 'Aren't they both the same?' Eddington replied firmly, 'No.' 'Well, you see,' Chandra remembered, 'that sort of shows his attitude. Somehow, he felt that nature must conform to what he thought was right.'[22] So did most other astrophysicists.

On the train back to Cambridge that Friday evening, Chandra recalled Milne saying, 'I feel in my bones that Eddington is right.'[23] Chandra snapped back angrily, 'I wish you felt it elsewhere.' Privately, Milne was elated because his own theory of stellar structure was at odds with the concept of stars disappearing into nothingness. Something had to prevent complete collapse. He too had preconceived notions. Visiting Chandra in 1934, he had begun to accept Chandra's results. 'But when Eddington said that the formula [for the upper limit of a white dwarf's mass] was wrong, Milne was all aglow.'[24] Soon after, Milne wrote that regardless of whether Chandra's theory was right, he would ignore it.[25]

Arriving back at Trinity well past midnight, Chandra could not face his room. Instead, he went to the senior common room and stood in front of the great fireplace, gazing into the flames. Alone with his thoughts, he reviewed the day's events. Until 6.45 that evening, he had been full of elation, bursting with self-confidence. He had revealed in all its rigour a momentous scientific discovery. In the red glow of the coals he saw Eddington's face, and his mind turned to the events that had all but obliterated his world. It had been 'a totally unexpected occurrence, that came close to destroying my scientific confidence'.[26] Had it all been for nothing – the sacrifices his family had made in sending him from India to Cambridge, the tearful farewells at the dockside in Bombay? He remembered his elation when he had made his shipboard discovery.

Less than a year later he had heard that his beloved mother had died. Perhaps, he thought to himself, he was reading too much into it; but just two days after he received the news was the first time he met Eddington. Despite his deep grief, which he could share with no one, he could not postpone that meeting.[27] He had been shown into Eddington's study in his house next to the observatory. The great man sat at his desk with his back to french windows that looked out onto a beautiful garden. The room was wreathed with pipe smoke and laced with the sweet smell of apples. Eddington was always eating apples, which he chewed up entirely, core and all.[28] Though Chandra felt empty inside, he had performed well.

Now he stood in front of the fire, repeating to himself T. S. Eliot's famous lines as he remembered them: 'This is how the world ends. This is how the world ends: Not with a bang but with a whimper.'[29] He had hit rock-bottom, but he would not concede. Sometimes Chandra and Milne sarcastically referred to Eddington as the Devil incarnate.[30] What happened on 11 January 1935 was Chandra's first experience of that manifestation of him.

Chandra never forgot that moment of humiliation when he had confronted the devilish titan of astrophysics. He was convinced he was right. Yet what he could not understand was why Eddington should have attacked and humiliated him in such a vicious and unbridled manner. It was indeed extraordinary that a nineteen-year-old Indian youth had managed to make a discovery that had eluded the great minds of European astrophysics. To understand how that came about, we have to step back in time to India not long after the end of the Victorian age, when the Raj was beginning to show signs of weakness and young Indians dared to step forward and claim their heritage.

2

A Journey Between Two Worlds

For Chandra and for many other ambitious young Indians of his generation, science offered the best way to break through the seemingly insuperable barriers erected by the British Raj. He wanted to express his patriotism by showing 'that Indians could be accomplished in a way which the outside world [could] recognise'. 'Certainly one of the earliest motives that I had', he said later, 'was to show the world what an Indian could do.'[1] To the end of his life, Chandra never forgot the day – 27 April 1920, when he was just nine – that his mother told him 'that a famous Indian mathematician, Ramanujan', had died the day before, at the age of thirty-two.[2]

Srinivasa Ramanujan's story was like a fairy tale. He had been a lowly clerk in the Madras Accountant's General Office. Then, in January 1913, he sent off a letter to G. H. Hardy, Eddington's colleague at Cambridge, outlining some mathematical ideas and suggesting there were more to come. Hardy was intrigued and impressed. Six months later, Ramanujan had been lifted from his hand-to-mouth existence and transported to Trinity College, Cambridge, where his only commitment was to do mathematics. In March 1918 he was elected a Fellow of the Royal Society, a singular honour in the British scientific world. That autumn he became a Fellow of Trinity College, able to devote himself entirely to research for the next three years. Later, Ramanujan would be recognised as one of the century's most original mathematicians. It

was Hardy who had brought all this about. For Chandra, his coun-
tryman's rags-to-riches tale inspired him and gave him the hope
that he too could 'break [his] bonds of intellectual confinement
and perhaps soar the way Ramanujan had'.[3]

Chandra was born on 19 October 1910, in Lahore. Now in
Pakistan, in those days it was the capital of the Punjab province of
British India. His father had been posted there as the assistant audi-
tor general for the North West Railways. As the first son of the
family, Chandra was given his grandfather's name, Chandrasekhar,
which means 'moon' in Sanskrit. The family moved on to Lucknow,
and then, when Chandra was eight, returned to their roots in south
India and settled in Madras.

In contrast to Ramanujan's early poverty, Chandra grew up in a
freethinking, upper-class Tamil Brahmin household. There were
always servants around. The only responsibility for him and his
three brothers and six sisters was to study.[4] Vidya, one of his sisters,
recalls that 'our home was really more like an educational institution. People seemed to be reading or debating or conversing about
interesting subjects all the time.'[5] The emphasis on the importance of learning was typical of Brahmins, for whom knowledge
counted for far more than wealth.

Ever since the 1850s, when India was established as the jewel in
Britain's colonial crown, the establishment of a British education
system had been a cornerstone of colonial policy. As Lalitha,
Chandra's widow, points out, the ultimate aim was 'not so much to
educate Indians as to have an army of cheap clerks'.[6] Nevertheless,
the result was the development of an elite cadre of young Indian
men, mostly Brahmins. For Indians, says Lalitha, education provided
'a window to the West'. In fact, it was this very 'exposure to Western
culture that led to the [Indian] renaissance', a national awakening in
all spheres of creativity but most notably in science, sparked by the
political movement to bring about India's independence.

Lalitha is the custodian of Chandra's memory. Now in her
nineties, she is a tiny, bird-like woman with an oval face, wavy
grey hair and round-rimmed glasses perched firmly on her delicate
nose. But in her youth, when she wore her hair smoothed into a
long plait, she was a beauty. Her exuberance, ready smile and
assertive speech still belie her years. Chandra's story, she insists, is

inextricably entangled with what was happening in India when he was growing up. 'The reason for the emergence of scientists like Ramanujan, [Satyendra Nath] Bose and Chandra was the renaissance in India that commenced in 1910,' she says emphatically. 'You must start here. The country was ready for something to happen. It was a crucial time. Mohandas K. Gandhi was about to return from South Africa.'

In 1910, the year that Chandra was born, Gandhi had been effectively in exile for more than two decades. After a conventional childhood, a friend had persuaded him that he should go to England to study law and become a barrister. For Indians, travelling 'over the water' meant becoming a virtual outcast. In later years Gandhi wrote of his terrible loneliness abroad in words that Chandra would echo three decades later: 'I would continually think of my home and country. My mother's love always haunted me. At night the tears would stream down my cheeks, and home memories of all sorts made sleep out of the question. Everything was strange.'[7]

Gandhi mastered several European languages, learnt to dance and studied the violin. He read widely, probing Christianity, Hinduism, Theosophy and the emerging ideological movement of pacifism. But when he returned to India in 1891, he found his chances for employment almost non-existent. He decided to try life in South Africa. He spent the next two decades in Durban and Johannesburg, working as a lawyer, fighting the discrimination suffered by Indian settlers. There he discovered his life's work: to bring about radical change in India by the methods of non-violent civil disobedience he had used so effectively in South Africa.

The movement towards Indian independence had begun on 28 December 1885 with the first meeting of the Indian National Congress. But it was not until after the Great War of 1914 to 1918 that the struggle began in earnest. India's role in the war inspired a new sense of pride and self-confidence. Half a million Indians travelled overseas to fight. What they saw in Europe fired their imaginations.

When Gandhi returned from South Africa in 1915, he exchanged his life as a successful middle-class lawyer for that of an ascetic. He gave up his possessions and wore the simple homespun garments of an Indian peasant. Soon people were calling him the 'Mahatma', the Great Soul or Holy One. After the war, along with the inspirational poet Lokamanya Tilak, Gandhi led a resurgence of Indian nationalism combined with the rediscovery of traditional Indian mores and the first moves towards breaking down the caste system. His method was pacifist non-cooperation. Initially the representatives of the British Raj dismissed him. In 1917, the then Secretary of State, Edwin Samuel Montagu, described him contemptuously as a fringe figure who 'dresses like a coolie, forswears all personal advancement, lives practically on air and is a pure visionary'.[8] It was an accurate description but a fantastic underestimate.

The British responded to the growing clamour for independence with a series of increasingly desperate laws, culminating in the detested Rowlatt Act, which made wartime restrictions on civil rights permanent. In the infamous massacre at the Sikh capital of Amritsar in the Punjab, a British Army contingent opened fire on an unarmed crowd of protesters, killing hundreds. That was the last straw. Multitudes who had patiently endured the yoke of imperialism flocked to the nationalist cause.

Growing up in the southern city of Madras, young Chandra was at first insulated from the growing foment of political unrest around him. In his family the emphasis was very much on the importance of education and, specifically, on science. The focus on learning was the legacy of Chandra's grandfather, Ramanathan Chandrasekhar, who had been professor of mathematics at Ankitham Venkata College in Vizagapatam (now Visakhapatnam), about 360 miles along the coast north of Madras. He died the year Chandra was born. But he left a library full of books on mathematics, including some he had written himself. Chandra treasured them throughout his life. Ramanathan also fathered ten children, two of whom were exceptionally brilliant.

Chandra's father, Chandrasekhara Subrahmanya Ayyar, was the eldest of Ramanathan's children. An exemplary student at his father's college, and then at Presidency College, he took the All India Examination and entered British government service in the Indian Audits and Accounts Office. There he worked in the railways section and travelled extensively to different parts of

India. In those days it was a tremendous achievement to enter the Indian civil service, which offered great opportunities for advancement. But his children looked down on him. 'The standards by which Chandrasekhar's family measured success were by intellectual and cultural achievements,' Chandra's younger brother, Balakrishnan, wrote.[9] They considered him 'a relative failure' because he had chosen the civil service route. It was a harsh judgement on a hard-working father.

Nevertheless, it was he who had built up the family's high standard of living. Ayyar, as he was known, was a strikingly tall, powerfully built man with a handsome round face, piercing eyes and a powerful, rather authoritarian personality. At home he dressed South Indian style in a shirt and *veshti* (a garment rather like a sarong). But for work he wore a Western shirt, tie, jacket and trousers. Like most Indians in British government service, he always wore a tightly wrapped turban to emphasise his reluctance to abandon the old ways completely. He was a brilliant exponent of Karnatic music, which he played expertly on the violin and for which he also developed a notation.

But for all his achievements, from a very young age he found himself eternally in the shadow of his younger brother, Chandrasekhara Venkata Raman. Raman was the star of the family. In photographs, while Chandra's father has a gentle cast to his face, Raman has a look of ferocious intelligence and no-nonsense determination. He raced through his education at an extraordinary speed, graduating from Presidency College at the age of seventeen with a degree in physics and finishing his MA when he was nineteen. Dogged by ill health, he gave up his studies in physics and, like Ayyar, joined the Indian Audits and Accounts Office. The brothers were posted to Calcutta, where they shared a flat, along with their wives. But Raman refused to give up his physics research and continued working on it in his spare time at the University of Calcutta. In 1917 he was offered a chair or professorship there and gave up his lucrative civil service posting to take up research full time. In 1930 he would receive the Nobel Prize for Physics. According to Chandra's cousin Sivaraj Ramaseshan, there was bitter sibling rivalry between Ayyar and Raman. Not only did Raman seem to do everything better than

Ayyar, but he had left the security of the civil service and eventually achieved world-wide fame. When the two brothers lived
together in Calcutta, the atmosphere between them was brittle.
Ramaseshan has studied their correspondence. 'There is no doubt
that C. S. Ayyar disliked C. V. Raman,' he says.[10]

Ayyar was a typical Indian father of his day. He showed little
affection towards his children. He assigned professions to his sons
and husbands to his daughters. But beneath his crusty exterior, he
cared a great deal about Chandra's emotional well-being, as is
clear from their correspondence. Yet their relationship was far
from smooth. Lalitha recalled that the 'children were disappointed
in Ayyar and, as grown-ups, did little to hide their frustrations
with him. They were closer to their mother.'

Sitalakshmi Balakrishna, Chandra's mother, was an extraordinary woman. Her marriage to Ayyar had been arranged when she
was fourteen, and she came from a typical Indian household with
an overbearing mother-in-law and an unhappy aunt. Over the
years she bore ten children, yet she always continued to study.
With Ayyar's help she learnt English, and went on to translate
Ibsen's A Doll's House into Tamil. It became the standard text for
high-school students.

In those days, middle-class families in India began their children's education at home. Early in the morning, before he went to
the office, Ayyar gave Chandra instruction on most subjects. His
mother taught him Tamil. Ayyar was particularly concerned that
Chandra should speak English perfectly; English was the key to
everything, from entering the Indian civil service to attaining
every bright young Indian's dream – study in England. Chandra's
parents were aware from early on that here was an extraordinarily
brilliant child. He made lightning progress in mathematics. He was
often to be found sitting alone in the library, devouring his grandfather's mathematics texts. By the time he finished high school, at
the age of fifteen, he had read them all. In later years, all ten children were to look back on their individualistic and free-ranging
early education with gratitude. As Chandra's brother Balakrishnan
put it, it prepared them for school 'without being cramped in the
early years by a load of educational lumber in the head'.[11]

The family did not really put down roots in Madras until 1923, when Chandra was thirteen. That December, Ayyar laid the foundation stone for Chandra Vilas, which was to be the family home. It was at 46 Edward Elliots Road in Mylapore, an affluent suburb of Madras, a world away from the squalor of neighbouring Triplicane where Ramanujan had grown up. In Chandra's day the house was an elegant, whitewashed two-storey mansion, built in the neo-colonial style traditional in Madras at that time, with long shady verandas, roofed wooden balconies and decorative eaves over the windows. The garden was planted with huge-leafed mango and coconut trees. It was a cool and inspiring place to read and talk, a haven from the intense light and heat of the day.

Ayyar had taken on the expense of providing a home for his family. But he was all too aware of the consequences. His family would no longer be with him when he was posted away from Madras. 'A lonely life for earning a livelihood faced me,' he recalled in his autobiography.[12] By the time Chandra Vilas was completed in 1924, Chandra was already in his last year at the Hindu High School in Triplicane. Those were carefree days. The three eldest brothers – Chandra, Visvanathan, two years his junior, and Balakrishnan, four years younger – were inseparable. At first they used to take the tram to school together. Later they cycled. Today, Edward Elliots Road has become Dr Radhakrishnan Salai, a noisy, traffic-filled thoroughfare and a far cry from the idyllic setting in which Chandra grew up. Chandra Vilas still stands, though it is much reduced and run down, tucked along an alleyway off the main road. Still, one can imagine Chandra and his brothers riding their bicycles along Edward Elliots Road to school nearly eighty years ago.

From 1925 onwards, Chandra's bicycle ride took him along the Marina beach to the splendid pink-domed halls of Presidency College. He was fifteen, two years younger than everyone else, but he had already acquired a reputation for brilliance. 'Chandrasekhar was born great and lucky – lucky at least to start with. From his early years he was recognised by all the people he came into contact with, including all the members of his family, as a mathematical prodigy or even genius,'[13] recalled his brother Balakrishnan. At Presidency he became famous for answering every question correctly in examinations, even ones that were not required. He usually scored more than 100 per cent. He read voraciously, devouring mathematics and physics texts at a single reading. His tastes in literature ranged from Robert Louis Stevenson to Thomas Hardy and Shakespeare, and he also learnt German. One of his college friends, S. R. Kaiwar, recalled that he was 'lightning-quick in comprehending, [and] could read a hundred pages in an hour quite easily. [He] was interested in everything around him.' Chandra had time to enjoy a normal boyhood. 'We laughed and joked a lot,' remembered Kaiwar. 'He wasn't by any means a prude.'[14] But he was prone to fits of gloom, sometimes withdrawing to his room to sit for hours in the dark.

Family life helped greatly to nurture the young man's brilliance. Conversations over dinner were not merely intellectual but focused on science. As Chandra later wrote, 'The atmosphere of science was always at home.'[15] By now his Uncle Raman had become one of India's most eminent scientists. One day, Chandra overheard Raman talking to his father about Ramanujan. Chandra was fired. He yearned to follow in his footsteps and become a mathematician.

After completing his first two years at Presidency College, Chandra had to choose what to study for his honours degree. He immediately opted for mathematics. But his father had other ideas – as far as he could see, there was no future in mathematics. He wanted Chandra to play safe, as he had, and join the civil service. Knowing he was determined to be a scientist, he pressured him to study physics, which at least had more practical applications. For Chandra physics was acceptable, but not the civil service. Unusually in Indian society, his mother supported him. 'You should do what you like. Don't listen to him; don't be intimidated,'[16] she told him. In the end, Chandra took his father's advice and in later years he was grateful. He wrote to Balakrishnan from Cambridge: 'I am ever so thankful to Babuji for having guided me properly in making the proper choice.'[17]† After all, Chandra continued, his physics research was so highly mathematical that 'my "first love" is not killed'.

Chandra often went to the Marina beach to think and dream.

† *Babu* is an affectionate Tamil term for Father; *ji* indicates respect.

His hero Ramanujan had also strolled there to escape the hot, dusty streets of Triplicane. Chandra was consumed with the urge to make his mark on the world as a scientist, to discover something that would transform our understanding. 'When I was in 5th form and later, I used to go to the beach and pray God – actually prostrate on the ground (I blush to tell you so) – to mould my life like that of an Einstein or a Riemann,' he later admitted to Balakrishnan.[18]

Chandra, Balakrishnan and Visvanathan would sprawl on the sand, joke as boys do, and listen to Chandra holding forth on mathematics, science or literature. In those days the Marina was a highly fashionable resort. In the evening, automobiles sputtered along side by side with the increasingly rare horse-drawn shuttered carriages in which Muslim ladies could take the night air without being seen. The beach is one of the longest and broadest in the world, stretching southward several miles from Fort St George to the sixteenth-century Portuguese town of San Thomé. From the venerable pink-domed clock tower of Presidency College, it is a long walk across the burning sand to the shimmering blue waters of the Bay of Bengal.

Established in 1840, Presidency College became the nucleus of the University of Madras, founded seventeen years later. In Chandra's day, as now, it was the best college in South India. The lecturers, most of whom were British, encouraged students to study rather than learn by rote, which was the norm at other Indian colleges. Chandra still complained that the system was exam-oriented and stressed the solving of problems instead of the understanding of science. One of the teachers, H. Parameswaran, had done his Ph.D. at Cambridge. He was, Chandra recalled, 'quite a competent experimental physicist [who] understood what research was'.[19] Chandra himself certainly did. His career at Presidency College was enormously successful. 'Everybody knows Chandrasekhar, he is a very famous man,' Balakrishnan wrote excitedly to him in November 1930. Today there are photographs of Chandra proudly displayed in the hallways of the physics department, alongside Bohr, Einstein, Parameswaran and Raman. In his study, the present head of the department has four photographs on his desk: two of Chandra, and one each of Raman and Einstein.

Most physicists and physics students struggle through physics texts equation by equation. Chandra could read them almost like novels, effortlessly grasping the essence and the details. As a teenager, Balakrishnan remembered, he had already absorbed some of the most daunting works in the Chandra Vilas library, such as Salmon's *Conic Sections*, Hardy's *Pure Mathematics*, Boole's *Differential Equations* and Burnside and Paton's *Theory of Equations*.

But his studies did not exclude all else. As a student in India in the late 1920s, he could not fail to be drawn into the struggle for Indian independence, even though his contribution was primarily to make his academic mark, to show that an Indian could succeed in the Westerners' world of science every bit as well as a Westerner. Chandra did not go so far as to join the Indian National Congress. But when Jawaharlal Nehru, its president, came to Madras in 1928 to speak at a mass rally on the Marina, Chandra was among the students who joined the crowd to listen to him, in defiance of the college principal's injunction not to do so. In later years, Chandra remembered Nehru in a way that so many others had, as 'a complex product of intellectual brilliance, physical attractiveness and stamina, human sensitivity and charm, which made him at once the hero to India's youth'.[20]

That summer, Raman offered his precocious nephew a priceless opportunity. He gave him a job in his laboratory in Calcutta, providing Chandra with first-hand experience of real scientific research. It was an extraordinarily exciting time. The previous February, Raman had discovered what quickly became known as the Raman effect, a way to glimpse a molecule's structure from the way it interacts with light. Chandra was swept along in the euphoria over this ground-breaking discovery. No doubt he hoped that he too would achieve something similar one day. But his own experiments were disastrous. Broken equipment began to pile up. It became clear to everyone that his true calling was theoretical physics, especially when he began to give brilliant seminars to the others in the laboratory, lucidly explaining his uncle's discovery.

That spring, Raman had paid a visit to Chandra's home in Mylapore, bringing with him a book with the tantalising title *The Internal Constitution of the Stars*, by Arthur Stanley Eddington. In it Eddington took the reader to the very frontiers of astrophysics. He laid out the key issues and ended with one of his unforgettable flourishes: 'It is reasonable to hope that in a not too distant future we shall be competent to understand so simple a thing as a star.'[21] Chandra was spellbound.

A titan of science, Eddington had single-handedly established the field of modern astrophysics. By clarifying the relationship between a star's mass, brightness and temperature, he began to make sense of its life journey from its birth, amid a vast cloud of interstellar gas and dust, to its death as a cold inert rock; he had worked out methods of analysing the movement and distribution of the stars; and he had also made vital contributions to our ideas about stellar composition and energy generation, presciently suggesting that stars shine by converting hydrogen into helium. Most importantly, however, Eddington was the first to apply general relativity to astronomy, having introduced the theory to the English-speaking world in popular books still read to this day.

Eddington's research papers were masterpieces of elegant prose. With the minimum of mathematics, of which he was a master, he derived results from first principles, linking the basic laws of physics to experimental data, always with a bon mot or superb metaphor. In *The Internal Constitution of the Stars* he shows how, although at first sight understanding the stars seems hopeless, it is not. Stars have a gravitational field and emit light which 'manages to struggle to the surface and begin its journey across space. From these two clues alone a chain of deduction can start', based on universal laws of nature.[22] Eddington had a charmingly vivid and utterly unforgettable way of bringing astrophysics alive. Describing a photon of light, he wrote:

> You may picture a photon of radiation, barging first one way, then another, like a man in a rioting mob – absorbed by an atom and flung out again in a new direction. In this way a photon in the Sun will wander aimlessly round in the interior for a million years or more until, just by accident, it finds itself at the exit of the maze – shoots through – and makes a bee-line across space to the Oakridge reflector, where Professor Shapley photographs it.[23]

The American astrophysicist Henry Norris Russell wrote of *The Internal Constitution of the Stars* that, 'The book itself is a work of art . . . the product of a great teacher as well as a great investigator.'[24] It was not only a journey through almost everything that was known about the stars; it also set out the key problems that remained to be solved. Eddington was an inspiration to everyone from professional scientists to blazingly ambitious and idealistic young men like Chandra.

Working in Raman's Calcutta laboratory, Chandra leaped at the opportunity to borrow his uncle's copy and read it closely. He was fascinated by Eddington's wide-ranging discussion of how atoms emit and absorb radiation according to quantum theory, and how stars can be described in mathematical terms as spheres of gas. Eddington's discussion of quantum theory helped Chandra to understand Raman's discovery.

But then he came across the classic *Atomic Structure and Spectral Lines*, by the great German physicist Arnold Sommerfeld. By pure chance, that September 1928 Sommerfeld showed up in Madras to give a lecture. Full of youthful bravado and confidence, Chandra found out where he was staying and went and knocked on his door. What followed changed the course of his life.

Born in 1868, Sommerfeld was involved in virtually every phase of theoretical physics, right up to his death in 1951. He was immensely talented both as a researcher and a teacher. His famous institute in Munich produced many of the pioneers in quantum physics, from Hans Bethe and Wolfgang Pauli to Werner Heisenberg, his prize student. As Heisenberg often said, if anyone deserved a Nobel prize it was Sommerfeld – though inexplicably he was passed over by the Nobel committee, one of its few serious omissions in physics. Sommerfeld's *Atomic Structure and Spectral Lines* first appeared in 1919. It was regularly updated and became the bible for researchers. Chandra read the English translation of the 1924 edition.

Chandra was totally unaware of the discoveries that had been made in atomic physics over the previous two years, discoveries that had utterly changed scientists' views of matter at the microscopic level. An extraordinary new quantum world was unfolding in which electrons and light behaved in astonishing ways. The old images of electrons as particles and light as waves no longer held; both turned out to be waves and particles at the same time – a bizarre and unimaginable conclusion known as the wave–particle duality.[25] From this, scientists extrapolated all manner of quantum weirdness. Electrons, they concluded, were spread across space and time, capable of being in several places at once. Stranger still, electrons appeared sometimes to be able to 'sense' and react to an experiment carried out on another electron far away, as if by ESP. The old certainties of classical physics – that scientists could calculate exactly the position and velocity of an electron and plot its path through time and space – had been overthrown. There was now a completely new way of looking at these properties in this volatile new world. In his clipped, precise English, Sommerfeld gently explained it all. He left Chandra with copies of two papers he had published earlier that year in which he showed the results of applying quantum theory to a gas of trillions upon trillions upon trillions of electrons squashed tightly together – in a metal, for example.[26]† It seemed that the quantum nature of electrons unexpectedly produced an outward pressure, which could hold the metal together, helping to explain why metals were not crushed by the massive inward pressure of gravity.

Eager to learn everything he could, Chandra sought out the seminal articles that Sommerfeld had recommended. Chief among these were works by the brilliant young physicists Wolfgang Pauli, at Hamburg University, and Werner Heisenberg, at Leipzig, both of whom had started out as students of Sommerfeld and had gone on to become the instigators of the new quantum physics. Both were no more than a decade older than Chandra but were already legendary names. Chandra was particularly inspired by the

† The sheer size of stars, together with their densities, the number of atoms they contain, and their distances from the Earth and the Sun can only be expressed in huge numbers. A mathematical shorthand is to use powers of 10: 10^2 is a hundred, i.e. one followed by two zeros (100), 10^3 a thousand (1000), and so on. Similarly, 10^{-1} is one-tenth, i.e. one preceded by a decimal point (0.1); 10^{-2} a hundredth, or one preceded by a decimal point and two zeros (0.001); and so on. However, I shall often give large numbers in words, where 10^9 is a billion and 10^{12} a trillion.

research that underlay Sommerfeld's work. He pored over articles by Paul (P. A. M.) Dirac, who worked in Cambridge, and Enrico Fermi, in Rome, who had shown the dramatic effects produced by the wave–particle duality when electrons are cramped together in a gas.

The way physicists try to understand a concept is to use it to solve a problem. This is what Chandra did. Although only seventeen, he had already homed in on the area in quantum physics he wanted to pursue – the quantum properties of a gas of electrons – which was soon to lead him to his ground-breaking discovery about white dwarf stars. From his reading of Eddington's book, he explored how light, electrons and atoms interact in a star. He combined this with what he had learned from Sommerfeld, and within a week or two he had written his first paper. It was published in the *Indian Journal of Physics* in 1928.[27]

The precocious youth was convinced that his results were important enough to merit publication in the *Proceedings of the Royal Society*. He was aware that a paper for the *Proceedings* had to be submitted by a Fellow of the Society. As it happened, he had recently read Ralph Fowler's newly published *Statistical Mechanics* and noticed that Fowler was a Fellow. Boldly, he decided to take a chance and send him a copy of his newly written paper. It was January 1929.

That same month there was a meeting of the Indian Science Congress (modelled on the British Association for the Advancement of Science) in Madras. Raman, who had by now achieved worldwide fame, presided. The meeting was packed to the rafters. Chandra, a fresh-faced eighteen-year-old, rose to address the venerable gathering and presented a summary of his paper. It was his first professional presentation at a scientific meeting. Professor Parameswaran announced that the young man was only in his second year at college and added proudly that, 'he had written the paper without any guidance or advice from anyone'.[28] There was thunderous applause.

Chandra's brother-in-law, A. S. Ganesan, who was a physicist, had left his copies of the *Monthly Notices of the Royal Astronomical Society*, running from 1925 to 1929, in the well-stocked library of Chandra Vilas, the family home. Glancing through them in early spring 1929, Chandra came across a paper in which Fowler proposed that quantum physics might offer a solution to Eddington's paradox that a white dwarf might collapse completely, instead of ending up as an inert rock. He was excited to see that it used the new way of studying a gas of electrons which he had learned from Sommerfeld and had already applied successfully in his first two published papers. 'Right there, there was something which I could do. So that is how I started, you see,' he later recalled.[29] It was this that convinced Chandra that Cambridge was the place for him. That was where the action was.

That June, Chandra sent his father a letter outlining his work in progress. At eighteen, he was working on five papers at the same time.[30] In one he elaborated on the paper he had sent to Fowler and which he had submitted to the *Indian Journal of Physics*.[31] He waited impatiently, hoping desperately that Fowler would reply. Finally he heard from him.

When the paper from the unknown Indian lad landed on Fowler's desk, he read it with growing interest. He made a few suggestions, to which Chandra agreed. Chandra immediately withdrew the paper he had just submitted to the *Indian Journal* and incorporated its results into the one he had just sent to Fowler. Fowler was hugely impressed. To Chandra's great excitement and pride, his work came out 'as a fairly big paper' in the October issue of the *Proceedings of the Royal Society*.[32] Chandra wrote yet another paper which, he told his father, he passed to a professor at Presidency College 'for correction. But he indefinitely postponed it. So I sent the paper to the Phil. Mag.' (the *Philosophical Magazine*).[33] In Chandra's view, the paper needed no corrections. He was right, and it was published the following year. Chandra also sent off a letter to Eddington, but sadly lost Eddington's reply. 'After losing Eddington's letter, I am very careful about my letters from Fowler,' he wrote to his father.[34]

In October 1929, Presidency College was abuzz. Werner Heisenberg had arrived to give a lecture, as Sommerfeld had the previous year. Everyone was poring over Heisenberg's amazing research results, particularly the new atomic physics – dubbed quantum mechanics – that he had discovered in 1925, at the age of twenty-four. Two years later he had discovered his famous uncertainty principle.

Léon Rosenfeld, Bohr's close colleague, wrote of him, 'A wonderful combination of profound intuition and formal virtuosity inspired Heisenberg to conceptions of striking brilliance.'[35] Heisenberg was fond of recalling that he learned physics backwards – first quantum physics, then classical physics; he had, he wrote, 'a very irregular education and training in physics'.[36] This hybrid education was no doubt the source of his daring, no-holds-barred approach to research. He had joined Sommerfeld's class in theoretical physics in the autumn of 1920, when Sommerfeld was discussing atomic physics, an area then in enormous turmoil, and immediately began writing ground-breaking papers. By 1927, he was a professor at Leipzig University and was already being spoken of as a future Nobel prize contender.

The staff at Presidency College suggested that their foremost physics student, Chandra, should give a lecture on Heisenberg's work and introduce the great man. Chandra's family 'had our own way of looking at the matter', wrote Balakrishnan. 'We told one another that the professors realised their inadequacy to talk on the work of Heisenberg and left it to Anna (as we used to call him, the word in Tamil meaning "elder brother").'[37]

Chandra was thrilled to be able to spend time in the company of the virtuoso physicist. The two men were about equal in height, but there the physical resemblance ended. While Chandra was slight and dark-skinned, Heisenberg had short fair hair, fair skin and the robust build of someone who regularly trekked and skied. People who met him in later life recalled his rock-hard handshake, youthful looks and purposeful walk. Chandra borrowed a car and showed Heisenberg around Madras. The two young men spent the entire day together. 'On one day by merely talking to him I could learn a world of physics. At night, we drove along the Marina, and he was telling me about America. On the whole his stay in Madras was extremely profitable. He has suggested one or two problems also,' Chandra wrote with enormous pride to his father.[38]

A month later, Chandra sent off another paper to Fowler entitled 'On the probability method in the new statistics', based on a problem Heisenberg had suggested to him. But the paper was never published. Chandra himself may have withdrawn it after

Heisenberg pointed out an error to him. 'What days they were!', he recalled 'wistfully', writing to Balakrishnan many years later.[39]

In January 1930, a few months after Heisenberg's visit, Chandra was invited to attend a meeting of the Indian Science Congress Association in Allahabad. There he met the distinguished Indian scientist Meghnad Saha. Saha congratulated Chandra on the paper he had recently had published in the *Physical Review*, describing it as 'very suggestive'.[40] Chandra reported excitedly in a letter that Saha had also invited him to tea, and then to a small luncheon, so that Chandra could meet Saha's group. The young man felt as if he had truly arrived.

But on the train journey back to Madras, he was rudely reminded that he was still just an Indian under the Raj. With his father working for the railway, Chandra always enjoyed free first-class travel. That day he found himself sharing a carriage with an English couple. The train had barely left the station when the woman began complaining loudly at having to share the compartment with an Indian. The only consolation, she added, was that at least he was wearing Western clothes. Infuriated, Chandra stormed out of the carriage and returned in a South Indian *veshti* and shirt. When the conductor came through, the woman demanded that Chandra be moved to a second-class compartment. Chandra suggested that she move to a second-class compartment instead. With supreme arrogance, she pulled the emergency cord, then threatened to pull it again when the train restarted. In the end, Chandra won the battle. 'After a while they moved to another compartment – second class or something, I don't know,' he wrote triumphantly.[41]

That month, Chandra wrote to tell his father the strictly confidential news that the government was planning to offer him 'a scholarship *straight without advertising for it* . . . as an exception they do it for me as a very special case'. The money would enable him to study in England. In exchange, Chandra was to return to Presidency College and take up a professorship there in theoretical physics. The staff at Presidency had high hopes for his future. The principal, P. F. Fyson, was the very man who not long before had fined Chandra for attending the rally on the beach where Nehru had spoken. Now, he asked Chandra if he thought he could become a Fellow of the Royal Society by 1934. 'I told him that if I could succeed [by] 1940, I shall consider myself lucky. I explained to him how Dirac and G. P. Thomson both of them are not yet Fellows,' Chandra wrote to his father.

Meanwhile, Chandra's final examinations were looming and he had to put his calculations aside. From working at the cutting edge of physics, he reverted to undergraduate status. Around this time, Chandra's mother became very ill. He was devoted to her and contemplated abandoning his plans to study abroad. 'You must go,' she insisted. 'You must pursue your own ideals to the utmost.'[42] She advised him strongly not to accept advice or help from his Uncle Raman. There had been a rift in the family ever since the two brothers and their wives shared a flat in Calcutta, many years before. Raman had impudently criticised Sitalakshmi's looks and lack of formal education, and contrasted her with his own wife, whom he had married after a courtship rather than in an arranged marriage, as Chandra's parents had done.[43] Apart from such thoughtless cruelty, Chandra's family also disliked his arrogance and self-importance.

To complicate matters further, Chandra had fallen in love. Lalitha Doraiswamy was vivacious, outgoing and extremely determined – essential qualities for a woman who had set her sights on entering the male-dominated world of physics.[44] Part of the Indian renaissance was a burgeoning women's movement. It was thanks to this that she received an education and, as she proudly explains, met her husband. Both Chandra and she came from well-known families. Lalitha's aunt, Subbalakshmi Iyer, had become a widow at the very young age of twelve. This was almost invariably a disaster for a Brahmin girl, who would be hidden away, shunted off to do menial work and denied a second chance of marriage and motherhood.[45] Subbalakshmi was lucky. Instead of punishing their daughter, her liberal-minded parents sent her to college. She graduated in 1911, outshining all the men in her class. For a Brahmin widow to win such plaudits made headlines across India. Subbalakshmi went on to found a home for child widows and she lobbied the government to build a college for women. Queen Mary's College stands opposite the Marina, very near Presidency College.

All this greatly affected Lalitha's family. Instead of being married off at an early age, the daughters were encouraged to complete high school and university, then to marry husbands of their choice.[46] To this day, love marriages are rare in Indian culture.

Lalitha was born just four days before Chandra, on 15 October 1910, in Triplicane. Her mother, Savitri Doraiswamy, was as formidable as Savitri's sister Subbalakshmi. When Lalitha was ten, her father, Captain Doraiswamy, died suddenly, probably as a result of trauma suffered in the Great War. Left alone, Savitri continued her education in Hindu tradition and culture and, with the help of her husband's wartime pension, educated her children. She even saved enough money to buy a house in the suburb of Mylapore, just behind Chandra Vilas. A wall separated the two properties.

In the honours course in physics at Presidency College, Lalitha was one year below Chandra. The main lecture room there is steeply banked, with twenty rows of fixed wooden tables with benches. As one of the few women, Lalitha was permitted to sit in the front row. She always sat on the far right-hand side. Chandra shyly followed her around. It was love at first sight – although, as Lalitha remembers, 'Indian society makes it difficult to actually begin conversations'. There is only one entrance, so Lalitha must have seen Chandra every day as he came in and quietly slipped into the seat behind hers.

Eventually, Lalitha took the initiative and asked Chandra if she could borrow his laboratory notebooks – 'and that's the way we formally met'. Another chance for them to speak was on school trips. At colloquia they sat as close as possible but did not converse. Lalitha remembers how impressed she was when Chandra demonstrated his deep understanding of physics in lectures he gave to much older and more experienced scientists. He seemed to know everything. Shyly and protectively, he would cycle home along Edward Elliots Road behind Lalitha, and the two would turn off into their respective driveways. Smiling, Lalitha recalls one of Chandra's first approaches, when he gave her a rose to pin on her sari at an annual party.

One of his first gifts to her was an inscribed copy of the 1924 English edition of Sommerfeld's *Atomic Structure and Spectral Lines*, which had been such an inspiration for him. He remembered that he gave it as 'a gift to my girl friend, who later became my wife'.[47] They were soulmates, sharing a passionate interest in science. For a long time nobody suspected there was anything between them. Just before he was due to leave on his momentous voyage to England, Chandra called on Lalitha at home to give her a list of books she had asked for. Suddenly they found themselves alone and unchaperoned. Speechless, they stood and stared at each other. The spell was broken by a family member bringing tea.

By the time he left India, Chandra had read widely and deeply in mathematics and physics. He had developed a sophisticated scientific taste and method of learning. The day before he boarded ship for Europe, he wrote a letter to his brother Balakrishnan from Bombay. Balakrishnan had asked him what he ought to read and whether he should take up a career in physics. Chandra advised him to begin with the 1930 edition of Sommerfeld's *Atomic Structure and Spectral Lines*. 'Of course you may not fully understand it on first reading,' he wrote. 'Never mind that. Read say the 1st chapter through – simply as if it were prose. Along with the book you can take E. N. C. Andrade's *Structure of the Atom*.' He also recommended Arthur Hass's *Wave Mechanics and Quantum Theory*, 'the one which is at present at home', and Richard C. Tolman's *Statistical Mechanics*. 'But before all this,' he continued, 'you will have to know calculus very well.' For this he suggested Horace Lamb's *Infinitesimal Calculus* '(it is at home)' and also advised Balakrishnan to 'work through Daniel Alexander Murray's *Elementary Course on Differential Equations*'.[48] By 'working through' a text, Chandra meant not only reading it but doing all the practice problems, a challenging task indeed. Balakrishnan replied that 'he will begin with the physics books' and mentioned the subject no more. The mathematics books were dauntingly detailed, while the physics texts took the reader into the strange new world of the atom. For Chandra to be so well acquainted with these very difficult texts at such a young age was quite extraordinary.

Chandra spent his last week in India in Bombay, taking care of last-minute details and giving a lecture at the Royal Institute of Science. Then came the tearful farewells at the dockside. Chandra's beloved mother was too ill to make the trip to see him off. His Uncle Raman wrote that great things were expected of him: 'We are all looking forward to your doing for Indian physics what Ramanujan did for Indian mathematics.'[49] This was high praise, especially from his famous uncle; the recently deceased Ramanujan had been one of India's greatest mathematicians. Everyone stayed on board until the last minute. No doubt Chandra breathed a sigh of relief when, in the late afternoon of 31 July 1930, the *Lloyd Triestino* slipped out of her moorings into the Arabian Sea. Bursting with youthful optimism, he was eagerly looking forward to England, to studying with Fowler and spending time in the company of Eddington. It would be an incredible adventure.

3

Rival Giants of Astrophysics

Arthur Stanley Eddington once wrote, in reference to himself, that 'Human personalities are not measurable by symbols any more than you can extract the square root of a sonnet.'[1] Eddington's portraits reveal little of the man. His stiff pose, impenetrable gaze, domed forehead, long nose and unsmiling lips are reminiscent of the prism-like expression of a previous Cambridge man, Sir Isaac Newton. Chandra perceived Eddington as the quintessential product of an Edwardian England in which everyone knew their place in the grand scheme of things and was confident of their right to the life of privilege they enjoyed:

> He was a man who was very distinguished, in the sense that one felt when one talked to him that one was talking to someone really substantial. The British, particularly in earlier times, can be very nice and kind, but at the same time, an element in their behaviour makes it very clear that [they feel] they're on a different level. There's no snobbery involved in it. It sort of comes naturally to them. Eddington was that sort of man.[2]

We know little of Eddington's private life. He usually destroyed his correspondence as soon as the business at hand was completed. In the autumn of 1944, as he lay dying, he began to dispose of his personal letters. His spinster sister Winifred, four years older and with whom he had lived for most of his life, destroyed much of what remained. In his will he bequeathed to the Royal Astronomical Society a tallboy – a large standing cabinet – filled with papers. The venerable Colonel 'Chubby' Stratton examined them, pronounced them 'merely of biographical interest' and threw them away.[3] The only biography of Eddington is more like a hagiography. It was written by a former student of his, Allie Vibert Douglas, and contains no personal information of any sort. Winifred commissioned the biography, from which we can deduce that there was no personal correspondence left.

Eddington was a study in contrasts. Students recall him as so painfully shy and arrogant that he was almost unapproachable, and as a man who expected great reverence. As a teacher he was appallingly dull, in the worst Cambridge style. Yet among his peers, particularly when dining at high table, he could be the life of the party. In a notebook full of random thoughts and memories, Chandra listed twenty-five stories he had heard Eddington tell at high table.

One reveals Eddington's distinctly offbeat sense of humour. At a gathering one night, a solemn discussion was under way on major technological discoveries. 'Eddington sat back,' recalls Chandra,

> puffed on his trademark pipe, and said that 'as far as he was concerned, zips were the only genuine discovery'. [He then] commented that the use of zips in women's dresses might lead to awkward situations. For example at an evening party one may inadvertently unzip a woman's evening dress and the dress may fall to the ground to every one's embarrassment . . . Eddington was amused at the spectacle he imagined but the staid prudes next to him did not even smile.[4]

Eddington was a highly cultivated man. The books he wrote, both technical and popular, bristle with quotations from French, German and Italian literature, cited in the original languages. He was an expert chess player and a whiz at *The Times* crossword puzzle, which he could complete in the time it took to move his pen across the grid. Despite his dull classroom lecturing style, he was a highly regarded public speaker.

The American physicist W. H. Williams, with whom Eddington shared an office in Berkeley, California, for a couple of months in 1924, wrote that 'he suffered from British uncommunicativeness in most severe form'. Eventually, however, the two discovered a mutual love of golf. Thereafter they played together twice a week at the Claremont Club, 'perhaps the worst golf ever seen on that course'. He also remembered that Eddington was an admirer of *Alice in Wonderland* and a composer of entertaining doggerel.[5]

The eminent American astronomer Harlow Shapley recalled that when Eddington visited Harvard University for its tercentenary celebrations in 1936, he 'divided his interests between galaxies and the Red Sox', the legendary Boston baseball team who were in town at the time.[6] He dominated the world of astrophysics. Shapley told Chandra that he had sent a circular to the leading American astronomers that same year, asking them to rank their colleagues to decide who should be awarded an honorary degree. 'Eddington was the first in every single list received!', he told him.[7]

Eddington was indisputably the greatest astrophysicist in the world and had developed the field of astrophysics virtually single-handedly. All other scientists were in awe of him. He wielded enormous power, not so much because of his position, but from the sheer brilliance of his work and from his network of highly influential colleagues around the world. He was an intimidating figure who cast a spell (as Milne was later to write of him) purely through force of intellect. 'Eddington had the gift, characteristic of genius, of being able to reach correct conclusions by arguments at first sight dubious', wrote Thomas Cowling, a student of Milne who many times witnessed Eddington in action at the RAS.[8] Questioning any of his work was a risky business which involved crossing not only Eddington himself, but devoted colleagues such as the powerful American astrophysicist Henry Norris Russell, who wholeheartedly supported Eddington's wildest theories and also controlled appointments in the United States.

Eddington was born on 20 December 1882, in the picturesque

town of Kendal among the hills of the beautiful Lake District.
Contrary to Chandra's assessment of him, his background was far
from privileged. His family were Quakers. His father, the head-
master of a famous old Quaker school, died in a typhoid epidemic
when Eddington was just two years old. His mother, who must
have been a formidable woman, moved to Weston-super-Mare, on
the coast near Bristol, taking her two young children with her.
There, he and Winifred grew up in some poverty. Even as a child,
Eddington was interested in astronomy and used to survey the
night sky through a three-inch telescope which had been lent to
him. He also had a great fascination for large numbers, and a
prodigious memory. One day his mother went out for the afternoon, leaving him alone. When she returned she asked how he
had spent his time. 'I've been counting all the words in the Bible,'
the precocious child replied. He actually succeeded in getting
through the Book of Genesis.[9] At night he counted the stars in the
sky.

Whereas Chandra came from an affluent background,
Eddington had to apply for every available scholarship. He went to
a small school in Weston-super-Mare, but as a day pupil, not a
boarder, to save money. He was popular at school. Besides mathematics, his other great love was English literature. He was fond of
recounting that he had once won a contest for the best 'jabber-
wocky' poem – 'Stand by the hedge and sound like a turnip' was
his impenetrable winning phrase, according to Chandra's recollection of the story.[10] He also loved cycling. A brilliant youth, he
won a scholarship to Owens College (shortly to become
Manchester University) when he was not yet sixteen, officially too
young to enter. A series of scholarships enabled him to study
physics and also some mathematics, mechanics, English history
and Latin there. He graduated with first-class honours in 1902, at
the age of nineteen. Another scholarship took him to Trinity
College, Cambridge, where he studied mathematics. In his second
year he came first in the fiercely competitive mathematics examinations and was awarded the rank of Senior Wrangler, effectively
guaranteeing him a research career. He was the first person ever to
achieve the rank in such a short time.[11]

By 1906 he had decided to focus on his childhood love –

astronomy. That year the Astronomer Royal offered him the position of Chief Assistant at the Royal Observatory in Greenwich, where he quickly realised how extraordinarily little was known about the stars. Astronomers and astrophysicists put forward a ragbag of confused hypotheses but seemed to have little clear idea about what stars were, how they moved and their life cycle. Here was a challenge – to bring order to and advance the field of astrophysics. He began by looking into the movement of stars. He won a Fellowship at Trinity College in 1907 and produced such distinguished work that in 1913, when George Darwin, one of Charles Darwin's sons, died, he was offered the prestigious position left vacant, that of Plumian Professor of Astronomy and Experimental Philosophy at Cambridge. The following year he was appointed director of the Cambridge Observatory. The post provided him with a comfortable home in the observatory house, and he brought his mother and sister to live there with him. The Great War broke out that same year.

Eddington lived in the sort of world for which the term 'dreaming spires' was coined. Oxbridge dons formed an almost monkish community. They lived in college for their entire adult lives and devoted themselves single-mindedly to their studies. G. H. Hardy and John Littlewood, Eddington's colleagues at Trinity, were of that sort. The most famous team in twentieth-century mathematics, they produced superb papers on number theory. One would dash off a letter to the other with the words 'just had an idea', and shortly afterwards a paper would be in the post.[12] Such men played their cards close to their chest. In middle age, Littlewood was often seen in the company of an attractive, much younger woman. He always introduced her as his niece. There were a few raised eyebrows, but in proper Cambridge tradition no one uttered a word. In his eighties, in the Combination Room (the staff common room at Trinity), Littlewood finally let out the news that she was his daughter, from a long and secretive liaison with a married woman. 'Next day he was depressed that nobody had blinked an eyelid!'[13] Aggrieved though he may have been, such sang-froid was totally in keeping with the Cambridge ethos.

In Eddington's day, the dons who lived at college spent most of their time in the company of other dons. Still in force was the late

Victorian ideal of 'romantic friendship within exclusively male communities, a phenomenon so normal and respected throughout the period . . . that the greatest care must be taken to avoid slick and dismissive judgements'.[14] Decades were to pass before a man could safely confess that his relationship with another man was anything more than friendship.

G. H. Hardy embodied the ethos. With his high cheekbones, ice-clear eyes, narrow nose, austere demeanour and look of perpetual youth, colleagues routinely described him as the quintessential 'beautiful' man.[15] He was a member of the Cambridge Conversazione Society, an elite secret society better known as the Apostles, because when it was established in 1820 there were twelve members. Its luminaries had included Alfred, Lord Tennyson, Bertrand Russell and James Clerk Maxwell, some of the most brilliant minds that Cambridge ever produced. By Hardy's time the society had such a strongly homosexual atmosphere that one member wryly observed, 'even the womanisers pretend to be sods, lest they shouldn't be thought respectable'.[16] In the late 1890s its members included the nucleus of what would become the Bloomsbury literary circle, as well as the economist John Maynard Keynes. In the 1930s it was notorious as a breeding ground for Communist agents, many of whom were seduced and then recruited by Guy Burgess. Homosexuality was taken for granted among its members. Littlewood, speaking at a time when homosexuality was not only taboo but illegal, referred to Hardy as 'a non-practising homosexual'.[17] The pioneer of digital computing, Alan Turing, was more straightforward. He spoke of him as 'just another English intellectual homosexual atheist'.[18]

And what of Eddington? Women found him charming. Unlike most of his Cambridge colleagues, he went out of his way to be courteous to them. One of the early female students of science, a young woman named Cecilia Payne, recalled how rude the eminent physicist Ernest Rutherford was to her. He often began his lectures, in which she was usually the sole woman in the audience, by looking her straight in the eye and uttering the word, 'Ladies', followed by a long pause, '. . . and gentlemen.' Rutherford's daughter, Eileen, a friend of hers, told Payne that Rutherford, prickling with macho arrogance, had said to her, 'She isn't interested in you, my dear. She's just interested in me.'[19] Payne was so offended that she gave up physics and switched to astronomy, a field that had attracted her ever since hearing Eddington lecture. Eddington immediately set her a problem on the structure of stars and was very helpful to her.

Many years later, she told Chandra that 'My interest in astronomy originated with a lecture by Eddington on his results verifying relativity. At that time I was doing the natural science Tripos, but because of Eddington I changed to the mathematics Tripos in order that I could become Eddington's student . . . I do not mind telling you that, in fact, I fell in love with him.'[20]

But although women found Eddington attractive, there is no evidence that he found them so. His biographer, Allie Vibert Douglas, noted that Eddington never wanted to get married. Apart from his mother and sister, 'his interest in women was simply and solely as acquaintances or, in the case of the very few women astronomers in various countries, as friendly colleagues'.[21] There was, she records, only one person with whom Eddington 'formed a lasting friendship [and with whom he] could throw off all the hesitant diffidence which formed an almost impenetrable barrier to intimacy with others.'[22] This was a man whom Douglas refers to as Eddington's constant companion and 'most intimate friend', Charles Trimble.

Trimble's acquaintances from Trinity College recalled that Eddington was the 'strongest, certainly a most rewarding influence on his life'.[23] Both came from working-class backgrounds and met at Trinity, where working-class students were thin on the ground; perhaps that was part of the initial attraction between them. Born in Bath in 1883, Trimble studied at Christ's Hospital, a private school which offered an excellent education to boys from impoverished backgrounds. At Trinity College, he graduated fourth in his class (Eddington was first). While Eddington went in for research, Trimble entered the civil service. In 1910 he took up a post as a mathematics tutor at his old school, Christ's Hospital. But he was plagued by mental problems and suffered several breakdowns.

Former students of Trimble's remembered that he tried to instil in them a love of literature as well as mathematics. A photograph of him taken in 1923, when he and Eddington were at the height of their friendship, shows a pensive man with gentle features and immaculately combed dark hair. By that time Eddington, in contrast, had acquired the supercilious, worldly look, his hair already greying, of a man who lived the rough-and-tumble of international science.

Eddington usually cycled alone, but when he did not, he made a note of his company. In three separate entries in his cycling diary he notes that he made trips of several days 'with Trimble'.[24] The two went on regular walking trips together and enjoyed glissading down icy and snowy slopes. On these jaunts Eddington abandoned his Quaker sobriety and cheerfully indulged in alcohol, tobacco and visits to the theatre and the cinema. Trimble recalled one glissading episode where the two men tumbled through the snow and ended up with their clothes torn and wet. Eddington refused to take his shirt off, and insisted that Trimble take a needle and thread and mend it right there while he was wearing it.[25]

It could well be that the two were more than simply friends. But in those days it was downright dangerous even to be suspected of homosexuality. Everyone was aware of the penalties of getting caught. The Oscar Wilde affair of 1895 still cast its shadow. If Eddington was a practising homosexual he would have had to have been extremely careful in his liaisons. Had word got out, he would have been ruined. Robert Kanigel, Ramanujan's biographer, wrote of the fragile balance that Hardy maintained in his life: 'He was a friend of many in Cambridge, an intimate of few.'[26] In Eddington's case, he had just two intimates, Trimble and Winifred. If indeed he was homosexual, the strain of maintaining secrecy must have been horrendous.

In his working mode, Eddington was a loner. Few students attended his classroom lectures, which were usually stilted and dull. They were more likely to glimpse the real Eddington in his public lectures, which were stylish as well as highly informative. Entering his study, visitors were struck by the quantities of papers and books all over the floor, littering the sofa and tables, and covering his desk. He sat right in front of french windows which opened onto the Cambridge Observatory's beautifully manicured lawn. If students interrupted him in mid-thought, their reception was usually off-putting, verging on the comical. Several recalled walking into his study for an appointment, at the precise time he had arranged to see them. He would look up in surprise, usher them in and politely listen to them, all the time with a quizzical look on his face, as if he was wondering who on earth this person was and why they were there, meanwhile chewing up apple after apple. Once the Norwegian astrophysicist Svein Rosseland dropped by to see Eddington for an arranged appointment. At the time, Eddington was obsessed with the idea that he could deduce the existence of elementary particles from mathematics alone. Rosseland knocked, again and again. There was no response. Then Eddington shouted, 'Oh, it's you. Come in. I have just found the neutron.'[27]

When Eddington took up his position at the Royal Observatory in Greenwich in 1906 and began his research into the stars, astrophysics was in its infancy. Astrophysicists assume that stars – the Sun, for example – are essentially balls of gas, made up of atoms.[28] But very little was known about their structure until the Danish physicist Niels Bohr proposed a theory of the atom in 1913. He depicted it as a minuscule solar system with a positively charged nucleus in the middle, representing the Sun, surrounded by enough electrons to give it a net charge of zero. The simplest and lightest atom is hydrogen, which has a nucleus and one electron circling it like a planet.[29] Helium, the next lightest, has two electrons.

Inside stars the temperature is unimaginably high, in the millions of degrees kelvin.† The intense heat energises the electrons orbiting the nuclei of the atoms so much that they tear themselves free. As a result the inside of a star is made up of a huge number of swirling electrons, and heavier nuclei lumbering along more slowly. In the 1920s astrophysicists investigating a star's structure focused on the electrons there. At the time, no one understood

† In science, temperatures are measured in kelvin, a unit of temperature named after the nineteenth-century British scientist William Thomson, Lord Kelvin. Kelvin temperatures can be converted into Celsius by subtracting 273. The internal temperatures of stars are in the millions of degrees kelvin, and at that level the difference between the two scales is negligible. In this book all temperatures are in kelvin.

much at all about the nuclei – although Eddington, for one, was already beginning to speculate about whether they might be the fuel that a star burns to make it shine.[30]

To study the structure of a star, astrophysicists assumed that a star could be regarded as an idealised gas known as a 'perfect gas'. The concept of the perfect gas was first developed in the nineteenth century, when scientists discovered that there is a simple relationship between the pressure, volume and temperature of a very thin gas of about the same density as the air we breathe. No matter what chemicals it is made up of, if the temperature remains the same and the pressure on it increases, it will decrease in volume. If it is kept at a constant pressure, in a pressure cooker for example, then when the temperature increases the gas will increase in volume – in other words, it will expand as it gets hotter. The mathematical relationship between the pressure, volume and temperature of a gas is the 'perfect gas law'.[31]

Through his research, Eddington came across Henry Norris Russell. In his thirties, Russell was a professor at Princeton University. As an 'Old Princetonian' he sported high-laced boots, starched collars and expensively tailored suits. He even cultivated a British accent, acquired during a brief period at Cambridge. A prudish man with little sense of humour, he loathed the oft-repeated quip that he was a world authority on Eros, the asteroid he happened to have studied for his Ph.D. thesis. He wielded a huge amount of power on the American scene. Up-and-coming astrophysicists feared running up against him. Many remember him as selfish, overbearing and opinionated. Russell bore a lifelong grudge against Eddington's colleague and arch-rival, the astrophysicist James Jeans, because Jeans had been recruited to Princeton in a more senior position and with a much higher salary. But despite his personal failings and eccentricities, Russell was a superb astrophysicist. His students populated nearly every major astronomical institution.

Russell was looking into the most fundamental problem of all – the life cycle of stars, how they are born, how they live and how they die. One of Eddington's first projects was to take up one aspect of Russell's work by investigating Cepheid variable stars.

The brightness of Cepheids fluctuates, from maximum to minimum then back to maximum; the period of fluctuation can range from a few hours to many days. By 1908, over 1700 of them had been found, and a definite relationship had been established between a Cepheid's period of fluctuation and its luminosity.[32] Eddington was able to deduce a relationship between its fluctuation and its density and was pleased to find that his result, obtained mathematically, more or less fitted astronomers' observations.[33]

He then began to wonder how stars became Cepheids and what happened to them once they had exhausted their energy, ceased to shine and stopped pulsating. He found his first clue in a lecture Russell gave at the Royal Astronomical Society in 1913, the year Eddington became Plumian Professor. That day, Russell was the last speaker in a long day of presentations. Everyone was dozing off by the time he presented his last slide. Eddington snapped to attention.

In the nineteenth century, scientists discovered that the universe is crisscrossed by different sorts of light, some visible to the human eye, others not.[34] At one end of the radiation spectrum (see Figure 2 in the astronomical picture section) are radio waves, with very long wavelengths – several metres long – and very short frequency. At the other is very-short-wavelength radiation – ultraviolet light, X-rays and gamma rays, the shortest of all. The only part of the spectrum visible to the eye is a small 'window' between ultraviolet and infrared. From the Earth, the universe looks quiescent and unchanging. But telescopes mounted on orbiting satellites high above the Earth's atmospheric blanket use instruments sensitive to parts of the spectrum invisible to our eyes. They reveal a startlingly violent universe, with gamma rays exploding and X-rays shooting out.

Every atom in a star emits light of a particular frequency, like the sound emitted when you strike a tuning fork. Starlight is a hodgepodge of frequencies, made up of all the light emitted by a huge number of atoms of many different elements. If millions of tuning forks of different frequencies were all struck at once, it would produce a continuum of sound in which it would be impossible to pick out any individual frequency. In the same way, starlight appears to be a continuous spectrum of light.

Astrophysicists unravel the different frequencies using an instrument called a spectroscope set in the eyepiece of a telescope. A single beam of starlight enters, and a fan of light of different frequencies emerges. When astrophysicists shine this spectrum onto a photographic plate, each frequency appears as a separate 'spectral line'. Spectral lines are like fingerprints; each chemical element has its own unique spectral lines, like DNA.

In the late nineteenth century, astronomers at the Harvard Observatory, backed by an army of underpaid female workers, engaged in the epic task of classifying the light from 500,000 stars. This involved teasing out similarities in the characteristics of light from each star as displayed on stacks of glass photographic plates. They then divided the stars into a number of spectral classes by grouping together those with similar spectral lines in their light. They identified the spectral classes with the letters O, B, A, F, G, K and M. The brightest stars were O and the dimmest M; O-type stars were the hottest and M-types the coolest. Stars in spectral class O had a surface temperature of about 50,000 degrees kelvin and those in M about 3000 degrees. These temperatures were worked out by measuring the wavelengths of the light emitted by the star.[35]

Russell's brilliant idea was to take certain of these stars and plot them on a graph, with brightness along one axis and temperature along the other. The result was a fascinating visual representation of the heavens.[36] By chance, a Danish photochemist-turned-astronomer, Ejnar Hertzsprung, had been doing similar work, so the diagram became known as the Hertzsprung–Russell (HR) diagram (see Figure 1(a) in the astronomical picture section). This was what grabbed Eddington's attention during Russell's 1913 talk at the Royal Astronomical Society meeting.

That day, Russell presented an HR diagram on which were plotted 300 stars whose distances from the Earth were accurately known at the time. To his own amazement, he found that they were not randomly scattered. Most were in a diagonal strip, running from hot, bright O-type stars at the upper left to cool, dimmer M-type stars at the lower right – stars about the same size as the Sun. He called this strip the 'main sequence'. The other large grouping of stars lay on a horizontal line jutting out from the main sequence at the top of the diagram, all very big and more or less of the same brightness. Following a suggestion of Hertzsprung's, he classified stars on the main sequence as 'dwarfs' and those on the horizontal strip as 'giants'. From this he proposed that there were 'two great classes of stars'– giants (bright stars ten to a hundred times bigger than the Sun) and dwarfs (about the Sun's size and of decreasing brightness).[37]

Some years later, Russell recalled that he invented the HR diagram 'because it represented the phenomena I wanted to describe in such a way that I could get them on a page'.[38] But it turned out to be much more than merely an economical way to present a lot of information at a lecture. It soon became apparent that studying stellar structure and the life and death of stars required astrophysicists to understand why stars populated certain regions of the HR diagram and not others, and how they 'trek' from one part of it to another in the course of their evolution. Eddington was one of the first to appreciate the diagram's importance and to try to fathom its cosmic message. His research was at the heart of the debate that would lead to a firm knowledge of the chemical constitution of stars and to an understanding of what makes them shine.

One star presented a particularly tantalising puzzle. It was completely outside Russell's neat layout. This was o^2 Eridani B, the dim companion of o^2 Eridani A.[39] The two stars revolve around each other in an endless dance known as a binary system. o^2 Eridani B did not fit the known pattern: it was neither a giant nor a dwarf. It was quite hot – about 10,000 degrees kelvin – but very dim and so lay well below the main sequence. At first Russell classified it as a dwarf star, while Hertzsprung used the term 'dark white star' to refer to it. It soon became known as a white dwarf. This was the first white dwarf to be identified.

Hertzsprung and Russell assumed that o^2 Eridani B was a freak because it fell so awkwardly outside the scheme of things, and they tried to ignore it. But then Russell remembered a conversation he had had in 1910 with Edward C. Pickering, at the time America's most influential astronomer and Director of the Harvard Observatory. 'It is just such discrepancies which lead to the increase of our knowledge,' Pickering had told him.[40] It was a salutary reminder. Twenty-five years later, Eddington and Chandra were to fight bitterly over the ultimate fate of these puzzling white dwarfs, which by then astrophysicists believed were burnt-out stars. Then, in 1914, an astronomer made some staggering new observations on another white dwarf, Sirius B, the star that was to change Chandra's and Eddington's lives and the course of astrophysics.

Sirius B is the companion of Sirius A, the brightest star in the night sky, twenty-six times as luminous as the Sun. Sirius A had been studied extensively since 1844, when the German astronomer and mathematician Friedrich Wilhelm Bessel worked out that it was 54 trillion miles from Earth. To astronomers of the mid-nineteenth century, this was unimaginably huge. The Earth is a mere 7926 miles across, while the Sun is 93 million miles from us and a billion miles from Pluto, the most distant major planet of our solar system. Sirius A is many trillions of miles farther away, yet still blazes so brightly that it all but obscures its companion star. The fact that it can be observed with a telescope shows how extraordinarily bright it is.

Nineteenth-century astronomers began to suspect that Sirius A had a companion star when they noticed that it wobbled instead of following a smooth path. In 1862 an amateur astronomer and lens-maker called Alvan Clark spotted it for the first time. Astronomers worked out that the next time the two stars would be far enough apart to make successful observations of Sirius B would be in another fifty years. In 1914, the year after Eddington encountered the HR diagram, Walter S. Adams, America's leading observational astronomer, was ready. Working at Mount Wilson Observatory in California, he measured Sirius B's brightness and spectral class and discovered that it too was in spectral class A, like o^2 Eridani B, and slightly cooler, at about 8000 degrees kelvin. o^2 Eridani B was not a freak after all;[41] Sirius B was equally strange. How could these stars be so hot yet radiate so little light? Surely there was some mistake. Or, as Pickering had cautioned, maybe not.

From the time it took Sirius A and Sirius B to orbit each other, the degree of irregularity of the orbit and the distance of the two from the Earth, astronomers were able to calculate the mass of

Sirius B.[42] It turned out to be about the same as the Sun, some 2 billion trillion trillion grams or about 10^{33} grams.[43]

They also calculated the radius to be some 11,280 miles, making it about thirty times smaller than the Sun and a little more than three times the size of the Earth. In other words, Sirius B packed the whole mass of the Sun into a volume not a great deal larger than the Earth's. That made its average density a stunning 61,000 grams per cubic centimetre. Transferred to the Earth, a single teaspoonful of this white dwarf matter would weigh nearly six tons, as much as a good-sized elephant. White dwarfs are so extremely dense that astrophysicists suspected that they would have to find some new version of the perfect gas law to describe them, though hopefully not drastically new. Eddington's comment on all this as it stood in 1914 was 'absurd'.[44] He decided to set aside the issue of white dwarfs and turn his mind to the physical structure of dwarfs and giants. What puzzled him was why stars should cluster in certain areas of the HR diagram and not others.

This bore on one of the most important questions that had yet to be answered: what makes stars shine? Eddington was certain that the theory formulated late in the nineteenth century by the German polymath Hermann von Helmholtz and the British scientist William Thomson (Lord Kelvin, of the temperature scale) was completely wrong. According to this theory, stars shone because the particles in the gas that made up the star were being squeezed together by gravity, which increased their temperature until they emitted radiation. But if this theory was right, it gave for the Sun an age of just 20 million years.[45] In 1917, radioactive dating estimated that the age of the Earth was around 2 billion years. How could the Earth possibly be older than the Sun?

Eddington proposed an alternative theory. Perhaps there was a slow process in which protons and electrons annihilated each other, thus creating light. There was such an immense number of electrons in a star that the energy supply they provided would be 'almost inexhaustible'.[46] The problem was that if the electrons and protons carried on destroying each other, the star itself would eventually disappear. Added to which, one would expect very dense stars, where the protons and electrons were packed together and jostling furiously, to be extremely bright. But in that case, why were very dense stars like white dwarfs so dim? Were there unknown processes that somehow affected the ongoing annihilation?

Eddington was sure that it had to be the nuclei of the atoms in a star that produced its energy, as opposed to the electrons, and that there had to be vast reserves of energy, enough to keep a star shining for billions of years. Then, in 1920, Francis Aston, a forty-three-year-old chemist at the Cavendish Laboratory in Cambridge, made a startling discovery. When he measured the mass of the nucleus of the helium atom (also known as an alpha particle) and compared it with the sum of the masses of the particles that made it up, the two figures were not the same. The mass of the helium nucleus was smaller. The missing mass was minuscule, just eight-tenths of 1 per cent.[47]

Aston reasoned that if a helium nucleus was forged from its separate particles, the vanishing mass would reappear as a huge quantity of energy, as predicted in Einstein's famous equation, $E = mc^2$. In this equation E is energy, m is mass and c is the velocity of light (i.e. 186,000 miles per second). c^2 – the velocity of light multiplied by itself – is a vast number. Thus a minuscule amount of mass would translate into an enormous amount of energy. Aston described his conclusion vividly: 'To change the hydrogen in a glass of water into helium would release enough energy to drive the "Queen Mary" across the Atlantic at full speed.'[48]

Eddington immediately realised that this could explain why stars continued to burn for so long. Perhaps what made them shine was the fusion of hydrogen, forming helium, by processes implied in Aston's measurements. 'What is possible in the Cavendish Laboratory may not be too difficult in the Sun,'[49] he pointed out. It was an inspired hypothesis, but it could be no more than that, for at the time scientists knew very little about the nucleus. But they did know that for fusion to happen, the temperature inside stars would have to be extraordinarily high – high enough to give protons sufficient energy to overcome their immense mutual electrical repulsion and slam into each other with enough force to stick together. Some astrophysicists argued that the temperature inside stars might not be hot enough. To this Eddington replied fiercely, 'We tell them to go and find a *hotter* place.'[50]

Eddington had by now embarked on an ambitious quest to work out the basic pattern at the root of all stars. He called it his 'standard model', and it embodied his developing ideas on the constitution of the stars. As he put it, 'It is as though nature had a standard model before her in forming the stars, and (except for occasional lapses of vigilance) would not tolerate much deviation.'[51]

His aim was to describe what went on in the interior of a star entirely through mathematical means. In 1917, when he first put forward his standard model, it applied only to giant stars, which have a density low enough for them to be studied by assuming that they obeyed the perfect gas law. The temperature inside a star is upwards of 10 million degrees kelvin, which means that the radiation there is in the form of very-high-energy, high-frequency X-rays. As the X-rays collide with the atoms in the star, they strip the electrons from the atoms, beginning with those in the outer rings which are less tightly bound than the interior ones. Knocking out an electron depletes the amount and strength of the light inside the star. The detached ('free') electrons swirl around until they are captured by another atom, only to be knocked out again, further depleting the star's store of light. Eddington described it all thus:

> The inside of a star is a hurly-burly of atoms, electrons and [radiation] . . . Try to picture the tumult! Dishevelled atoms tear along at 50 miles a second with only a few tatters left of their elaborate cloaks of electrons torn from them in the scrimmage. The lost electrons are speeding a hundred times faster to find new resting places. Look out! there is nearly a collision as an electron approaches an atomic nucleus; but putting on speed it sweeps round it in a sharp curve . . . Then comes a worse slip than usual; the electron is fairly caught and attached to the atom, and its career of freedom is at an end. But only for an instant. Barely has the atom arranged the new scalp on its girdle when a quantum of [light] runs

into it. With a great explosion the electron is off again for further adventure.[52]

In order to explore a star's temperature and the amount of radiation it emits, Eddington had to specify the average number of free electrons per nucleus, which astrophysicists call the 'mean molecular weight'. Astronomers made the assumption – perfectly reasonable at the time – that stars are made up of the same elements as the Earth is: that they contain almost no hydrogen or helium but are composed mainly of oxygen, iron, sodium, silicon, potassium, magnesium, aluminium and calcium.[53] Taking all this into account, plus the fact that not all atoms might be stripped of their electrons, Eddington adopted a mean molecular weight of 2.1. He proposed that the mixture of chemical elements was the same in all stars and that the mean molecular weight therefore applied in all cases.

Another issue that Eddington had to consider in modelling a star was its opacity – the extent to which its chemical composition prevents radiation from escaping from inside it. If a star is not opaque enough, too much radiation will escape and it will cool down too quickly. But if it is too opaque then radiation cannot easily escape and it may explode prematurely. Eddington began by taking the mathematical formula for opacity used in the laboratory, which he called the 'physical value'. He then altered it to fit the conditions inside a star, calling this version the 'astronomical value'. Taking a mean molecular weight of 2.1, he calculated a figure for the 'astronomical value' that he assumed was valid for every star.[54]

From this he made the important discovery of how to relate the mass of a star to its brightness – the more massive a giant star, the brighter it is. He called this the mass–luminosity relationship. Amazingly, it turned out to be valid not only for low-density giant stars, which could be studied using the perfect gas law, but also for dwarf stars, even though they were much denser. This meant that if astronomers could measure the luminosity of a star, they could work out its mass. It also provided firm theoretical backing for observations that more massive stars are brighter than less massive ones, and explained why, for example, Sirius A was so extraordinarily bright. The one sort of star that it was not valid for was the strangely anomalous white dwarfs.

Eddington was amazed at his own result. 'In making the comparisons for these dense stars [i.e. dwarfs – though not white dwarfs], the writer had no anticipation that they would agree to the curve [i.e. to his theory],' he wrote.[55] It completely disproved Russell's theory that dwarf stars are too dense to obey the perfect gas law.

But how was this possible? Eddington's inspired guess was that since the atoms inside stars are almost completely stripped of their electrons, they are some billion times smaller in volume than terrestrial ones and occupy much less space. This was why the matter in the interior of a star would behave like a perfect gas even if it was very dense. In ordinary dwarf stars, matter may be as dense as platinum yet still behave like a perfect gas – because even platinum is still a long way from the maximum density possible.

Eddington's mass–luminosity equation agreed perfectly with astronomers' observations. But the fit was achieved at high cost. In order to make his calculations work out, Eddington had to assume that the astronomical value for opacity was ten times the physical value. This became known as the 'opacity discrepancy'.[56]

It was a fundamental flaw in Eddington's standard model, as he readily admitted. The only way to make the astronomical and physical values match was to postulate a huge amount of hydrogen inside stars.[57] Comparing the mathematical equations for the physical and astronomical values made this crystal clear. But to do this would make the standard model untenable because, according to his mathematics, to increase the amount of hydrogen would decrease a star's radiation pressure, which was a key ingredient of the model.

As it happened, the Indian physicist Meghnad Saha had already made a discovery that threatened Eddington's standard model. In 1920 he had found a way to link the temperature at the surface of a star with the chemical elements that made up its surface layers. From this he deduced that there was a million times more hydrogen than any other chemical element in the Sun's atmosphere. Could this mean there might be hydrogen in the interior of stars, as well as on the surface? And if so, why should there be so much hydrogen in stars when there was so little on Earth? Eddington fervently hoped that further refinements to Saha's theory would explain away his finding, but none ever did.

Then, in 1925, Cecilia Payne, Eddington's former student who was by then a twenty-five year old Ph.D. student at the Harvard Observatory, used an improved version of Saha's theory to re-examine spectra from the Sun and proved conclusively that there really was a huge amount of hydrogen in the Sun's atmosphere. The irascible Russell, an ardent supporter of Eddington and in awe of him, wrote to Payne that this was utterly 'impossible'.[58] Fearing his wrath, Payne was forced to follow his line and toned down her remarks, declaring that the quantity of hydrogen 'is probably not real'.[59] When she visited Cambridge in September 1925 and told Eddington her result, he replied curtly, 'Well, that is on the stars, but you don't know that it is in the stars.'[60]

Eventually independent estimates forced Russell to accept Payne's conclusion.[61] By 1932, the weight of evidence was over-whelming. The twenty-seven-year-old Danish astrophysicist Bengt Strömgren, who was to become a friend and colleague of Chandra, proposed that at least a third of a star was made up of hydrogen. This brought the astronomical and physical values for opacity into agreement, thus removing the opacity discrepancy.[62] Eddington gave in.[63]

Alongside his work on stars, Eddington was also doing pio-neering work on relativity theory. Einstein had proposed his general theory of relativity in 1915, just after Eddington became Plumian Professor and Director of the Cambridge Observatory. But the world was at war, and Einstein's revolutionary theory did not immediately cross the Channel. Through his astronomical studies, Eddington came into contact with the distinguished Dutch astronomer Willem de Sitter, professor of astronomy at Leiden University in neutral Holland, who was examining the astronomical implications of Einstein's theory. In 1917 de Sitter sent Eddington a copy of Einstein's latest paper on general rela-tivity. It was almost certainly the first copy to reach Britain. Eddington immediately saw the significance of Einstein's work. The following year the Physical Society commissioned him to write the first report on it in English. His masterly 'Report on the relativity theory of gravitation' brought the theory to the attention of many British scientists and established Eddington as the leading exponent of relativity in the English language.

Meanwhile, the war intensified. Conscription had been introduced in 1916. As a Quaker, Eddington could have declared himself a conscientious objector, though it would have been considered an unpardonable disgrace to abstain from participating in a war when youthful patriots all over the country were volunteering to go and fight for their king and country. Instead, several major scientists at Cambridge prevailed on the government to exempt him on the basis of his scientific achievements. The Home Office appealed against the exemption. At the first hearing, on 14 June 1918, Eddington declared himself a conscientious objector. He saw it as no shame, he said, to join his Quaker friends in internment camps in Northern Ireland, peeling potatoes. In the end the Astronomer Royal intervened. He proposed that Eddington should be exempted on condition that he lead a British scientific team to observe the total eclipse of the Sun on 29 May 1919, to verify Einstein's general theory of relativity. The board granted him a year's reprieve to do so.

Soon afterwards, the Armistice was signed. But Eddington prepared for the journey nonetheless. The aim was to test a prediction of the general theory of relativity: that light would bend near a massive star like the Sun. At that time observations of stars close to the Sun could only be made during a total eclipse.

It was a great adventure. Early in March 1919, two expeditions sailed from England. They parted company at Lisbon. One went on to Sobral, in Brazil. The other – led by Eddington – set off for the tiny Portuguese island of Principe, off the west coast of Africa. By mid-May, Eddington and his colleagues had their instruments in place and took some test photographs. But on the day of the eclipse, disaster struck. The eclipse was due at two in the afternoon, but that morning there was heavy rain, which threatened to ruin all their plans.

'The rain stopped about noon,' Eddington recorded, 'and about 1.30 . . . we began to get a glimpse of the Sun. We had to carry out our photographs in faith. I did not see the eclipse, being too busy changing plates, except for one glance to make sure it had begun and another half-way through to see how much cloud there was.'[64]

He and his colleagues photographed stars that appeared in the sky close to the edge of the eclipsed Sun. Later they carefully measured the positions of the stars and compared them with their positions when the Sun was in another part of the sky. Eddington took a leading part in the observations and also made the critical micrometer measurements on the best photographs. To his and everyone's delight, the positions of the stars differed more or less by the amount predicted by the general theory of relativity. He never forgot the moment when he measured the first plate which 'gave a result agreeing with Einstein'.[65] In later years he referred to this as the greatest moment of his life. The *New York Times* headlined the results of the expedition in jocular fashion: 'Stars not where they seem or were calculated to be, but nobody need worry', for Einstein knew where they are.[66] The public particularly appreciated the irony that a British scientist had verified a 'German theory' not long after the end of the Great War.

In celebration, Eddington penned a small epic describing the dramatic events on Principe in a parody of Edward Fitzgerald's popular *Rubáiyát of Omar Khayyám*. One verse reads:

> *And this I know; whether Einstein is right*
> *Or all his theories are exploded quite,*
> *One glimpse of stars amid the Darkness caught*
> *Better than hours of toil by Candle-light.*[67]

The expedition made his name. Eddington became hugely famous as a world expert on relativity theory. He much enjoyed regaling high table with the tale of how, some years later, he bumped into a renowned physicist (whom he chose not to name) at the Royal Society. 'Well! Professor Eddington,' the physicist exclaimed in jocular tones, 'you must be one of the three people in the world who understands "Relativity".' Eddington savoured the amusing exchange that followed: 'I said, "Oh! I don't know", to which he said, "Don't be modest, Eddington." To which I replied, "On the contrary! I am wondering who the third person is!"'[68]

There was one topic that brought together all the different

strands of Eddington's work. That was the puzzling matter of those mysterious, hot yet dim white dwarfs.

Astrophysicists surmised that a star is probably formed in the following way. Deep in interstellar space, in regions rich in gas, particles of gas begin to coalesce, drawn together by their mutual gravitational attraction. Some compress more and more and change from a diffuse, spread-out mixture to become more compact and well defined. As the particles jostle about, following the perfect gas law, they exert an outward pressure which partly counterbalances the inward tug of gravity. But gravity is more powerful and presses the gas particles more and more tightly together, making them hotter and hotter until eventually they begin to glow. The light they radiate creates another outward pressure, called radiation pressure.

In everyday life, radiation pressure is minute: the light of a torch or a car's headlight does not exert any pressure that we can feel. A star, however, is so vast and powerful that radiation streams out of it with the force of a hurricane. But what exactly is the role of radiation pressure in creating a star that will not collapse? In 1917 Eddington made a brilliant suggestion. The key, he advanced, was balance. A stable star is formed when the star's gravitational pressure, squeezing inwards, is in perfect balance with the pressure of the gas itself plus the radiation pressure of the light emitted by the particles in the gas. Eddington's hypothesis marked a major step forward in our understanding of the stars and provided the cornerstone of modern astrophysics.

But what happens at the end of a star's life? White dwarfs, Eddington suspected, must be superannuated stars. If he could solve the puzzle of why they were both hot and dim, that would surely cast light on the final destiny of stars. He worked out the probable chain of events. As a star ages it grows dimmer. Its radiation pressure decreases, and the inward crush of gravity, now counterbalanced predominantly by the pressure of the gas itself, squeezes the star to make it smaller. But this led to a paradox. As Eddington pointed out, when enormously dense and compact stars – like white dwarfs – are crushed by gravity, they become denser still, far denser than even the hardest rock on Earth. But in order to cool down, he reasoned, they would need to become less dense, until they were no denser than inert terrestrial rocks. Instead they seemed to be stuck, eternally shrinking and compressing. Although they grew dimmer they could not cool down. But could they carry on in this way for ever? What happened eventually? Could it be that they collapsed to nothing? As Eddington put it:

> The star seemed to have got itself into an awkward fix. Ultimately its store of subatomic energy would give out and the star would then want to cool down. But could it? The enormous density was made possible by the high temperature which shattered the atoms. If it cooled it would presumably revert to terrestrial density. But that meant that the star must expand to say 5000 times its present bulk. But the expansion requires energy – doing work against gravity; and the star appeared to have no more energy available. What on earth was the star to do if it was continually losing heat, but had not enough energy to get cold?[69]

It was a tantalising mystery. In *The Internal Constitution of the Stars* he played down the issue. Nevertheless, white dwarfs seemed an awkward anomaly in the enormous progress that was being made – largely by Eddington – in understanding the structure of the stars.

Another mystery that Eddington wanted to crack was how a white dwarf could possibly be so small yet so dense. In 1914 he had described the incredibly high density of Sirius as 'absurd'. But ten years later, when he developed his mass–luminosity relationship, he discovered that the perfect gas law was valid even at the high densities inside dwarf stars such as the Sun. This led him to wonder whether, at the far higher temperatures inside white dwarfs, every atom might be completely stripped of its electrons in the same way, permitting even more atoms to be packed in.[70] In January 1924, he wrote to his friend, the astronomer Walter S. Adams, who had carried out the key observations of Sirius B in 1914, that he had 'recently been entertaining the wild idea' that the 'incredible' density of Sirius B 'may just be possible'.[71] He went on to suggest an ingenious way to check whether white dwarfs really could be so dense by using Einstein's general theory of relativity to verify the smallness of the radius.

General relativity predicts that the intense gravitational field of a white dwarf will affect the frequency and wavelength of the light it emits. Gravity on the surface of Sirius B is 100,000 times stronger than on the surface of the Earth, so strong that it stretches the wavelengths of the light emitted by the atoms there. To us, seeing them from Earth, they appear longer. Light of longer wavelengths is towards the red end of the spectrum, so the light has a reddish tinge. This is called gravitational redshift.

The difference in the wavelength of light emitted by an atom on Sirius B and of the same light from an atom on the Earth turns out to depend on the mass of Sirius B divided by its radius. Sirius B's mass and radius had already been calculated. (Assuming that Sirius B was 0.85 times the mass of the Sun, astrophysicists had worked out its radius, based on its luminosity and temperature.) From these Eddington calculated its redshift.

He then suggested that Adams make a direct measurement of the redshift of Sirius B and compare it with the figure he had calculated.[72] The measurement was exceptionally delicate. Sirius A was so extremely bright that it strongly impinged on the spectra of Sirius B, making it difficult to determine which spectral lines belonged to Sirius A and which to Sirius B. Adams, however, felt up to the task and produced a result very close to Eddington's figure. Confirming the gravitational redshift also verified his figure for the apparently infeasible smallness of the radius – which in turn confirmed the supposedly absurd density. Astronomers were sure that the mass was right. Eddington thus furnished firm evidence that Sirius B really was extraordinarily dense, as unlikely as it seemed. He for one was convinced by his own figures. His experiment also provided further proof of the general theory of relativity, in that Sirius B's gravitational field really did alter the light the star emitted, just as the theory said it should. Eddington was doubly delighted.

As he was writing *The Internal Constitution of the Stars*, Eddington realised that general relativity revealed something extraordinary about the structure of space around extremely dense objects such as white dwarfs. The supergiant Betelgeuse has a radius of about a hundred million miles, twice the size of the

Earth's orbit around the Sun, but it is a million times less dense than the Sun. What would happen if such a vast object were as dense as the Sun or, even, as a white dwarf? 'The force of gravitation would be so great that light would be unable to escape from it, the rays falling back to the star like a stone to the Earth . . . The mass would produce so much curvature of [space] that space would close up round the star, leaving us outside (i.e. nowhere),' he wrote.[73]

The concept that there might be 'dark stars' with a gravitational attraction so powerful that none of their light can escape was first argued by the peripatetic English natural philosopher John Michell, in 1784, some hundred years after Isaac Newton came up with his theory of gravity. In 1796, the French mathematician and scientist Pierre-Simon de Laplace popularised the idea – though without mentioning Michell. Eddington, however, was not aware of any of this. Colleagues recall him contemptuously dismissing the history of science as irrelevant to working scientists.[74]

According to the perfect gas law, white dwarfs ought to lack the energy necessary to expand and should therefore be unable to die in the way that stars were supposed to, by turning into cold inert rocks. But instead of picking up on his own fanciful prediction – that they might be crushed to a point of infinite density and disappear into some niche in the fabric of the universe – Eddington suggested bypassing this 'curious problem', adding that it was 'not necessarily fatal'.[75] Someone would have to return to it, however, because 'white dwarfs are probably very abundant'.[76]

They are indeed. In 1926, when Eddington was writing, only four were known with certainty. By 1938, about twenty had been found. Today, hundreds have been located in a small observable region close to the Sun, the only place where it is possible to see them because they are so dim. In this region white dwarfs account for about 9 per cent of all observable stars, and those are only the ones accessible with our telescopes. Astronomers suspect that they are the most abundant species of star in the heavens, lending support to the theory that the main way stars die is by becoming a white dwarf.

As early as 1926, Eddington suspected this too. The problem

was that, as he wrote in his preface to *The Internal Constitution of the Stars*, according to the perfect gas law they ought to collapse, a conclusion which was simply unacceptable as far as he was concerned. It was an impossible conundrum, a stumbling block that impeded the enormous progress he was making in reasoning out what went on in the interior of stars and laying the foundations of modern astrophysics. But that same year Eddington's colleague Ralph Fowler had the brilliant idea of applying the intriguing new quantum physics – on which Fowler was an expert – to stars. His aim was to clear up once and for all the knotty paradox that Eddington had uncovered – to prove that stars do not collapse but expire peacefully as inert rocks, thus re-establishing the harmony and beauty of the universe.

4

Stellar Buffoonery

Ralph Fowler was 'a big man, both physically and in the world of science; he was a hard hitter, in every sense of that phrase; he was a man of winning charm, who inspired friendships'; thus Arthur Milne described him.[1] Born in 1889, Fowler came from a privileged background and grew up fully expecting success in life. At Winchester College he was particularly renowned as an athlete. Throughout his life he continued to be a keen sportsman: he played golf, cricket and rugby, all with distinction, and was a formidable rock-climber and bridge player. An affable, easygoing man with a famously hearty laugh and open smile, he was much liked.

Quite apart from his own stellar achievements in the worlds of physics and astrophysics, he was at the heart of a circle of brilliant and influential scientists. It was he who introduced the brilliant and eccentric Paul Dirac, who was then his student, to quantum theory and later put Dirac, Heisenberg and Bohr in touch with one another. In 1921 he married Eileen, the only child of his close friend and colleague, the legendary Nobel-prizewinning physicist, Ernest Rutherford, by then Lord Rutherford, who had discovered the nucleus of the atom. The couple had four children.

Milne described Fowler with affectionate humour as 'the kind of man you can still remain friendly with, even when he has sold you a motor-bike; it is not possible to say more' and added that he was a 'prince amongst men'.[2] After his first meeting with him, in

September 1930, Chandra wrote home excitedly: 'Mr. Fowler – strong (big) healthy, middle-aged man, quite happy, full of the joy of life. He says "Splendid" very often.'[3] At that first meeting, Chandra was so nervous that he 'fell down a whole flight of stairs'. 'I can still hear Fowler's admonishing tone: "steady, steady!"'[4]

After finishing at Winchester, in 1908, Fowler entered Trinity College, Cambridge, where he studied pure mathematics. He happened to specialise in the solution of certain equations that dealt with the properties of gaseous spheres and were thus of great interest to astrophysicists. He was awarded a Fellowship at Trinity in 1914. When the Great War broke out, Fowler was commissioned in the Royal Marine Artillery. He was often to be seen at Cambridge wearing his gown over his military uniform. He took part in the disastrous Gallipoli campaign, where he was mentioned in dispatches and was wounded so severely in the shoulder that he was removed from combat. This turned out to be fortuitous, for as a result he met the eminent physiologist Archibald Vivian Hill, then a Captain in the Cambridgeshire Regiment.

With his sterling leadership qualities, broad scientific background and extensive contacts with key players at Oxbridge, Hill was the kind of scientist to whom governments naturally turn in times of war. He was asked to form an Anti-Aircraft Experimental Section within the Munitions Inventions Department. Its main task was to improve the effectiveness of anti-aircraft gunnery against German zeppelins and later against the powerful twin-engined Gotha bombers. Sighting anti-aircraft gunnery required new specifications. At high altitudes, conditions such as wind speed, temperature and pressure had to be taken into account.[5] Hill set about assembling a group of mathematicians and physicists and turned to Hardy, another Fellow of Trinity, for advice. Hardy responded that, 'although he was ready to have his body shot at he was not prepared to prostitute his brains for purposes of war'.[6] But he had no such compunctions about Fowler's brains and he heartily recommended him, as well as Milne, a student of his, then in his second year at Trinity. That was how Fowler met Milne. The two quickly became firm friends as well as colleagues.

Hill's group was based on HMS *Excellent*, docked at Whale Island, off Portsmouth. Unofficially they called themselves 'Hill's

Brigands', referring to their unorthodox methods in matters both social and scientific. It was a heady time. Fowler and Milne flew out from Farnborough aerodrome in the powerful new F.E. 2D biplanes, powered by 250 h.p. Rolls-Royce engines. They persuaded the Royal Flying Corps pilots to fly as high as possible so that they could check the atmospheric conditions, hanging over the wing to take readings of temperature and pressure. In an open cockpit and with no parachute or radio, this was not only thrillingly exciting but downright dangerous. The highly experienced combat flyers tried to scare their passengers by horsing around, performing daredevil nosedives and stalling turns. Fowler and Milne loved every minute of it.

Milne began with an equation for low-level trajectories of anti-aircraft shells and expanded it to apply to much higher elevations. His result is still in use. Fowler's contributions included the meticulous recording and writing-up of data from aircraft runs and shellburst tests, building good working relationships with naval officers, inspecting coastal AA facilities and making many visits to the lines in France to encourage scientists in the field.

Hill's effect on both men was enormous. Milne always remembered Hill for teaching him how to do research. As a result of their wartime work, both Fowler and Milne changed their research interests from mathematics, first to upper-atmosphere physics and then to astrophysics. As we have seen, in 1916, when conscription was introduced, Eddington was excused on the grounds that, as a distinguished scientist, it was not in England's long-term interests for him to serve in the army and possibly be killed.[7]

Fowler's forte was not so much in thinking up original ideas as in applying someone else's to solve a particular problem; his fatal flaw, however, was that he stopped short, rather than pushing an argument through to its logical conclusion. The prime example is his application of quantum statistics to a gas of electrons in order to solve Eddington's paradox.

A few months after Eddington published *The Internal Constitution of the Stars*, in 1926, Fowler proposed that quantum physics might offer a solution to the apparent paradox that a white dwarf might collapse completely, instead of ending up as an inert rock. Fowler was well aware that when a gas of electrons is incredibly dense, and certainly when it is millions of times denser than any terrestrial matter – in a star, for example – it no longer behaves as a perfect gas. Most astrophysicists assumed that the perfect gas law would have to be modified to take this into account. Fowler, however, realised that the situation called for more sweeping measures and that such a dense gas of electrons would have to be studied using the laws of quantum physics, which rule that when electrons congregate, they try to stay as far apart from one another as possible.

The need for a new law of nature emerged with Bohr's model of the atom as a solar system in which electrons circle the nucleus like planets round the Sun. Crucially, they can only occupy certain orbits.[8] Bohr went on to show how electrons are distributed like beads on what can be visualised as rings. The simplest atom – hydrogen – has one electron and one ring. Helium consists of two electrons which complete this ring, making helium inert, i.e. unable to react with any other chemical element. Then comes lithium, with three electrons threaded on two rings, followed by a succession of chemical elements each with an added electron in the second ring, culminating in neon, in which the second ring is completed with eight electrons; neon is also inert. Bohr's model gave the satisfying result that the number of electrons in an element determines its chemical activity, thus making sense of the periodic table of elements. Einstein declared that it was 'an enormous achievement'.[9]

But Bohr could not explain why there should be a specific number of electrons in each ring. Then, in 1925 the flamboyant twenty-four-year old Viennese physicist Wolfgang Pauli, a former student of Sommerfeld, came up with the answer. The missing ingredient was another mysterious ingredient of the quantum world: electron 'spin'.

Spin is a quantum property of an electron and quite impossible to imagine in terms of everyday experience, in the same way that it is impossible to imagine an electron as a wave and a particle at the same time. Physicists coined the term 'spin' because its mathematical representation is the same as if the electron were a spinning top. Spin turned out to be the missing quantity needed to explain why the rings filled up as they did. The full explanation required a new law of nature: that no two electrons in an atom can have the same properties. This has become known as the Pauli exclusion principle.[10]

Scientists can study a very low density gas of electrons using the theory of the perfect gas, based on classical, or Newtonian, physics. But for a highly dense gas of electrons such as exists in a star, laws of quantum physics such as Pauli's exclusion principle come into play. In a star, given that no two electrons can have the same properties, one by one the various possible positions, velocities and spins of electrons – their quantum states – become occupied. Pauli's exclusion principle rules that each quantum state can contain at the most only one electron with a certain position, velocity and spin, akin to the restriction on how many electrons can occupy any ring in an atom. Thus these quantum states begin to fill up rapidly, starting with the low-velocity states that can accommodate electrons which are barely moving.

When a star begins to run out of fuel and starts to darken, the radiation pressure pushing outwards decreases and the inward thrust of gravity begins to prevail. The star starts to contract. Deep inside, the many trillions of electrons are squashed tighter and tighter together, like people trapped in a crowd. The quantum states fill up, from the centre of the crowd, where electrons are milling slowly, to the fringes, where they are swirling more quickly. The remaining electrons scurry around trying to find a place to settle. Eventually the collection of electrons cannot be squeezed any tighter because they cannot infringe on one other's 'territory' or state. Collectively they resist any further compression by exerting a powerful outward pressure. This is far stronger than their mutual electrical repulsion, which is almost non-existent, having been neutralised by the positive charges of the nuclei of the atoms from which they were stripped.

Eventually the star goes completely dark and there is no more radiation pressure. The outward pressure exerted by the electrons is all that is left to counter the enormous inward push of gravity. When the electrons begin to resist being squeezed together, they are in what is called a degenerate state; the outward pressure they exert is known as degeneracy pressure. The result of this degeneracy pressure is that the star becomes rock-hard, so hard that it cannot be crushed by gravity. It can now die in peace as a cold, dark rock.[11]

Thus, abstract though it seemed, incorporating quantum physics into astrophysics appeared to return the subject to its classical purity. It resolved Eddington's paradox, ensuring that stars expired gracefully and supporting the age-old vision of a quiescent universe.

Fowler had demonstrated that a theory which had emerged from studying the micro-cosmos could also be applied to some of the largest objects known – stars. He framed his work in daunting mathematical detail. To allow for the fact that white dwarfs might be far denser than anyone had thought, he assumed a density of 100,000 grams per cubic centimetre, twice the amount used by Eddington. But at that very high density, electrons within a white dwarf still do not move at speeds approaching that of light, so it was not necessary to incorporate relativity theory. Fowler knew this but chose not to proceed any further. His goal had been to resolve Eddington's paradox, and he had satisfied himself that he had succeeded in doing so. In his usual forceful and authoritative tone, the ex-Royal Marine declared that applying quantum theory 'clears up Eddington's question in a convincing manner, and I am content to leave the matter so'.[12] Eddington, too, was content.

Thanks to their wartime adventures, together with Fowler's famously affable personality, Milne and Fowler were to remain firm friends for the whole of their careers. Milne's relationship with Eddington, however, was far more troubled.

When Milne and Fowler first met, Milne was just twenty, Fowler's junior by some seven years. Small and slight, he was a warm, likeable man with a round, boyish face and bottle-bottom glasses. His poor eyesight had excluded him from active military service, so he eagerly accepted the invitation to join Hill's Brigands. He came from a more modest background than Fowler. His parents were schoolteachers, and he had won a scholarship to Trinity College entirely through his own brilliance – in fact, he scored the highest grade that had ever been awarded in the examination. A colleague of his wrote shortly after his death, 'He could crack open an intractable ballistic problem, explain a forgotten theorem of Euclid, recite an ode from Horace or a speech from

Shakespeare, and even crack a bawdy joke, all in the same lunch-eon hour.'[13]

Milne got on well with his colleagues and was extremely popu-lar. He was hugely charming and very dedicated to his work. Chandra's friend William McCrea wrote that Milne 'could be said positively to exude vitality. His features were often set in intense concentration but in conversation they became wonderfully mobile and expressive. They became overspread with impish delight at any witticism that appealed to him.' He added that 'he was essentially a religious man; faith and courage gained the vic-tory in each succeeding struggle of his life'.[14]

At the end of the war, Milne was twenty-two and reluctant to return to Cambridge to complete his undergraduate studies. So Fowler and Hill set up an arrangement to enable him to submit original work that would hopefully qualify him for a Fellowship at Trinity. Amazingly, in a single year Milne completed no fewer than three theses: one on a mathematics problem, one on the propagation of sound waves in the atmosphere and one on the properties of the Earth's atmosphere at great heights. The last two were based on his wartime research, which Fowler had convinced him was outstanding. He was duly elected.

But it was not enough to satisfy Milne. Chandra recalls that 'the fact that he never went through the "Tripos mill" produced in him an "inferiority complex" (those were the words he used), and a feeling that he had never acquired that foundation which is the "stuff of science".'[15] This was extraordinary. If anything, Milne's mathematical rigour was excessive. Eddington's papers were a model mixture of a great deal of physics with just enough mathe-matics to back it up. He once criticised Milne for taking 'thirty-six pages of technicalities' to make a single point.[16] Milne was deter-mined to prove at every possible opportunity that he really did know some mathematics. He always retained 'traces of the under-graduate attitude' towards most senior scientists.[17]

One night at high table in Trinity, Hugh Newall, director of the solar laboratory, mentioned that he was looking for an assistant director. Suddenly he remembered that Milne, who was sitting right next to him, had studied atmospheric problems during the war. Why not work on the solar atmosphere?[18] As a result, Milne spent ten years, from 1919 to 1929, researching the Sun's outer layers. It was the most successful period of his career; he was to be awarded the Gold Medal of the Royal Astronomical Society on 11 January 1935 for his achievements during this decade.

But tragedies accumulated in Milne's life. In 1921, when his father died, he seriously considered leaving academia to support his family. Then, in April 1924, he had an attack of epidemic encephalitis, prevalent in England at the time. Cambridge friends looked after him. He recovered by July but he was never quite the same. Milne later left Cambridge to become Beyer Professor of Applied Mathematics at the University of Manchester. There he pursued a vigorous and fruitful academic career. He also met and married Margaret Scott Campbell, who had been brought up there by her uncle, a professor of history. Everything seemed on track. Then, in 1928, he was offered the position of Rouse Ball Professor in Mathematics and a Fellowship of Wadham College at Oxford University. At the time, compared to Cambridge, Oxford was a desert in mathematics. Several others had turned the position down. Milne accepted. But establishing a mathematical presence took up a great deal of his time, and he lost contact with colleagues and friends in Cambridge.

At Oxford, Milne's personal life began to decline. Margaret had been happy in Manchester, taking part in academic life along with her husband. But Oxford was different. For one thing, women were not permitted to dine in college.[19] Milne began to feel isolated and out of touch with developments in atomic physics and quantum mechanics, subjects which were beginning to have a bearing on research into the atmospheres of stars. So he decided to change fields and work on creating models of the interiors of stars, which did not require such knowledge.

Milne had been invited to present the prestigious Bakerian Lecture of the Royal Society, on 6 June 1929. It seemed the ideal opportunity to introduce his new ideas, which he considered nothing less than revolutionary. Everyone present expected him to speak on his ground-breaking work on stellar atmospheres. But he had set his sights much higher. Three years earlier, in his own Bakerian Lecture, Eddington had presented new work that brought the study of interstellar matter – gas and dust in regions of space far from any star – into the mainstream of astrophysics research. Milne's great idea was to extend his work on the atmospheres of stars to look into their interiors, rather than working from within the star to its surface as in Eddington's standard model. But before presenting his lecture, Milne decided to visit Eddington and discuss his research with him. He should have guessed that there would be ferocious arguments. Eddington was totally convinced that a star's surface temperature could only be understood by examining its internal structure.

When the two met, one fine May day, Eddington totally demolished Milne's theories. There were only a few weeks to go before 6 June. With an almost superhuman effort, Milne wrote another lecture from scratch and delivered it as planned. It was a tour de force that not only consolidated his research on stellar atmospheres but also put forward original discoveries. It would be referred to by researchers in the field for many years.[20] Yet Milne was dissatisfied. He told friends that he did not consider his lecture worthy of the occasion.

A few months later, at the beginning of November, he had another chat with Eddington. He insisted that an important result such as Eddington's mass–luminosity relation could not be deduced just from balancing the outward push of its radiation and gas pressures against the inward tug of gravity. It would also be necessary to look into the structure of a star, and in particular the source of its energy. The life of a star was far more complex than Eddington made it out to be, he declared excitedly. We can imagine Eddington sitting, puffing on his pipe, intermittently chewing up apples, trying to restrain himself. Was not the mass–luminosity relation astounding? Did it not reveal that, amazingly, the perfect gas law, which everyone had thought applied only to less dense stars such as giants, extended to the incredible densities inside dwarf stars, and thus to all stars, except in the one case of white dwarfs? Yet Milne dared sit there and tell him that he, Eddington, had just been having a good day, having been lucky enough to assume from the start that stars were perfect gases, whereas all along Eddington had known perfectly well that white dwarfs were not perfect gases, as Fowler had shown three years earlier. Milne obviously did not have the slightest idea about building models of stars! Thus once again Eddington got the upper hand.

Nonetheless, Milne decided to present a shortened version of his ideas at the Royal Astronomical Society meeting on 8 November. Eddington dismissed Milne's work out of hand. 'It is difficult to discuss this paper,' he wrote superciliously. 'Professor Milne did not enter into detail as to why he arrives at results so widely different from my own; and my interest in the rest of the paper is dimmed because it would be absurd to pretend that I think there is the remotest chance of his being right.'[21] Eddington continued to pick away at Milne in a succession of short papers which he published the following January.

Milne did not take Eddington's criticisms well. He wrote to his brother that Eddington

> uses very strong language about my work – he calls it sophistry, mysticism, suggestions without foundation. He's getting very pontifical these days, and is touchy about his old theory, which is rotten to the core. It's astounding that such a complete hoodwinking of the scientific world should have gone on for so long. His theorems about stars are mostly baseless conjectures.[22]

In his writing, Milne referred to giants and dwarfs as 'centrally condensed stars' and white dwarfs as 'collapsed configurations'.[23] He built on the work of Fowler and Edmund Stoner, a physicist at Leeds University. Stoner had worked out that when a white dwarf began to collapse, the pressure would be so great that the electrons there would move at speeds close to that of light. On the basis of this he devised an equation, taking special relativity into account, which would allow a white dwarf to become rock-hard provided it was below a certain mass. Milne argued that although the outer layers of all stars behave as a perfect gas, working inwards one comes to a point where density and pressure increase to such an extent that the perfect gas law fails. If one is to calculate the star's internal pressure and temperature, one has to take quantum theory into account, using first Fowler's equation of state and, further into the star, Stoner's equation, which includes the effects of relativity. It may be necessary to find yet more equations to describe other exotic gases that will have to be explored as one moves still further into the interior of the star, where pressures increase enormously. What Milne's contention added up to was that all stars are made up of an envelope of a perfect gas surrounding a series of rock-hard cores, each of which prevents the star from collapsing beyond a certain pressure. These cores – being rock-hard – were also called degenerate cores.

Like Eddington, Milne balked at following this line of reasoning to its logical conclusion – that white dwarfs could continue to dwindle for ever, becoming smaller and smaller and denser and denser, towards an infinitesimally small size and an infinitely high density. This would be a singularity, the word used when the mathematics of a theory of physics comes up with the result 'infinity'. Traditionally, infinity is anathema to science – a sure sign that the theory is breaking down. But singularities had been cropping up regularly in quantum physics, and physicists, less hidebound than astrophysicists at that time, set about finding ways to take them on board and eliminate them. But astrophysicists such as Milne simply refused to deal with them, which often made astrophysics a laughing stock for physicists. To avoid creating a singularity, Milne made the ad hoc hypothesis, with no supporting arguments, that right at the centre of a star there had to be some sort of medium with incredibly high incompressibility.

In a critique published in the 29 March 1930 issue of *Nature*, Eddington poked fun at Milne's scenario. Milne, wrote Eddington, contended that 'physically a star cannot "contract indefinitely"; sooner or later it must find a (contracted) configuration of equilibrium'. By the same argument, Eddington responded, 'Humpty Dumpty's position on the wall was not unstable; ultimately he found a new configuration of equilibrium.'[24] Even if Humpty Dumpty fell off the wall a thousand times, he would always put himself back together again and be back on the wall. Similarly, according to Eddington, Milne argued that no matter how far a star collapsed, it would never collapse completely. There would always be a core inside a core inside a core, like an infinite nest of Russian dolls.

Milne wrote to Herbert Dingle, a well-placed figure in the

astrophysics community, to lodge a complaint about Eddington's caustic comments at RAS meetings: 'I cannot tell you what spiritual turmoil this has caused me. I have to face being accused of talking absolute nonsense before the whole R.A.S., just because I arrived at different results.'[25] Eddington, too, was reaching the end of his tether. 'I almost despair of extricating my main controversy with Prof. Milne from the tangle in which it has become involved,' he wrote.[26] Things were getting messy. Milne felt himself a martyr to science, a misunderstood genius. But he would go down fighting. Posterity would bear him out.

Still smarting from Eddington's criticisms, he began the most detailed exposition yet of his ideas on the constitution of stars. It was published late in November 1930. This paper, he explained, 'brings forward considerations which compel a drastic revision of our views as to the structure of the stars'.[27] Eddington's standard model would have to be replaced. While Eddington took as his basic assumption the concept of the star as a perfect gas, Milne's was that every star must have an incompressible core of some sort.

Nevertheless, for all the bitterness of their disputes, there was one thing on which they all concurred. As Jeans put it, 'Eddington, Milne and myself are of course all agreed that actually the density at the centre of a star must be finite.'[28] The concept of infinite density was unthinkable.

For Eddington, speculation about stars was only one part of his work. He was becoming more and more involved in his quest for one grand overriding theory that would cover everything from atoms to stars. This was his attempt to fuse quantum theory with the special and general theories of relativity, in pursuit of the Holy Grail of scientific endeavour, the quest that has driven scientists ever since science began – to find a so-called theory of everything. He called it his 'fundamental theory'. It was to absorb him so much that it would cloud his judgement on a number of matters – not least of which was the fate of white dwarfs.

Eddington had begun his search in 1928, inspired by Dirac's new theory of the electron. This was based on both the special theory of relativity and quantum theory, though it did not include general relativity. One of its weirder implications was that there had to be electrons with negative energy. But this was unthinkable – a physical impossibility, like negative time. Heisenberg complained to Pauli that the theory was the 'saddest chapter in modern physics'.[29]

Eddington was not particularly disturbed by the strange implications of Dirac's theory, but its mathematical structure struck him as odd. Until then, everyone had thought that the appropriate mathematical form for relativity was tensors – elaborated versions of vectors in the sense that they can be built up from them. Central to classical physics, vectors are utterly unlike ordinary numbers and do not obey the same rules. With ordinary numbers, one plus one makes two, but with vectors it could be zero, one or two, or numbers in between. But to create his new theory, Dirac had evolved an equation based on an esoteric form of numbers unfamiliar even to most physicists, neither vectors nor tensors, called spinors. What made Dirac's theory groundbreaking was that spin – one of the quantum properties of the electron – emerged from it. Eddington had a deep understanding of the mathematical basis of physics, and could see that Dirac had tapped into something far more profound than simply using a new form of mathematics to provide a basis for spin.

Eddington had a strong mystical streak, derived perhaps from his strong Quaker convictions. It took the form of an awe of nature that went beyond science. Mysticism offered an escape from the closed logical system of physics, but he always made it clear that physics must not be brought in to support any mystical view. Mysticism was a religious experience. Yet behind his imperturbable mask he agonised about the spiritual world and struggled with the dichotomy between the two worlds, both equally invisible, of science and of the spirit.

Eddington was also fascinated by 'seen' and 'unseen' worlds and the connections between them. His first book on the fundamental theory, *The Nature of the Physical World*, begins with a brilliant comparison of these two worlds.[30] In it he describes two tables, both exactly the same. We can place books and papers on both. One is an ordinary, everyday table. The other is a table as perceived through the eyes of a scientist: mostly empty space crisscrossed with electric and magnetic fields, with electrons buzzing around in such a way that our books and papers do not fall through. Although both tables are real and substantial, the concept of 'substance' requires redefining for the scientific table. What was the link between these seen and unseen worlds?

Eddington was convinced that the answer was mathematics. That was the key that would open the door between the two worlds. Dirac's equation was based on four numbers (the four dimensions of space and time – length, width, depth and time). Eddington used these as the basis on which to create sixteen new and enlarged spinors which he called 'E-numbers'.[31] From them he intended to construct a new mathematical system large enough to bring together all the parts of the universe, both large-scale, as described by the special and general theories of relativity, and at the quantum level.

Next, Eddington focused on seven numbers which scientists regard as 'primitive constants of physics', numbers which remain the same wherever you go in the universe. Much like the music of the spheres, 'we may look on the universe as a symphony' played on these numbers, as on the notes of a musical scale, he explained.[32] These numbers are so fundamental that they must also underlie the scientific theories of civilisations everywhere in the cosmos, or so scientists believe. The primitive constants that Eddington isolated were: the electrical charge of the electron – a minuscule figure, 10^{-10} in terms of size (measured in centimetres), mass (measured in grams) and time (measured in seconds); the Planck constant, a measure of scale in the atomic world and smaller still, 10^{-27}; the speed of light, a huge figure, 10^{10} centimetres per second (measured in the same units as the other constants); the masses of the proton (10^{-24}) and the electron (ten thousand times smaller than that of the proton, 10^{-28}); Newton's universal gravitational constant (10^{-8}); and the cosmological constant of the general theory of relativity (a kind of anti-gravity which allows the universe to expand, and unimaginably minuscule, 10^{-55}).

Eddington's ultimate goal was to use his new mathematics to deduce the number of electrons and protons in the universe, then to compare his result with the results that scientists obtained from experiments. As a first step, he deduced certain combinations of the seven primitive constants from his E-numbers, then compared them with the actual measured figures. He was particularly fascinated by something called the fine-structure constant, which can be constructed from the charge of the electron, the Planck constant and the speed of light.

When scientists studied the spectral lines of the light emitted by atoms, they discovered that some are made up of even narrower, very closely spaced lines: in other words, these lines have a fine structure. Sommerfeld explained this brilliantly by incorporating relativity theory into Bohr's theory of the atom, which pictures atoms as minuscule solar systems. It turned out that the spacings between the components of a spectral line depend on the fine-structure constant. Early experiments suggested that this figure was the ratio of two integers, or whole numbers: 1/137.

Integers lie at the heart of the quantum theory of the atom. They also played an important role in ancient Greek thought. The mathematician Pythagoras and his followers believed that through whole numbers they could tune in to the music of the cosmos. Even Sommerfeld, apparently a down-to-earth German physicist, rhapsodised that the 'language of spectra' expressed a 'true "music of the spheres" in chords of integral relationships'.[33] The number 137 itself has mystical connotations. In the Kabbalah, numerical quantities assigned to certain key phrases referring to Yahweh add up to 137. Sadly, subsequent measurements showed that the fine-structure constant is actually not 1/137, but 1/137.036. But Eddington refused to be deterred. He also idiosyncratically insisted on referring to the fine-structure constant as 137, not 1/137.

A key element of Eddington's argument was that particles can never be considered in isolation, and that there are no systems in nature in which the interactions between particles can be ignored. According to the general theory of relativity, it is gravity – the gravitational tug between particles – that gives each particle its mass and sculpts space and time. Besides being a measure of the fine structure of spectral lines, the fine-structure constant is also a measure of how strongly two electrons interact. For Eddington this was further support for his contention that there is no such thing as an isolated particle in quantum theory either.

As particles cannot be considered in isolation, Eddington argued, any theory of the electron has to deal with at least two electrons. Applying his new mathematics, each could be described using 16 E-numbers. Multiplying 16 by 16 gives a total of 256 different ways in which electrons can combine into a single system. He then showed that, of these 256 ways, only 136 are actually possible; 120 are not. He gave this the mathematical signature of 256 = 136 + 120. Thus, like pulling a rabbit out of a hat, he was able to produce the number 136 from purely mathematical reasoning.

Now, 136 was not 137, but for Eddington it was close enough. He was convinced that the additional 'one' 'will not be long in turning up'.[34] The obsessive pursuit of 137 literally took over his life. Russell recalled meeting him at a conference in Stockholm. They were in the cloakroom, about to hang up their coats. Eddington insisted on hanging his hat on peg 137.[35] Said Dirac, 'He [Eddington] first proved for 136 and when experiment raised to 137, he gave proof of that!'[36]

Nevertheless, he considered his result encouraging. To Eddington his fundamental theory was more than just physics. He saw himself as a spinner of grand philosophical systems that extended to all facets of science and the mind and to the essence of being. Every now and then he needed to confirm his intuitions by comparing them with data which scientists could obtain from experiments – such as the rate of expansion of the universe. But in general he preferred to draw his conclusions from thought and mathematical reasoning alone.

Eddington was determined to calculate the number of electrons and protons in the universe. He did so using quantum and relativity theories to make the point that his fundamental theory was a union of all these, then calculated it again using his own new mathematics, based on the 16 E-numbers. First he took the universe as governed by the general theory of relativity, and deduced an equation containing the number of electrons and protons in it. Next he imagined this universe as a giant atom made up of electrons and protons, and used quantum statistics to arrive at a second equation. From these two equations he isolated the number of electrons and protons, which added up to the massive figure of 1.5×10^{79}. He used this to determine the rate at which the galaxies are receding from us, i.e. the rate of expansion of the universe. To his delight, the result was fairly close to the rate measured by astronomers.[37] The problem was that his figure was a fraction, and a fractional number of particles is an impossibility – added to which he was determined to tie in his result with the fine-structure constant, which he insisted was a whole number.

At this point he turned to his E-numbers once more and worked out the number of electrons and protons in the universe again. He wrote the result as 136×2^{256}, a figure which held an aesthetic and mystical aura for him. He pointed out that 2^{256} is almost exactly the same as 10^{79}, and that the number he had deduced compared perfectly with the figure obtained from experiments. It seemed to prove that it was no accident that the universe contains 10^{79} electrons and protons. (The key part of the calculation was the huge figure on the right – 10^{79} or 2^{256} – which is not much affected whether it is multiplied by 1.5 or 136.) Through mathematics, Eddington had found the key to understanding the structure of the universe – or so he felt.

Unfortunately, the equation that Eddington deduced from quantum theory turned out be wrong. To get the result he wanted, he had to force general relativity and quantum theory to match up. He did so by using a false figure.[38] Colleagues were beginning to detect, in the books and papers Eddington was writing on his fundamental theory, a dramatic shift from the succinctness and clarity of his earlier writings on astrophysics. Chandra was later to quote one acerbic critic: 'His [Eddington's] papers are very clear up to a point and then at the critical moment they become obscure, to become clear again after important results have been deduced.'[39]

Eddington was not alone in his quest to deduce the fine-structure constant from theory. Pauli obsessed over it as well and he too was fascinated by the multifaceted connotations of the mystical number 137.[40] In 1958 he was in hospital in Zurich, undergoing treatment for cancer. When the physicist Victor Weisskopf, who had been his assistant in the early 1930s, visited him, Pauli suddenly blurted out, 'Viki, I'm never going to get out of this room alive!' 'Why?' asked Weisskopf, puzzled. 'Look at the number of this room,' Pauli replied grimly. It was number 137. And indeed, Pauli died there.[41]

But this was still a long way from the obsessive behaviour of

Eddington, who was prepared to bend basic laws of physics to achieve his ends. Pauli dutifully read almost everything that Eddington produced, but he had nothing but scathing comments. 'I consider Eddington's "136-work" as complete nonsense: more exactly for romantic poets and not for physicists,' he wrote.[42]

But despite the growing criticism surrounding Eddington's work on a unified theory, his reputation at the dawn of 1930 was at its height. His model of the constitution of the stars was at the hub of astrophysics; his ideas on cosmology were considered intriguing; and his philosophical meditations on the universe were best-sellers, widely discussed. He had managed to tie his calculation of the number of electrons and protons in the universe to the rate of its expansion, as well as deducing it from his E-numbers. And, in his work on the fate of stars, Fowler had cleared up the disturbing paradox he had uncovered. As far as Eddington was concerned, everything was settled. That year he received the final accolade – a knighthood. It was also the year that Chandra arrived at his door.

5

Into the Crucibles of Nature

Chandra left India on a high. As the city of Bombay receded into the distance, he may have thought back over the achievements of his nineteen short years of life. It had been a golden youth. Besides completing his university training, he had already published five research papers and favourably impressed two of the world's greatest scientists, Heisenberg and Sommerfeld. The two leading Indian physicists – Raman and Saha – were full of high hopes for him. Raman proclaimed, 'This young man shows all the signs of being a genius. He will surely leave an indelible mark on physics.'[1] He was guaranteed a professorship at Presidency College when he returned. During the voyage he would think about Fowler's white dwarf theory, as well as remembering Lalitha, the young woman he was leaving behind. He was on his way. The world of physics was at his feet. He had good reason to believe that no one could resist his charms or be anything less than deeply impressed by him.

For the first few days the sea was so rough that the ship had to sail at half speed. His Uncle Raman's advice had been to 'keep cheerful and remain on deck if possible. You will certainly have a very pleasant voyage once you get near Aden.'[2] This proved to be the case. As soon as the weather had improved, Chandra separated himself from the other Indians. He kept pretty much aloof, to the annoyance of his countrymen, who were eager to take full advantage of everything the West had to offer.

Chandra settled into a deckchair and pulled another alongside on which to pile up his books and papers. He had with him Fowler's 1926 paper, 'On dense matter' (in which Fowler laid out his solution to Eddington's paradox), the American Nobel laureate Arthur Holly Compton's book *X-rays and Relativity*, and Eddington's and Sommerfeld's books, as well as his own published papers. The sea was calm, without even a ripple. The vast sky was intensely blue and the air smelt invigoratingly salty.

Those were idyllic days. Chandra had found peace after the hectic pace of the previous weeks. He had nothing to do but think and calculate. He knew precisely what to work on – the problem he had been forced to set aside some months earlier, of 'making Fowler's conclusions more precise'.[3] He had already begun by tying them in with current ideas on how to calculate a star's internal temperature and pressure from its surface temperature, mass and radius as measured by astronomers. Eddington had given formulae for doing so in *The Internal Constitution of the Stars*. Yet no one, not even Eddington, had thought of applying these methods to Fowler's new theory of white dwarf stars. Why not? It was so straightforward. Perhaps Eddington had become complacent now that, as far as he was concerned, Fowler had straightened out the heavens for him.

Some years later, Chandra would recall what happened next as 'so simple and elementary that anyone could do it'.[4] Quickly, he calculated the density at the centre of a star as required by Fowler's result and found that, for the white dwarf star Sirius B, it was *a million grams per cubic centimetre* – a million times denser than water. Chandra pondered his finding. 'Interesting,' he must have thought. 'Very interesting'. But what made this huge number so interesting? We can imagine his mind drifting away, with the beauty of the sea before him.

Suddenly he remembered the story his Uncle Raman often told him, of how he had made his most famous discovery. The inspiration had come to him on a sea voyage to Europe in 1921. As he pondered the sparkling blue opalescence of the Mediterranean, it occurred to him that the sparkle must be caused by collisions between sunlight and water molecules.[5] From this he worked out a way to unravel the structure of molecules by examining how they alter the frequency of the light striking them – the famous Raman effect.

Turning his gaze from the azure sea, perhaps Chandra glanced at the deckchair next to him, on which lay Compton's *X-rays and Relativity*. This was where he had learned about Einstein's theory of special relativity, about what happens to objects moving extraordinarily fast, at speeds comparable to that of light, an area wholly beyond the bounds of classical, Newtonian physics. The speed of light plays a central role in Einstein's theory: it is generally taken as 186,000 miles per second and is always the same, regardless of the motion of the source. It is the cosmic speed limit; nothing can move faster. This new way of peering into nature – relativity theory – also reveals that a particle's mass changes with its speed. It is not constant. This can lead to surprises, as Chandra was about to discover sitting in his deckchair deep in thought. In a flash of inspiration, for the sake of argument, for completeness, 'I asked myself what the velocities of the electrons would be at the centre of the white dwarfs'.[6] It might be interesting, he thought, given the extremely high density he had just calculated. It turned out to be extraordinarily fast, more than half the velocity of light.

Chandra was stunned, though he was well aware that there could be unexpected consequences when relativity is introduced into an equation. Astonishingly, Fowler had forgotten to apply special relativity. He had taken for granted – mistakenly, as it turned out – that the pressure inside a white dwarf star was no greater than a hundred thousand grams per cubic centimetre, and he therefore assumed that the electrons in a white dwarf moved slowly enough for their motion to be studied using Newtonian physics. The upshot was that his result had been only approximate, not exact as everyone had thought.

Chandra's moment of inspiration was more than serendipity. Equations spoke to him. He could lose himself in numbers and symbols, as if in a cocoon. He was a perfectionist – he could not stop until every detail was right. But he could also see the broader significance of his results. He had the vision to see the big picture. This was the key to his genius.

Before he went any further, Chandra had to make sure he was right, by working out why these astonishingly high speeds were correct. The clue lay in an astounding implication of quantum mechanics, which Heisenberg had discovered three years earlier, in the spring of 1927. To Chandra, of course, Heisenberg was far more than a name on a research paper. There was a face behind the name, a handshake, encouragement.

Dubbed the uncertainty principle, Heisenberg's discovery was that there is an inverse relationship between a particle's position and its velocity: we can measure very precisely where a particle is, but we cannot simultaneously measure with the same accuracy how fast it is moving, and vice versa. Any attempt to force the electrons in a star to occupy a precise location results in a frenzied retaliation in velocity. This is why electrons inside white dwarfs move at speeds close to that of light. As an expert on quantum theory, Fowler should have incorporated special relativity into his equations right from the start.

Fowler had explored non-relativistic degeneracy, in other words, degeneracy theory which does not take relativity into account. By including the theory of special relativity Chandra could explore what he called a relativistically degenerate gas of electrons.[7]

Chandra realised that a calculation using the full paraphernalia of special relativity was just too difficult and would have to be postponed. He decided instead on an approximation in which he assumed that the electrons within a white dwarf moved very close to the speed of light. He calculated the relationship between the pressure and density of a relativistically degenerate gas of electrons using as a basis an idealised quantum gas, ignoring the electrical interaction between particles. As Fowler had, he focused on white dwarf stars that have completely cooled off. Bringing in relativity, however, produced an astounding result: there was a limit to how much mass a white dwarf star could have. Chandra calculated it as a little less than the mass of the Sun.[8]

But what would happen to a white dwarf which, after burning up all its fuel, ended up with a mass that exceeded the limit which Chandra had just discovered? With nothing to brake its collapse, could it be that it collapsed for ever? In any case, it seemed that he had overthrown Fowler's theory and found a new solution to Eddington's problem – one that Eddington himself had toyed with but had dismissed as absurd.

The ship finally docked in Genoa, from where Chandra made the overland trip to the English Channel. He arrived in London on 19 August. But the proper documents had not been forwarded. Chandra arrived practically without notice. It was an irritating bureaucratic muddle arising out of the usual morass of Anglo-Indian misunderstanding. In desperation he wrote to Fowler, who was on vacation in Ireland. The kindly Fowler responded just in time to cut through the red tape. 'That was all there was to it!' as Chandra recalled, remembering Fowler's concern.[9]

Chandra arrived in Cambridge in buoyant mood and with high hopes. He had visions of being received there as he had been in India, where he was feted and recognised by all 'the people he came into contact with [as a] genius'.[10] But he was in for a shock. The precocious youth had had almost no competition in India, but in Cambridge he found himself up against some of the most brilliant minds in science. 'I had a shattering experience; to suddenly find myself in an environment where there were people like Eddington and Hardy, not to mention all the other well known names, is a very strong sobering experience,' he wrote.[11]

His self-confidence evaporated. For a while he was convinced that the only reason he had been admitted was 'purely due to the accident' that he had corresponded with Fowler for the past two years. 'Why I should have written then to Fowler God alone knows. I suppose that was because Fowler is to help two years later,' he wrote to his father that September, just after arriving in Cambridge.[12] He took it to be his karma.

By 2 October, Chandra had written two papers which he was looking forward to presenting to his thesis adviser. One, on the density of white dwarfs, was a pure application of Fowler's solution to Eddington's problem. The other was more original and ambitious. It concerned the maximum mass for white dwarfs that Chandra had just discovered. Fowler read both. He made some suggestions about the density paper, and offered to send it to Milne and then submit it for publication.

Fowler was noticeably less enthusiastic about the second paper. He suggested that Milne might take a look at it. Chandra was disappointed. He was convinced that this shorter, second paper 'was more important'.[13] Fowler reminded Chandra that, in 1930,

Edmund Stoner had also arrived at a limiting mass for white dwarfs by including relativity. Chandra dutifully cited Stoner's work in his 1931 paper, pointing out, once again, that his own results were based on a more realistic model of a star and so had far broader implications.[14]

Stoner's result applied only to a star which was the same density all the way through, whereas Chandra assumed the more likely scenario that a star became denser as one moved towards the centre. Moreover, Stoner was concerned only with the question of whether the temperature and pressure inside a white dwarf became unexpectedly high as the star's mass reached a certain critical point – Chandra's upper limit. Chandra, however, was prepared to follow the idea of a maximum mass to its logical conclusion – that a white dwarf of more than this mass might eventually collapse to nothing. Chandra could not have avoided seeing Stoner's footnote in which he thanked Eddington for conversations about his research.[15] He must have realised from this that Eddington had more than a passing interest in relativistic degeneracy.

In the end, Chandra recalled, 'Fowler and others were not doing anything about the paper I gave them in September.' He was tired of them simply commenting, 'Very interesting.'[16] Sir James Jeans, however, took notice. Chandra wrote excitedly to his father: 'I have proved that the MAX Mass a Dwarf star can have is just that of our Sun. This result I communicated to Jeans. He has written quite a nice letter and thinks the result "quite important".'[17] Thus encouraged, in mid-November Chandra sent his paper on the maximum mass to the American *Astrophysical Journal*. It was initially rejected, but finally published there in March 1931.[18]

And yet, despite the staggering implications of his discovery, Chandra had the distinct feeling that 'neither Fowler nor Milne thought of [it] as very important . . . Very few people were interested in what I was doing.' The young man's confidence was ebbing away. 'Somehow or other, I felt I didn't belong there. It seemed to me that there were far too many big people, far too many people doing important things, and what I was doing was insignificant in comparison. I suppose I was afraid.'[19]

Like all Indian students newly arrived in England, Chandra had to contend with loneliness, homesickness, bland food and problems with keeping to a vegetarian diet, not to mention worries about what was happening back home and the racial prejudice that was as prevalent in Cambridge as anywhere in England. Fiercely proud of his Indian heritage, Chandra strongly objected to any hint of discrimination. Babuji was concerned for him, no matter how uncaring a face he may have presented to his children. 'I am feeling very lonely here . . . suffocatingly lonely,' Chandra wrote to him.[20] Once term began, however, he became immersed in his studies and felt a little better. But he remained a loner.

Fowler, his thesis adviser, was frustratingly difficult to pin down. He seemed always to be absent from Cambridge or busy in his study at Trinity or in the library. Chandra would spot him going into the library and wait patiently outside, hoping that he would emerge and invite Chandra along to his office to talk. But he never did. Chandra soon gave up.

The young man's feelings towards his new home oscillated violently. Rubbing shoulders with people not much older than himself who had made revolutionary discoveries was a thrill. 'I saw Dirac yesterday,' he wrote to his father:

> A lean, meek shy young 'fellow' (F.R.S.) who goes slyly along the streets. He walks quite close to the walls (as if like a thief!), and is not at all healthy. A contrast to Mr. Fowler . . . Dr. Dirac is pale, thin, looks terribly overworked. Tomorrow his lectures begin. I may get introduced to him sometimes. The place is so big that it takes a long time to get to know these people.[21]

It was an apt description. But a month later, Chandra's excitement had palled. 'The formality of introduction is so great and even then it is not worth the trouble of getting introduced,' he wrote sourly to his father.[22] Compared with the warmth and openness he was used to from South India, the smug reserve of most of the English people he encountered was a severe culture shock – so much so that he withdrew into himself. It was grindingly difficult to get used to life alone, in this dull, grey country, far from the comfort and support of his family. Here, there was nothing to do but work.

Isolation was fine for him, he wrote to his brother. 'There is nobody to talk to, and mathematics has become my morphia. Dull and stupid, it helps me to forget everything else. I take some walk in the evening, and Cambridge is so small that you come back to the same place in ½ an hour.'[23] All day long he would work alone in his rooms. 'As it is, life is so lonely. Confined to one's study all day long and even during walks not to get rid of oneself! It's all so sad, so oppressive!' he complained bitterly to his father.[24] By November, he was writing that he felt like a 'single electron in deadly free space' – as if Babuji would have the faintest idea what his son was talking about. He waxed enthusiastic about the papers and ideas he was working on; he had become a 'physics animal', he said.[25] Three months earlier he had been immersed in Shakespeare, but 'now the diversion is quantum mechanics or stars to "chemi-luminescence" or "theory of valence"'.[26]

'It makes me sad to think you are feeling a bit lonely,' Babuji wrote to his son on New Year's Day, 1931. Chandra had been away for five months now. 'The feeling will come on, as the human ties press on the mind. Hereafter I will keep a diary.' A stickler for grammar and spelling, he added, 'Please also keep for reference an Oxford dictionary as your letters are full of *spelling* mistakes in English.' In reply, Chandra apologised for being an 'unlessoned boy'. He was writing, he added, on Dirac's recent relativistic electron theory.[27] Rather than astrophysics, he was trying to work on pure physics, as did Dirac, whom he idolised. But when he handed in the paper he had written, Fowler told him in no uncertain terms that it was unsatisfactory. 'One can laugh only bitterly!!' Chandra wrote to his father. 'Feeling lonely', he decided, was a weakness.[28] After this, there would be no further outward displays of such feelings.

Chandra was also attending Eddington's lectures on relativity. 'He is in contrast to Dirac,' he told Balakrishnan. 'Dirac is always serious – "a martyr to meditation" . . . But Eddington is always funny and humorous. On the whole the lectures this term are very useful and good.'[29] Chandra was in the minority in his opinion of Eddington's lectures: most students complained that they were very dull. But Chandra's innate brilliance led him to see many things differently.

There was one astrophysicist, at least, who took an interest in Chandra. From the beginning, Milne took Chandra under his wing, sometimes too much so. He made his first surprise visit to Chandra's rooms in October 1930, not long after the young Indian had arrived. But, as Chandra recalled, he talked almost exclusively 'about his own ideas as to how all stars must have degenerate cores'.[30] The young man was awestruck by the surprise visit, which made a permanent impression on him. Shortly afterwards, Milne wrote Chandra a letter, suggesting that his (Chandra's) results on the limiting mass of white dwarfs were 'particular cases of my analysis of "collapsed stars"'.[31] Chandra was piqued. He thought the exact opposite. 'Prof. Milne's results are in a sense a generalisation of mine own,' he wrote to his father on 14 November 1930. In point of fact, Milne's analysis was no more that a pastiche of Fowler's and Chandra's results, but, on the other hand, Chandra was wrong in claiming that Milne's theory was no more than an expanded version of his own. Chandra's letter was written from Burlington House, where he was attending the Royal Astronomical Society meeting at which Milne presented yet another paper critical of Eddington. 'Others (including me) believe that he has completely *destroyed* Eddington's view of the interior of stars,' Chandra told his father. He added that he still felt far too nervous to join in the discussion with Eddington, Jeans and the others.

By this time Chandra had begun to realise that his own work, on the limiting mass for white dwarfs, was being ignored for some reason. The only way to proceed seemed to be by melding his result into Milne's idea that all stars had an incompressible core. But that ran directly counter to Eddington's standard model, which assumed that most stars behaved like a perfect gas. The young man was about to enter a bruising battle with giants.

Over at the Royal Astronomical Society, the run-ins between Eddington, Jeans and Milne were becoming vicious. Eddington and his colleague and arch-rival, James Jeans, disagreed on just about everything, most notably what stars are made of and what their energy source was. Their disputes at the RAS had become famous for their vociferous tone. Like Eddington, Jeans had been fascinated by numbers from a very young age – at seven he was memorising logarithms. But unlike Eddington, he came from a privileged middle-class background. In 1905, when he was just twenty-eight, he was already a star lecturer at Trinity College. Woodrow Wilson, then president of Princeton University, lured him to what was at the time little more than a posh small-town college to take up a professorship. There Jeans enjoyed a massive salary, far greater than the pittance teaching staff received at Cambridge. By 1910 he was back in Cambridge. The holder of the Plumian Professorship was on his deathbed, and Jeans knew he was next in line. But as a result of some Machiavellian internal politics, the professorship went instead to Eddington, at the time a mere junior fellow.

Disgusted with academia, Jeans resigned. While in the United States he had married Charlotte Tiffany Mitchell, of the Tiffany jewellery empire. The two lived in regal splendour in their villa, Cleveland Lodge, in Dorking, just outside London, where Jeans speculated about the universe and turned to popular science writing. His book *The Mysterious Universe* was a best-seller. He was also a superb musician, and played Bach brilliantly on a magnificent organ – Chandra once visited Jeans and was mightily impressed. Compared with the regular denizens of high table, Chandra considered him an 'aristocrat', although most thought otherwise and were, recalled Chandra, 'sarcastic' about him. 'Anyone who can play all of Bach's organ compositions cannot be a trivial person,' was Chandra's opinion.[32]

On 9 January 1931 things came to a head at the RAS. By now the gloves were off. Tucked away at the back of the room, Chandra must have been aghast, seeing these mighty minds slugging it out. Launching the proceedings, Milne took a swipe at Jeans's hypothesis that, at high pressures, a star's interior became an exotic liquid in order to prevent complete collapse. He accused Jeans of trying to account for observed facts by 'introducing hypotheses which are extrapolations of the laws of Physics'. 'To be perfectly frank,' he added, 'I don't suppose that anyone really believes his theory.'[33] He then turned his fire on Eddington, reiterating his criticisms of the mass–luminosity relation and the fact that he completely ignored the issue of the source of a star's energy. They were all, of course, in agreement that the density of the core of a star had to be finite; infinite density was unthinkable and impossible as far as they were concerned. They were simply arguing about how and why that was true.

Then Eddington rose with a magisterial air. He brushed aside Jeans with the comment that any difference between the liquid- and gas-type theories was 'purely a matter of plausibility of hypotheses', and therefore not serious. Milne's case was different: it concerned the 'correctness of a mathematical or logical deduction'. Eddington went for Milne's Achilles' heel, his inferiority complex about his mathematical ability. Milne had argued that Fowler's theory of white dwarfs implied very high temperatures and pressures inside stars. In Fowler's theory, a dense gas of electrons which are moving at speeds much less than that of light, and so are non-relativistic, reach a point where they cannot be squeezed any more by gravity. The electrons retaliate by generating an outward non-relativistic degeneracy pressure, and the core becomes as hard as a rock. Milne had pointed out that the temperatures and pressures inside stars would be even higher if the electrons were moving at speeds comparable to that of light, which was what Chandra's theory of white dwarfs predicted. In other words, stars with incompressible cores must be extraordinarily hot and dense inside, and all the more so if the electrons in the star are moving at such high speeds that they are governed by the theory of relativity. Eddington disagreed. Such high temperatures and densities were simply impossible, he claimed. He reminded his audience that the opacity discrepancy – the nagging difference between the opacity measured in the laboratory and the opacity expected for Eddington's standard model of the stars – arose not only in his model, but in Milne's, and more severely too. The only way for Milne to avoid it would be to apply his model to other types of star; but if he did so, the internal pressure of the star might end up as a negative quantity, which could not possibly correspond to any real star. 'So the star must be wrong as well! . . . Professor Milne is between the devil and the deep sea – or rather between me and the deep sea,' he concluded triumphantly.[34]

Jeans waded in. 'As Prof. Milne has been so frank in the

expression of his views as to my theory, I feel that I can be equally frank about his work,' he snapped. 'In brief, I do not think it gets anywhere.' He then took a sideswipe at Eddington, saying, 'I don't think that Eddington's gaseous model is tenable any longer.' Then, as the *coup de grâce*, he concluded, 'It seems to me that [Milne] is not studying stars at all, but solutions of a differential equation. In saying this I am being as frank as Prof. Milne has been about my work.'[35]

W. G. M. Smart, a renowned celestial mechanician at Cambridge, reminded everyone that stellar models have to be simplified to be 'amenable to mathematical treatment . . . The stars themselves must be the final referees in a matter of this kind, and personally I think that the stars behave very well according to Sir Arthur Eddington.'[36] The great Cambridge mathematician and dandy, G. H. Hardy, had joined the RAS specifically for this sort of row. He came up with the showstopper:

> As a mathematician I don't care two straws what the stars are really like. I have a mild preference for Milne's theory, since it takes account of all solutions of the differential equations concerned. [But] I am particularly interested in Mr. Fowler's paper [which] is probably the only one of the collection which is of lasting value, for he is certainly right, whereas it is extremely likely that everyone else will be shown to be wrong.[37]

Milne sank deeper into gloom. In the end Hardy was proved to be right, at least as far as this particular argument went. Thomas Cowling, a former student of Milne's, recalled that the atmosphere was 'diamond cut diamond. I think everyone was absorbed, sitting at ringside seats.'[38]

As for Chandra, his recollection was that 'the controversies between Eddington, Jeans, and Milne led most onlookers to say "a plague on both of your houses!" Essentially this sentiment was expressed by P. W. Merrill in his [book] "Variable Stars" where he writes that the papers on stellar constitution are no more than "exercises in differential equations".'[39] Merrill's exact words were these:

Technical journals are filled with elaborate papers on conditions in the interiors of model gaseous spheres, but these discussions have, for the most part, the character of exercises in mathematical physics rather than astronomical investigations, and it is difficult to judge the degree of resemblance between the models and actual stars. Differential equations are like servants in livery: it is honourable to be able to command them, but they are 'yes' men, loyally giving support and amplification to the ideas entrusted to them by their master.[40]

It was beginning to seem all too likely that Eddington, Jeans and Milne were busy solving complicated equations to create models of stars that bore virtually no resemblance to the real thing.

February was a black month for Chandra. 'I take walks every day along the "Trumpington" – the long road which leads to London – alone every evening,' he wrote to his father. 'The road is usually very lonely. I go "down" 2 miles along the footpath and back, 4 miles walk which usually takes an hour or more.' He took a break from time to time, though he felt very guilty about it:

Occassionally on a Sunday – once in a month or sparser – I take a holiday. After breakfast I go on a long long walk across the fields. As for instance last Sunday – after a month of more or less uninterrupted steady work – I took a holiday. I am ashamed to confess that I needed it, but one wants a recreation once in a while.[41]

That day he strolled alone along the winding Madingley Road, through frozen fields.

A main cause of Chandra's gloom was that Milne was putting pressure on him to formulate his results so as to support Milne's own. Milne welcomed Chandra's theory because it revealed the unexpected properties of gases at the incredibly high densities in a star's interior. Milne expected to find more surprises as astrophysicists delved further into the interior. Yet he did not want to accept Chandra's concept of a maximum mass above which a white dwarf, after burning out its fuel and becoming dark, might collapse to nothingness. Milne wrote to Chandra in January 1931: 'Your conclusion in its present form arises from the curious properties of [relativistic degeneracy] but I think you have fallen into the Eddingtonian error of inferring physical consequences from what can be only an incomplete algebraic treatment.'[42] Among the 'curious properties' was the upper limit to the mass of a white dwarf.

Milne then raised an intriguing issue. Take the case of a white dwarf with a mass just under Chandra's upper limit. It is therefore stable and will not collapse any further: its electrons, moving at speeds close to that of light, can no longer be compressed by gravity; they generate an outward pressure – relativistic degeneracy pressure – to form a rock-hard core. But what would happen if just enough inert matter were sprinkled onto this white dwarf so as to push it over the maximum mass? Would it begin an infinite collapse? 'Until you have solved this question your publication of your paper would raise more questions than it would answer,' he told him.[43] In Milne's view, it was simply impossible for a star to shrivel into nothingness. To add to all his other worries, Chandra now had to ponder whether stars could be stable at higher densities. This problem would take a new turn the following year with the discovery of the neutron.

Four weeks later, Milne was urging Chandra to explore how the different possible conditions for the interiors of stars (perfect gas, non-relativistic and relativistic degenerate gases) might 'transform one into the other'.[44] Chandra was beginning to feel like a pawn in the bruising battle of egos Milne was engaged in. He must have wondered when his own turn would come. Whenever it did, he was determined not to suffer Milne's fate, although he was not sure how he would manage this.

At the RAS meeting on 13 March 1931, Eddington, Jeans and Milne returned to the attack. Eddington kicked off the proceedings with a paper in which he calculated the maximum pressures and densities of stars that obey the perfect gas law and of non-relativistically degenerate stars (i.e. low-density stars with incompressible cores, such as Fowler had dealt with in 1926).[45] The figures he came up with were a billion degrees kelvin for the maximum internal temperature of a star and 10 million grams per cubic centimetre for its maximum density. These figures were clearly much too high, he said, a result of considering ideal rather than real stars and 'not likely to be attained' in reality. Nevertheless, he wanted to present these results because he thought that they 'may be of use in curbing riotous speculations'.[46] He was referring, of course, to Milne's and Chandra's results, which included relativistically degenerate stars, and which predicted even higher pressures and temperatures.[47]

Eddington even expressed doubts about Fowler's result, which also implied infeasibly high internal temperatures, declaring that it was far from certain that the 'formula of Milne's [originally Fowler's] for the pressure of matter in a completely degenerate state is correct'.[48] Chandra was shocked. Not only was he sure that Fowler's argument – that a white dwarf can never collapse – was right as far as it went, but his own work was an extension of Fowler's, and based on it. Eddington was indirectly attacking him. In a letter to Stoner, Eddington went so far as to dismiss Fowler's results as 'conjectural and fishy'.[49]

Fortunately for Chandra, he had one close friend, Sarvadaman Chowla. Short and jovial, Chowla was just the sort of person that Chandra needed to help preserve his sanity. A graduate of Government College in Lahore, Chandra's birthplace, Chowla had gone to Cambridge to study number theory with the famous mathematician J. E. Littlewood. 'Life is so lonely here, that it is some consolation to have somebody,' Chandra wrote to his father.[50]

In May came news that Chandra's beloved mother had died. Chandra was devastated. 'Oh! My Heavens, little did I dream that when I left her at the station that it was my last farewell to my mother. I am alone in my grief and it paralyses me to stay in my rooms and shed tears,' he wrote.[51] For two weeks he could not work at all. Then, with a mighty effort to make work his consolation, he returned to his calculations. He worked six to seven hours a day – which he considered 'taking it easy'. From June he began actively collaborating with Milne and ratcheted up the hours to eight or nine a day – though he claimed Milne worked faster.

Concern about the family back home was a serious problem for any young Indian student abroad. Chandra worried about his father's state of mind and also had to encourage his brother Balakrishnan, whom Babuji had just put into medical college. A dreamy sort, thin and ascetic, Balakrishnan aspired to being a writer. Instead of applying himself to his medical studies, he spent his days in cafés, talking and reading. Recently he had begun to feel that he had dropped behind so badly that it was too late ever to catch up. Life was passing him by. Chandra exhorted him to do his best, in the process revealing a great deal about his own state of mind. 'Age does *not* matter. It is *never too late* to begin,' he admonished him with an older brother's sternness. Then his own self-doubts began to colour his remarks:

> Do you think that just because one read a little calculus and conics when his equals did not, read a little statistical mechanics which others of his age did not, happened to publish a few papers, Do you think that he is better than his friends, Do you imagine that he has in any case proved *his intrinsic merit*? You are woefully mistaken if you think so, and I will cry shame on him . . . I wish I could divulge to you the sorrows of my heart, and tell you how I feel at times that my heart will break by the oppression of my ignorance . . . I cannot repeat too often that nobody can be so dissatisfied with oneself, *as* I am of myself.[52]

No matter how much he achieved, he could never shake off his sense of worthlessness.

Chandra sympathised with his brother, forced to do medicine when he longed to be a writer. '*That* surely was not my feeling when I took physics in preference to mathematics,' he wrote to him.[53] In subsequent letters, he patiently counselled his younger brother to follow Babuji's advice and reminded him that, 'Chekhov – or was it one of the other great Russian writers? once said that every artist must undergo the training of a medical doctor to fully understand human suffering in the large.'[54] Chandra felt certain that Balakrishnan was destined to be one of the best of India's new wave of writers, and tried his hardest to convince him of this. Yet from Balakrishnan's letters one senses that he was still in awe and jealous of his brother.

Under pressure from Eddington, and with little support from his colleagues, Milne badly needed an ally. As a powerless graduate student, Chandra was the perfect candidate. As far as Milne was concerned, his calculation of an upper limit for a white dwarf was simply part of his own research programme: when a star contracted so much that its internal pressure exceeded the pressure allowed for by Fowler's theory, Chandra's came to the rescue. Milne now wanted Chandra to prove that some of Eddington's basic assumptions were wrong – that, in the case of white dwarfs, for example, Fowler's solution was valid right to the core. He wrote to Chandra that:

> From the point of view of general science it is important that where a mistake is made it should be pointed out, as courteously as possible, and as there has been more than enough of controversy between Eddington and myself, it is desirable that it should appear that the pointing out of the error is not merely the consequence of pre-existent antipathies between A.S.E. and myself.[55]

It seemed that Milne was lining up Chandra as his front man to draw some of Eddington's anger away from himself. Everything was piling up against Chandra.

Milne now saw errors everywhere in Eddington's work. He wrote to inform Herbert Dingle, who was on the editorial board of the *Monthly Notices of the Royal Astronomical Society*, that he had found a crucial flaw in Eddington's paper on the limits of temperature in a star which made the whole argument 'absolute nonsense'.[56] He even wrote it up for publication, then two days later had to withdraw it because he realised that Eddington's calculation was correct after all. He complained to Dingle that whatever he wrote on the blackboard during his lectures at the RAS was 'categorically ignored' by his audience, particularly by Eddington, who never left him in peace.[57] He intended to write lengthy replies to Eddington and insisted that Dingle should publish them. Milne submitted another paper to the *Monthly Notices of the Royal Astronomical Society*, but then had to withdraw it and pay the printer's costs. Possibly he had tried to publish the same paper twice. He wrote to Dingle: 'you may have felt afterwards that I had not played the game . . . Since members of the Council think the direct controversy has gone on long enough, I will cease writing controversial papers.'[58]

Milne was also pushing his impressionable young colleague to 'set about tackling some of the harder problems of this subject . . . You should tackle something more ambitious' than the fate of white dwarfs, he insisted. 'For example suppose you take my rotating star paper of 1922 or so and re-do it.'[59] Chandra, in fact, did just this for his Ph.D. thesis. But he was strong enough to resist taking Milne's advice any further than that.

By July 1931, Chandra's friend Chowla had completed his Ph.D. and left for India. 'So my only friend* [*rather acquaintance] here is gone,' Chandra wrote gloomily to his father, demoting Chowla. He added stoically that he had grown 'accustomed to being lonely'.[60] It was all becoming too much to bear. Chandra needed to get away from Cambridge, and from all the pressure there, at least for a while. The same month that Chowla left, he set off for Göttingen, in Germany, one of the centres of the new quantum mechanics. There he had arranged to study under the renowned German physicist Max Born.

Born, then forty-eight, was a very distinguished-looking man with fine, sensitive features. In the 1920s he had been a pioneer of quantum mechanics. He had done important work in electromagnetic theory, applied mathematics, astronomy, acoustics (in the German army during the Great War, when it was used for spotting enemy artillery batteries from the sound of their guns), relativity, crystallography and atomic physics. He was particularly renowned as the mentor of physicists of the calibre of Heisenberg and Pauli. Heisenberg had developed quantum mechanics while working at Born's institute in June 1925. Born's major discovery was the deep meaning of the wave function, a key and much misunderstood aspect of the Austrian physicist Erwin Schrödinger's wave mechanics. In 1926, Born realised that the wave function describes the probability of finding an electron in a certain region of space, whether it is in an atom or moving freely. Einstein, a friend and colleague of Schrödinger's at the University of Berlin, utterly repudiated this non-classical view of science according to which an electron's whereabouts cannot be predicted with absolute certainty, but only the probability of it being in a particular location, a theory linked to the electron's wave–particle duality. 'God does not play dice,' insisted the great man.[61]

For Chandra, Göttingen offered a thrilling opportunity to hobnob with the international physics community. Besides Born, he met the twenty-three-year-old Edward Teller, who enjoyed discussing astrophysics with him. They renewed their friendship between 1933 and 1934, when Teller visited Cambridge. Born in Budapest in 1908, Teller's scientific brilliance was recognised even when he was a boy. At the end of the Great War, the Austro-Hungarian Empire broke up and Hungary became independent for the first time in centuries. The result was political chaos. Teller decided to leave. He went to the University of Karlsruhe in Germany to study chemical engineering. But he soon discovered that his true love was physics and moved to Sommerfeld's institute in Munich. In 1928, he was hit by a tram and lost his right foot. He learned to walk with a prosthetic foot at amazing speed and went on to become an excellent skier. Perceiving that the real discoveries in atomic physics were being made in Leipzig, where Heisenberg had just opened an institute, Teller transferred there and completed his Ph.D. in 1930. At Göttingen he held a post-doctoral position. A handsome young man with floppy black hair, he was to become internationally known as 'the father of the H-bomb'.

A week after he arrived, Chandra wrote excitedly to his father that he was hoping to see Heisenberg shortly. 'I will see if he remembers me. Possibly not!'[62] He did. Heisenberg frequently dropped into the institute at Göttingen, though mainly to see the visitors, not the director, Born. Born was a complicated, rather tormented man and had made many enemies. Chandra too confided that he found Born 'very unpleasant'.[63] While on the continent, Chandra took the opportunity to visit the art museum in Kassel with its collection of Dutch masters; his favourite was Rubens. He went on to Berlin to see the National Gallery and the zoo. From there he went to Potsdam to visit Erwin Finlay-Freundlich, an astronomer of mixed German and Scottish descent and a close friend of Einstein. Freundlich invited Chandra to give a lecture on his recent work which, in his opinion, dealt the 'death blow to Eddington's theory'.[64] Chandra duly reported all this to his father.

Back among the dank fens of Cambridge, nothing had changed. Chandra found himself plunged once again into Milne's increasingly desperate struggle.

Once again the gloom descended. 'On Monday next, I am 21!' Chandra wrote to his father. 'I am almost ashamed to confess it. Years run apace, but nothing done! I wish I had been more concentrated, directed and disciplined in my work.'[65] Babuji replied immediately. How deeply he felt for his son, he wrote, so far away, and so lonely. Chandra, he said, should go to London more often. 'You need *not* save money from your scholarship . . . You have become an *adult* and should start to have your own experiences (emotional and otherwise) . . . I daydream of the near future when you will return to India, [to] take up your professorship (like your grandfather).'[66]

As well as handing out sagacious advice, Babuji was beginning to remind Chandra that it would soon be time to return home. As Amar Singh wrote in a study of Indian students in Britain, 'To the traditional three don'ts – wine, women and meat – one more seems to have been added, i.e. don't stay in Britain after completing your studies.'[67] In India, having had an English education brought enormous cachet. To be an 'English returned' ensured instant access to the highest levels of Indian society.

Things had fallen so flat for Chandra that he was reduced to asking Fowler for something to do. At the time, he was devouring a series of ground-breaking papers by Heisenberg and Pauli on the interactions between electrons and photons, which laid the foundations for a brand new subject, quantum electrodynamics. Hugely excited by these papers, he began to think he should shift to theoretical physics, desperate to 'attempt something "substantial"', he wrote to his father.[68]

Then he received a letter from Milne. Once again, Milne had found an error in Eddington's work, but once again it had turned out to be a false alarm. He had decided to give up corresponding with Eddington, who, he complained, 'always grossly misrepresents my arguments'.[69] Shortly afterwards, Milne paid another unannounced visit to Chandra, at 10.30 in the morning. 'We

discussed for a long time together and he left! ... my general
quality of work is still low! (too low in fact!),' wrote Chandra
gloomily of Milne's blunt assessment.[70]

Chandra pressed on with his calculations and, late in 1931,
discovered another astounding implication of relativistic degeneracy: a star which has a radiation pressure of more than 10 per
cent of its total pressure cannot develop a degenerate core to prevent it from collapsing completely. No matter how densely its
electrons are crowded together, the star's internal temperature
will always be so high that it will remain a perfect gas. Without
degeneracy pressure, a rock-hard incompressible core cannot form,
so there is nothing to prevent the burnt-out star from being
crushed by gravity into an infinitely dense, infinitely tiny point.
This was another way of demonstrating that such stars were liable
to carry on shrinking and shrinking and becoming denser and
denser for ever. But Milne forced Chandra to put this work aside.
'Milne refused to accept the result and the reason was simply that
it violated his thesis' – that all stars must develop degenerate
cores, or some sort of incompressible core, and will never disappear
into nothingness.[71]

Superficially, at least, 1932 started off on a high note. On
8 January Chandra presented a paper at the RAS on a topic guaranteed to be free of controversy. It extended Milne's research on
the Sun's atmosphere and had nothing to do with the internal
structure of stars. 'After I spoke, Sir A. S. Eddington made some
"fine" remarks about my work and Professor Milne was in his usual
raptures!' Chandra wrote to his father. 'The whole thing was quite
a success! But somehow the whole thing fell flat on me.' For some
reason, he felt, he was losing his 'original cheerful and proud view
of things. As one ages so one gets "sour" goes the saying!'[72] At
twenty-one, Chandra already felt he was getting old. In the year
since his arrival in Cambridge, he had published eleven long and
complex papers containing important conclusions and spent over
a month in Göttingen. Yet he was far from satisfied.

The following week, classes began. Chandra was in almost daily
contact with Dirac. 'He is *just* wonderful!' he wrote excitedly to
Babuji. 'His philosophical insight into the general formalism of
theoretical physics, his mathematical profundity to penetrate with

ease any region of unexplained physical or mathematical thought and with all this what humility!'[73] Dirac did the sort of pure physics that Chandra aspired to do. Fowler, however, encouraged him to continue working on astrophysics with Milne. Perhaps Fowler felt that Chandra was aiming too high. But Chandra struggled to avoid immersing himself completely in astrophysics; he wanted to consult Dirac. In Germany he had decided that what he really wanted to do was study with Bohr in Copenhagen, the world centre for quantum mechanics.

During Eddington's course on the structure of stars, Chandra had asked several incisive questions. Eddington 'was quite "floored"!', Chandra wrote to his father. 'Intend to continue the discussion this evening.' But it was not what he was hoping for: 'I met Sir A.S.E. but nothing much came out of it.'[74] Eddington asked Chandra to do some calculations for him, but Chandra said he was too busy with the work he was doing for Milne. He was taking a risk: it was poor strategy for an aspiring graduate student to refuse to help out with a calculation when an eminent scientist asked him to.

Chandra was beginning to worry that he was 'making a mess of the whole thing!' He was becoming more and more convinced that the answer was to change fields: 'My Astrophysical work has become largely a matter of numerical work and by such work one never learns anything.'[75] In his spare time he had begun feverishly reading up on quantum electrodynamics and on group theory. It was Heisenberg who first told Chandra about the latter when he visited India in 1929. Group theory is a beautiful and rather esoteric branch of mathematics which is used in tandem with the Pauli exclusion principle to find the correct mathematical equation to represent the behaviour of collections of electrons.[76]

Then Dirac and Fowler returned 'from a short visit to Copenhagen where, in residence, were Heisenberg, Pauli, Darwin, Kramers, Meitner, Kronig, Coster, an intellectual aristocracy of a breadth and magnitude hardly ever "written" in History,' Chandra breathlessly told Babuji.[77] Fired with excitement, he was more and more determined that he too would breathe that rarefied air and mix with these gods of science. It was bound to advance his work and his career. Perhaps he also thought that there he might be appreciated as he had been in India. He was finally able to consult Dirac, who was in full support of his plan.

By the end of June, Chandra was ready for his trip. In preparation he had been trying to solve problems using the newly learned group theory, but with no success. Then Dirac proposed a problem to him and he dropped everything to work on it. Still in his twenties, Dirac had just been elected to the coveted Lucasian Chair. It was the most prestigious professorship in Cambridge, and had been held by Newton. Nevertheless, Chandra could not wait to leave. 'Cambridge now – in spite of Dirac! – gets on my nerves and the same little ten feet square – my garret! – chokes me,' he wrote to Babuji.[78]

That August he was finally on his way. Bengt Strömgren, of the University of Copenhagen, met him at the station when he arrived. Two years older than Chandra, Strömgren was already a well-known astrophysicist. The son of the director of the Copenhagen Observatory, he had been a student of the stars since boyhood and had been trained in observational astronomy by his father, Elis Strömgren, himself an astronomer of some note. His work on the abundance of hydrogen in stars had changed Eddington's mind on the subject, forcing him to rethink his standard model. But he had also become an expert in quantum mechanics through spending time at Niels Bohr's institute. He was an astrophysicist who could speak the language of physicists.

For Chandra, the atmosphere of high intellectuality in Copenhagen was hugely inspiring. Bohr, a titan of physics, had won a 1922 Nobel prize for his discovery of the first modern theory of the atom. In prestige he was second only to Einstein, who by then was no longer working on cutting-edge issues concerning the quantum physics of atoms and nuclei. Bohr was an imposing figure. In his youth he had been an exceptional athlete and a world-class soccer player, and he still skied, cycled, sailed and played formidable table tennis. By the time Chandra met him he was in his late forties, a heavy-set man with unruly black hair smoothly combed back, a high, domed forehead, heavy features and large hands, one of which usually held his trademark pipe. He spoke in low tones, in heavily accented English. Listeners had to strain to hear him. His hesitant mode of speech reflected his unrelenting desire to look ever more deeply into the essence of quantum physics, which for him meant probing beyond science into thought itself. He liked to quote the words of the eighteenth-century German philosopher and poet Friedrich Schiller: 'only fullness leads to clarity/And truth lies in the abyss'. Einstein once described Bohr as being 'like an extremely sensitive child who moves about the world in a sort of trance'.[79]

'In Munich and Göttingen you learned to calculate . . . In Copenhagen you learned to think,' the young physicists who visited Copenhagen often said.[80] Ideas were bubbling there. The more bizarre an idea seemed, the better it was received. Everything was examined and turned over in group sessions, with no holds barred. There were no nuances or stage antics, no annoying English understatement. Everyone interacted directly with everyone else. The Germans in particular argued tooth and nail. After the debates there was none of the animosity that drove so many of the Oxbridge circle to despair. Students, postgraduates, visitors and established scientists all mixed. Ideas were thrown up, critically examined and debated daily in a common quest to understand the increasingly complex mysteries of the atom. Bohr's famous response to the wildest ideas that came up was, 'That's a crazy idea. The trouble with it is that it's not crazy enough!' Cambridge was dull in comparison.

But as far as Chandra was concerned, there was a problem: Bohr and his colleagues were not interested in astrophysics. Physicists and astrophysicists were poles apart. It was not just that physicists focused on what made stars shine while astrophysicists busied themselves building models of stellar structure. Their philosophies were diametrically opposite. In his astrophysics research, Eddington played down mathematics and insisted on using just enough of it to analyse data (though ironically his research on general relativity and the fundamental theory was strewn with abstruse equations). Physicists, on the other hand, used mathematics – a very hybrid and strange variety – as their launching pad and sole guide for their exploration of unknown territory full of ambiguities, counter-intuitive concepts and singularities. Chandra recalled Hans Bethe, the physicist who finally solved the problem of how stars shine, saying that the 'attitude of physicists towards the astronomers was one bordering on contempt'. Eddington was the 'sole exception', due to his 'enormous prestige'.[81]

'I didn't belong to the scientific community any more in Copenhagen than in Cambridge,' Chandra later recalled. 'For example, I remember that every Friday in Bohr's house there used to be a tea [after which Bohr] and others would go with Bohr to his study to work. I used to stay behind and play with [Bohr's little boys]. But the atmosphere was better [than Cambridge] and I made so many lasting friends.'[82] He even relaxed enough to join the boisterous young scientists in a 'harmless bit of girl-watching [and] soon developed a system of mentally classifying them, according to their looks, under three heads – Alpha, Beta and Gamma!'[83]

Inspired by the liberating atmosphere of Copenhagen, that September Chandra boldly submitted a paper to the *Zeitschrift für Astrophysik* which included the result that Milne had rejected so vehemently when he had proposed it the previous year: that stars whose radiation pressure is greater than 10 per cent of their total pressure cannot develop a rock-hard core and will eventually shrink to nothingness. The idea was to bypass the Cambridge circle, for the *Zeitschrift*'s editorial offices were in Potsdam. But as luck would have it, Milne happened to be passing through town when Chandra's paper arrived. The editor, Finlay-Freundlich, whom Chandra had met the previous year, asked Milne to look it over. The response was all too predictable. 'Milne wrote a long letter of some twelve pages to me arguing that my conclusion was fallacious and that I would do harm to my "growing" scientific reputation by publishing the paper,' Chandra recalled. 'But I stood my ground with the encouragement particularly of L. Rosenfeld. And the paper was eventually published. This experience left me sceptical, already in 1932, of Milne's objectivity.'[84]

Chandra recalled somewhat bitterly that he 'modified the paper a bit to mollify Milne',[85] reducing the impact of his conclusion with the caveat, 'We are unable to avoid the central singularity by appeal to Fermi–Dirac statistics alone.'[86] (Fermi–Dirac statistics is the quantum statistics used for electrons.) The key word was 'alone', suggesting that there might be some as yet undiscovered way, other than Fermi–Dirac statistics, of avoiding the seemingly inevitable conclusion that the star would eventually collapse to a singularity.

Put in the hopeless position of being forced to try to save Milne's theory, Chandra waffled about the possibility of *'incompressibility setting in late'*.[87] Maybe somehow, at the last minute, the star might develop an incompressible core. Nevertheless, even Milne had to concede that his theory was doomed. Chandra's latest discovery proved incontrovertibly that, in certain cases, stars did not need to have degenerate cores and thus, under certain conditions, may remain a perfect gas – exactly as Milne's arch-enemy, Eddington, postulated in his standard model. Chandra's paper had a totally devastating effect: 'My work did not alter [Milne's model] in a trivial way: it eliminated all of it!'[88]

Meanwhile, at Copenhagen the magic was beginning to wear off. 'Bohr is difficult to pin down for a conversation. I have had opportunities to speak to him on various occassions,' Chandra wrote to his father – once again misspelling 'occasions' – 'rather hearing him speak to me!'[89] The origin of this souring of relations was the quantum mechanics problem that Dirac had given Chandra, concerning the statistical mechanics of electrons and photons. Unthinkingly, Chandra wrote up his results and gave the paper to Bohr to submit to the *Proceedings of the Royal Society*, completely bypassing Dirac. He had mentioned Dirac's name in the acknowledgements, so the editor sent it to Dirac to look over. Dirac wrote directly to Bohr: 'It seems to me that the result of the paper must be wrong.' Chandra had overstepped the bounds of mathematics as it is applied to physics. It was an old habit of his; Heisenberg had caught him out on a similar point two years earlier. In essence, he had tried to deduce 'purely from mathematical considerations' – namely, group theory – that there were other possible kinds of statistics besides the known ones that apply to electrons and photons. Dirac added that if Chandra's paper were published, he would have to append a note stating his disagreement.[90]

Chandra insisted he was right and even claimed that Dirac had never read his paper. 'This means another long drawn out controversy,' he wrote dejectedly to his father.[91] What happened turned out to be rather embarrassing for all concerned. Chandra had

omitted the impact of spin on his calculations. Bohr and his colleagues too had failed to notice this error. 'Our only excuse for
overlooking so trivial a point may perhaps be that we were under
the impression that you, at any rate for a moment, had considered
the question worth investigating. But even that may be a mistake,'
Bohr wrote to Dirac.[92] He discussed the matter with Chandra.
Gloomily, Chandra agreed to withdraw the paper. Hugely down-
cast about what had happened, Chandra told Babuji that the
'Quantum mechanics paper is WRONG . . . 6 months of hard
work is all SMOKE.'[93]

That failure put an end to his thoughts of becoming a theoretical physicist. 'I have become a confounded pessimist – for
instance I hold the view that "things never get better, they always
get worse" – believing in nothing,' he wrote to Balakrishnan from
Copenhagen. The only thing to do seemed to be to concentrate
on astrophysics and deal with his backlog of work; but it would
mean more squabbles with Milne. On the bright side, he was
working hard on his Ph.D. thesis. He reassured Balakrishnan that
although in the West he had acquired 'new habits of thought', this
applied only to his intellectual life. If Balakrishnan were to ask
whether he had changed his ways regarding 'smoking, drinking,
dancing, going out with girls, eating meat (or even putting on a
hat!) then the answer is negative', he told him firmly.[94]

Chandra left Copenhagen on 17 March 1933. On his way back
to Cambridge he passed through Hamburg to see the mathematician Emil Artin. Artin was an important contributor to number
theory, a highly abstruse branch of mathematics. Chandra's friend,
Chowla, knew Artin and wrote that he was 'astonished that
[Chandra had] gone so far into the algebraic theory of numbers'.[95]
The visit with Artin went well. Chandra still dreamt of becoming
a mathematician and wrote excitedly to his father, 'Might study
with him in July instead of Milne!'[96] But by now everyone was
aware that Germany was on the path to war, so he had to abandon
his plan.

Back in Cambridge, Chandra looked into sources of funds to
extend his stay in England. His father was putting relentless pressure on him to return to India after he had completed his Ph.D.
'Your *mathematical* work *can be done in India*,' he wrote. 'There are

more things in this world to experience and realise, than *merely "intellection"*. There is the experience of love of a woman, the love of a child. These two in due course enrich a man's life [because] she is a complement to your mental and moral character.'[97] He urged Chandra to think about it. After all, he declared, a father knows best. But Chandra dreaded the thought of going back to his homeland.

Springtime brought a round of garden parties, including one in May at the Eddingtons', where Chandra met Russell. He had finally finished his thesis, for which he had chosen to elaborate and extend Milne's results on the shape of stars, taking the conventional route of focusing on a traditional problem area. He described his work to Russell, 'who was frightfully enthusiastic about it all'.[98] He asked Fowler if he wanted to see it. Perhaps wary of the fearsome detail involved, Fowler's response was, 'No, definitely no! Just hand it to the registrar.'[99]

Chandra's Ph.D. viva, with Fowler and Eddington as examiners, took place on 20 June in Eddington's study at the Cambridge Observatory. As Chandra later recalled, it was almost comic. Fowler roared up in his car fifteen minutes late. While Chandra and Fowler were in formal academic gowns, Eddington was in everyday clothes and slippers. He looked at them and muttered laconically, 'I cannot rise to the occasion.' The two professors then quizzed Chandra on his thesis and on astrophysics in general, interrupting Chandra's answers to argue ferociously with each other over obscure details. Forty minutes into the examination, Fowler suddenly looked at his watch, exclaimed, 'Goodness, I'm already late!' and abruptly left. As Chandra stood there, at a loss as to what to do, 'Eddington simply said, "That is all."'[100]

Shortly afterwards, Babuji wrote that he was lonely and wanted to come to London. This was the last thing Chandra needed. He wrote back quickly, listing all the drawbacks: the poor quality of English vegetarian food, the rice half-cooked and the curries frightfully hot, and that when his father wanted to come it would be winter and he would find the lack of sunshine disturbing.

With the situation in Germany growing steadily worse, many of the country's leading scientists, most notably those who were Jewish, were fleeing. Chandra's Uncle Raman had high hopes of expanding India's scientific community, declaring that 'Germany's loss is India's opportunity'.[101] Chandra disagreed: 'For one thing all the "great men" have (or can) found (or find) positions in Europe or America (Max Born for instance is coming to Cambridge),' he wrote to his father. 'Only the younger men – the Jews! – will at all think of going to the tropics and C.V.R. [Raman] by over looking them is shooting off his mark.'[102]

Determined, if at all possible, to stay on after completing his Ph.D., Chandra had decided to go through the arduous process of applying for a fellowship at Trinity. Fowler was positively discouraging about his chances of getting one, but he persevered nonetheless. After all, he had published more than ten papers as a graduate student, on almost every facet of astrophysics. He had received rave reviews when he lectured on his thesis research at the RAS. Milne declared it 'beautiful', while the formidable Russell congratulated him on the 'lucid presentation he has given of an intricate problem', adding that 'his work is going to be of great practical value'. Although Eddington had some doubts about the way in which Chandra had modelled certain stars, he too found it highly interesting.[103]

But despite the depth and breadth of his publications and accolades from the most important astrophysicists of the day, weighing heavily in the balance against him was the fact that he was Indian. Only one Indian had ever achieved such an honour: the great mathematician Ramanujan. Chandra was so convinced that he would not succeed that, on the day the fellowships were announced, he had already packed and was preparing to leave his lodgings, to go to Oxford and work with Milne for a term before returning to India. Purely out of curiosity he stopped by at Trinity to see who had won. 'I was astonished to find my name among the people who were elected,' he later recalled. 'Well, this has changed my life!' he told himself. 'I do not know what my future would have been, except to say that it would have been very different.'[104]

Chandra wrote rather movingly to his father, describing his many conflicting feelings on accepting the honour, which would mean that he would not be returning to India for several years:

I may as well mention (to be fair to myself) that this final decision has not been made without great pain, and indeed during the whole of Sir J.J's [Thomson's] speech last evening, and during all the time when so many were congratulating me there was only one picture before me – my mother in her silk saree tying with trembling hands a thread round my head praying God to look after me, and blessing me with all the force of love to "*go forward*". Indeed I always have this vision which has been a great source of inspiration – intensely saddening yet stimulating.[105]

Milne wrote to congratulate Chandra. Later that year, he wrote warmly to his young colleague, 'Perhaps you would do me the honour of dropping the title "Professor" in our future correspondence. It used to be a good rule at Trinity that once a man became a Fellow he dropped all titles with other members of the high table.'[106] At last, Chandra had arrived.

Babuji's reply to Chandra's letter was full of joy, pride and happiness, but mixed with sadness. 'Your decision is not news to me,' he confessed. 'What I feared, you have stated.' Nevertheless, he could not help but be happy for his son: 'What you have achieved is only my dreams or hopes come true, of my son – obtaining intellectual "laurels".' He was, he added, definitely going to come to Europe.[107]

To everyone's surprise, Babuji's trip was hugely successful. He set sail on 9 November 1933 and had such a wonderful time that he extended his leave of absence from the accounts office where he worked, and did not return until 9 July the following year. A violin player of some renown in India, he had arranged to give several private recitals. He performed in London, Paris, Vienna, Florence and Geneva, among other places, and one performance was broadcast on BBC radio. Father and son spent the Easter holidays together in Munich. Chandra was far happier than he had ever expected to see Babuji. They talked far into the night about many things, including literature. Chandra had to apologise for seeming perhaps too argumentative: 'It's just the excitement.'[108] Babuji crossed the Channel again in March. He performed in Aberdeen, Scotland – which he particularly enjoyed – and then in the North of England. In May he arrived in London, where he and Chandra spent more time together.

Shortly after he left, at the beginning of June, Chandra wrote him a moving letter. His father's trip, after all, had been sparked by his own decision to stay on in England. Chandra assured him that he had taken care to maintain the ethical and moral standards that his parents had instilled in him. 'I have learned to know you better during your stay in England and I am sincerely glad that you did finally come to Europe,' he wrote. 'Finally the kind and sympathetic way in which you understood me on the last night of your stay in 7 Worsley Road moved me very much and when I went to my room I prayed: "Oh! God, Let me be the Son of my Father".'[109]

That February, Chandra had published a short paper in *The Observatory* summarising the results he had originally dared publish only in a foreign journal, the *Zeitschrift für Astrophysik*, which effectively demolished Milne's work. Now that he was a Trinity Fellow, he felt more sure of himself. 'The general evidence then is in favour of Eddington's perfect gas hypothesis for ordinary stars,' he concluded, perhaps to draw Eddington's attention to the fact that he was switching camps and joining the winning side.[110] White dwarf stars, however, were an entirely different matter, as Eddington too agreed.

In July that year, B. P. Gerasimovich, Director of the Pulkovo Observatory in Leningrad, invited Chandra to the Soviet Union. The two had corresponded since Chandra's days at Presidency College, and Gerasimovich had expressed great interest in his progress.[111] Like other Western scientists who visited Russia in the early 1930s, Chandra was taken on a special tour which ensured that he would come away with the impression of a free and dynamic country; the darker side was concealed. Chandra was very impressed. As he told Balakrishnan, 'Russia gives the impression of a young man, full of ideals, who has such indomitable moral courage and indefatigable physical strength to go forward in spite of setbacks and who with his hopes derives consolations from his ideals during times of adversity.'[112] He did not realise it at the time, but this idealistic personification of Russia would come to fit himself all too well.

In Leningrad he met the Russian physicist Lev Landau and the

Armenian astrophysicist Viktor Ambartsumian. Ambartsumian invited Chandra to give a lecture there on his results on white dwarfs. Much impressed, he urged Chandra to write these results up in as broad and rigorous a manner as possible. Until now Chandra had lacked a firm mathematical basis for the astounding implication of his theory of the upper limit for the mass of a white dwarf. Ambartsumian suggested that he needed to deduce it from the mathematical formalism that astronomers used to examine the structure of spheres of gas no matter which gas law they obeyed – perfect, imperfect, classical or quantum.[113] If this exact version of his theory was right, it would be expected to yield Fowler's result for white dwarfs, which holds for electrons travelling at speeds much less than that of light. An important step in validating and clarifying a new result in theoretical physics is to demonstrate that it reverts back to previously accepted cases which, it would turn out, were valid only within certain limits. This is the case with Einstein's theory of special relativity: for speeds much less than that of light, it becomes Newton's theory of motion. Chandra thought the idea an excellent one and resolved to turn to it when he was back in Cambridge.

In Moscow, Chandra gave a lecture at Moscow University's Sternberg Astronomical Institute. In the audience was Yakov Zel'dovich, Russia's explosives expert who was then developing an interest in astrophysics. Landau may also have been present. Chandra's reception in the Soviet Union made a welcome change from Cambridge, where everyone either ignored his work or claimed it as a special case of their own.

On his return to Cambridge, Chandra turned his attention to personal matters. Something had been on his mind for a while. He wrote to Babuji, 'I wonder if you would excuse me if I made bold to be impatirent [impertinent] enough to ask what you thought of Lalitha – she wrote to me that you had met her.'[114] Babuji wrote straight back. He had met 'Miss D.L. . . . on *three* occasions,' he said. Lalitha had helped him with the sound quality of his performances of Indian music. His sense of propriety prevented him from saying anything very specific, but he was prepared to concede that his 'general impression is that she is a modest and quite reserved young lady'.[115]

Chandra's response stunned Babuji. Throwing aside Indian tradition, he did not bother to ask his father's permission but came straight to the point. 'It was very kind of you to have understood me and written to me of your conversations with Lalitha,' he replied. 'Actually now Lalitha and I have come to *definite* mutual understanding that we should get married on my return. L knows too that you are aware of the fact that L and I have been interested in each other for these years.'[116]

Babuji wrote back rather touchingly, 'On receipt of your letter, I could not exactly express the feelings which arose in me. I was incredibly more sad than happy – the reason which caused the sadness to arise in me was that your mother was not alive to welcome home your bride-elect.' He confessed that he had had some reservations. 'My impression of the alliance was not favourable' because of Lalitha's aunt, who had 'married a certain gentleman – who had already a wife and some children.' But of course, he went on, 'present-day morals are different from the past generation. All this flashed in my mind in a space of 2 minutes or less . . . I only pray to God that the union you propose should be blessed and happy.' He ended on a sentimental note, quoting from Alfred, Lord Tennyson's 'Idylls of the King' (Tennyson had been Babuji's favourite poet when he was a romantic young man of Chandra's age): 'I do hope and pray that "this maiden passion for a maid" will bring all happiness that a father can wish for his son.'[117]

At the end of October, Chandra submitted a paper on some work in progress, summarising three major papers he was preparing to publish in 1935.[118] It was notably more self-assured than the earlier articles he had written on the subject, and promised broader results than any before.[119] He wrote that he had deduced the precise equation governing the structure of spheres of gas made up of slow- and fast-moving electrons, as Ambartsumian had suggested. It was so straightforward, he added, that 'it is surprising that it has not been isolated by anyone previously'.[120]

In the months that followed, Eddington visited Chandra's rooms regularly, two or sometimes three times a week. He even arranged for Chandra to be given the fastest mechanical hand-calculator available. Something strange was going on – perhaps he smelled blood. Chandra had convinced him that the seemingly endless Eddington–Milne controversy was finally to be settled, and in Eddington's favour.

As he was to state in the summary of his paper at the RAS meeting on 11 January 1935, Chandra's thesis was that 'For a star of small mass the natural white dwarf stage is an initial step towards complete extinction. A star of large mass [greater than the upper limit for white dwarfs] cannot pass into the white dwarf stage, and one is left speculating on other possibilities.'[121] He put together his results in a graph which he exhibited at the RAS, which showed very clearly that as the mass of a white dwarf reaches the maximum mass, the star's radius becomes zero and the star shrinks to nothing.[122]

Chandra had already notified Milne of his results. Milne congratulated him: 'the concept of a "complete" equation of state [for relativistic degeneracy] is valuable and new to the subject formally', he wrote. But he was still unclear about Chandra's concept of the upper limit to the mass of a white dwarf. He repeated his query: 'If we add a piece of dust to the star' whose mass is just under the maximum mass, will it become unstable? To which he added, 'You may think me very conservative and stick-in-the-mud.'[123] Lack of stability was anathema to a physicist steeped in classical physics.

The battle lines were drawn. Chandra had shown Eddington and Milne the manuscripts of his first two 1935 papers. They were highly detailed, dense with mathematical formulae and extensive numerical results. In November 1934, Chandra invited Milne to be his guest at the Trinity Commemoration Dinner. 'During this visit, I almost persuaded him to concede that his work ... was essentially left without any basis,' he recalled. 'I was hopeful that my work would dissipate the Eddington–Milne controversy.'[124]

The energetic director of the Harvard Observatory, Harlow Shapley, a former student of Russell's, had also begun to take an interest in Chandra. He suggested that he spend some time at Harvard. Babuji was becoming more and more concerned. An invitation by someone of Shapley's stature had to be taken seriously. Chandra's return to India seemed to be receding ever further into the future.

The day was fast approaching of the all-important meeting at

which Chandra was to present his result. 'I am overpowered with
the work I have to do before January on my stellar structure,'
Chandra wrote to his father, adding, 'As the meeting day
approaches I have to work more and more and faster and faster.'
He was running out of time. As well as his calculations and new
results, there was his impending marriage to Lalitha and the need
to schedule a trip to India. 'I feel I am pretty deep in a mess,' he
wrote:

> I should feel much worse if it were not that I am now passing
> through an essentially successful period in my scientific life.
> When I am deeply absorbed in my work I often recall a
> statement which Bohr made when he heard of the
> engagement of Rosenfeld – 'It is a great pity that human
> beings cannot find all their satisfaction in scientific
> contemplativeness.'[125]

Chandra's brilliance and the originality and truth of his discovery were surely about to be recognised at long last, even by the
stern old men who populated astrophysics at Cambridge.
Everything should have been euphoric, yet Chandra could never
put aside the darkness in his mind. 'Involved in the puzzles of the
interior of stars, battered by differential equations, boxed by
numerical calculations, impeded by ignorance, rushed on by the
dawning of the New Year, I have at last emerged not indeed with
the anticipated joy with which I began [diving] into the Crucibles
of Nature, but burnt and smoking, dissatisfied, tired.'[126] Thus
Chandra wrote to Balakrishnan in his last letter before that fateful Friday, 11 January 1935.

6

Eddington's Discontents

On the morning of 12 January 1935, Chandra awoke into another world. The previous day he had been full of hopes and fears, apprehension and excitement, daring to imagine a glorious future for himself after he revealed his momentous discovery. He must have dreamt of being greeted with the kind of acclaim and recognition he had enjoyed five years earlier in India. Perhaps he hoped that, even though he was young and Indian, he would finally be accepted into the inner circles of Cambridge academia. Instead, Eddington had utterly humiliated him.

Eddington's rebuttal had been delivered with far more than his usual venom. For Chandra it had been a complete shock. The essence of the great man's argument, delivered with superlative arrogance, was that 'there is no such thing as relativistic degeneracy!'[1] Relativistic degeneracy was, of course, the bedrock of Chandra's theory. Without it, the situation reverted to Fowler's theory, which applied only to a dense gas of electrons moving much more slowly than light. In Fowler's theory there was no limit to the mass of a white dwarf. Such a star would always die a peaceful death, no matter how massive it was. Eddington was claiming that Chandra was totally wrong and his work worthless.

At the time, astrophysicists assumed that most stars turned into white dwarfs at the end of their lives, except for a few that blew themselves to pieces. All stars, they thought (mistakenly, as it turned out), went through the same lifecycle, zigzagging across the HR diagram. They began as brightly burning giant stars, then joined the main sequence as dwarfs, and continued to contract until they finally burnt off all their fuel and expired as white dwarfs – hot but dim, tiny but astoundingly dense. But here was Chandra, daring to suggest there might be another end to the life of the stars.

Eddington continued his rebuttal with the convoluted argument that, while he agreed with Chandra's mathematics, he considered the physics incorrect because it was based on a combination of two diametrically opposed theories – relativity and quantum theory which, as far as Eddington was concerned, did not allow for the possibility of electrons travelling at speeds comparable to that of light: 'and I do not regard the offspring of such a union as born in lawful wedlock'.[2] Chandra was aghast. Thoughts ran through his mind at a furious pace. Surely, the man must have his head in the sand! Had not their colleague at Cambridge, Dirac, been awarded a Nobel prize in 1933 for formulating just that – a relativistic version of quantum mechanics? What was going on? But worse was to come. Next, Eddington claimed that the Pauli exclusion principle, which was at the very basis of Chandra's theory of white dwarfs and of quantum theory itself, was not a general law of nature. As Pauli himself would have put it, 'Why, that's not even wrong!'

In view of this, Eddington argued that although Chandra's mathematics might be correct, the result – relativistic degeneracy – had nothing at all to do with the stars. It was just playing with numbers. This is what Eddington had meant by saying that 'there is no such thing as relativistic degeneracy'. But he was not finished yet. He added that Fowler's theory of white dwarf stars, which did not agree with relativity theory, as Chandra had discovered, actually did take relativity into account. In order to prove this Eddington resorted to arguments so convoluted and obscure that they would eventually wreck his reputation as a scientist.

Eddington, of course, was fully au fait with relativity theory, more so, perhaps, than anyone in the world other than Einstein. He was also deeply involved in his own effort to construct a theory that would combine general and special relativity with quantum theory – namely, his fundamental theory. So to refer to such a union as not 'lawful wedlock' was inconsistent, to say the least.

In his book *The Internal Constitution of the Stars*, Eddington had happily considered the implications of Einstein's general theory of relativity on the study of stars. According to the general theory – which, unlike special relativity, encompasses the effects of gravity – space (or, more properly, spacetime) is distorted by objects to a degree determined by the object's gravity. Spacetime has a structure, a fabric, and the extent to which that fabric is distorted depends on how massive the object is. In the same way that a stone placed on a rubber sheet will create a hollow, an object creates a well in spacetime; the heavier the object, the deeper it sinks. The hollow in spacetime in which a large object is is called a gravitational well. Thus the curvature or warping of spacetime around a massive body is a measure of the object's gravity – the stronger its gravity, the deeper the well. Far away from massive objects such as stars, spacetime is more or less flat.

Eddington had used the general theory of relativity to try to understand what happens when stars collapse under their own gravity. According to general relativity, he pointed out, a small dense star would create a far deeper gravitational well than would a star like the Sun. It would distort space around it so much that no light could escape, and it would disappear; it would, in his phrase, be 'nowhere'.

But to Eddington all this was mere fanciful speculation. So he pooh-poohed Chandra's argument that if a star burnt up all its fuel and still ended up with a mass exceeding the upper limit which Chandra had discovered, it would inevitably collapse to a singularity, infinitely small and infinitely dense. If that was so, then for Eddington the only conclusion was the apparently absurd scenario that 'the star has to go on radiating and radiating and contracting and contracting until, I suppose, it gets down to a few kilometres in radius, when gravity becomes strong enough to hold in the radiation, and the star can at last find peace'. But for Eddington this was unthinkable. 'Dr. Chandrasekhar has got this result before, but has rubbed it in in his last paper', he concluded dismissively.[3]

As Chandra recalled in later years, Eddington was the only one to realise, with his 'enormous physical insight', that the 'existence of a limiting mass implies black holes' and that this must occur 'once one accepts the physics. If [he] had accepted that, he would have been 40 years ahead of anybody else.'[4]

Yet despite the flimsiness of Eddington's arguments, the established scientists had chosen to support him against Chandra, the outsider. What must have been bitterest of all for Chandra was that he had full confidence in his own discovery. He knew in his bones that he was right. He would have to pull himself together. But who could he turn to for advice and support? Certainly not Babuji; he wouldn't understand. McCrea had disappeared after the session, and in any case seemed to be a blind follower of Eddington. Chandra's close friend Léon Rosenfeld came to mind. He dashed off a letter to him.

At thirty-one, only six years older than Chandra, Rosenfeld already had a sterling pedigree. A warm, easygoing Belgian physicist, always immaculately dressed, with glasses and a high, domed forehead, he had completed his Ph.D. at the University of Liège in 1926. He then spent two years in Paris with the maverick physicist Louis de Broglie – an outsider in the science community, perpetually at odds with the interpretation of quantum theory hammered out by Bohr and Heisenberg. Rosenfeld moved on to become an assistant to Wolfgang Pauli, and after 1930 began a lifelong collaboration with Bohr. Chandra had first met him when he visited Bohr's institute in Copenhagen in 1932. The two became firm friends, particularly after Chandra went to stay with him and his family in Liège. Rosenfeld knew how to handle himself in the intellectual rough-and-tumble of Copenhagen, and perhaps would have some helpful advice.

Chandra wrote immediately, telling him what had happened and asking him to pass the news on to Bohr. 'Eddington sprang a surprise on everyone,' he declared. Eddington, he explained, had dismissed Chandra's method for deriving his new equation (describing a gas of electrons governed by relativistic degeneracy, inside a white dwarf) as completely wrong, and worse still, had claimed that Chandra had used two fundamentals of physics – Pauli's exclusion principle and the theory of relativity –

Eddington at his writing table in 1931.

Portrait of Chandra as Fellow of Trinity College, Cambridge in 1934. (AIP Emilio Segrè Visual Archives)

Marina Beach in Madras, where Chandra and his brothers often went to escape the summer heat. (Author photo)

Ralph H. Fowler a 'big man in every sense of that phrase', circa 1930. (Courtesy Royal Society)

Chandra's favourite picture of his friend, Edward Arthur Milne, taken by Chandra at the McDonald Observatory, Texas, in 1939. (Courtesy Meg Weston Smith)

James Jeans lecturing at the Royal Astronomical Society, March 1935, in the meeting room where Eddington and Chandra had clashed two months earlier. (Courtesy Royal Astronomical Society)

G. H. Hardy, the 'quintessential beautiful man', in around 1930. (Master and Fellows of Trinity College, Cambridge)

Paul Dirac (on the left) and Werner Heisenberg (on the right) looking dapper during Heisenberg's visit to Cambridge in 1933. (AIP Emilio Segrè Visual Archives)

Arnold Sommerfeld in 1927, a year before he met Chandra in Madras. (AIP Emilio Segrè Visual Archives)

The Solvay Conference in 1927. Einstein is fifth from the right in the front row. In the middle row, Dirac and Compton are fifth and sixth from the left, directly behind Einstein. Born and Bohr are the last two on the right. Schrödinger is sixth from the right in the back row. Pauli is fourth from the right and Heisenberg is to his left, followed by Fowler (tall with a moustache). (AIP Emilio Segrè Visual Archives)

Lev Landau, George Gamow and Edward Teller (left to right) with Bohr's sons Aage and Ernest in the garden of Bohr's institute in 1931. (Niels Bohr Archive)

Taken at the 1939 Paris conference on novae, supernovae and white dwarfs. Left to right, front row: Stratton, Cecili Payne-Gaposchkin, Henry Norris Russell Amos Shaler, Eddington, Serge Gaposchkin. Back row: Gerard Kuiper (fourth from left), Bengt Strömgren (fifth), Chandra, Walter Baade (end o row). (Courtesy Yerkes Observatory)

Charles Trimble with students at Christ's Hospital School in 1923. (Courtesy Christ's Hospital)

C. Subrahmanyan Ayyar (Babuji), Chandra's caring father in 1917.

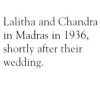

Lalitha and Chandra in Madras in 1936, shortly after their wedding.

Enrico Fermi (on the right) and Edward Teller (on the left) at the University of Chicago in 1951. (AIP Emilio Segrè Visual Archives)

J. Robert Oppenheimer (second row, second from the left, in jacket and tie), Richard Feynman to his left and Enrico Fermi (first row, third left) at a seminar at Los Alamos in 1944. (AIP Niels Bohr Library)

Stirling Colgate in Los Alamos in 2003. (Author photo)

Einstein and Eddington in Eddington's garden in Cambridge University in 1932. Eddington recalled that he once had invitations for dinner with Einstein in London and Manchester on the same evening. He continued, 'This is doubtless explainable by the Principle of Indeterminacy; still I hope on this occasion ψ [the wave function, i.e., Einstein's] will have a more condensed distribution'. (Winifred Eddington)

Chandra giving his lecture on 'The Series Paintings of Claude Monet and the Landscape of General Relativity' at Presidency College, Madras, in 1992. Lalitha is at his side. This is the same room where he courted her more than sixty years earlier. (Courtesy the Physics Department, Presidency College, Chennai)

A relaxed Chandra at the Raman Institute, Bangalore, in 1982. (Courtesy G. Srinivasan

incorrectly. Instead of feeling satisfied with the fruits of working twelve hours a day for the last four months, he was 'terribly worried'.[5] Could all his work have been for nothing? The next few months saw a flurry of letters between Chandra and Rosenfeld. Rosenfeld fully supported Chandra. Eddington's remarks, he declared, were 'utterly obscure'.[6]

One assertion of Eddington's particularly astounded Rosenfeld and Bohr: that Pauli's principle depended on considering electrons only as standing waves as opposed to progressive or travelling waves. If you attach one end of a rope to a door knob, then shake the other end, you can generate a wave that appears to move but in fact simply goes up and down – a standing wave. Progressive waves move forward, like those in the sea. To analyse electrons in an atom, you can imagine them to be confined and therefore at rest, like standing waves. Eddington declared that this was the correct way to apply the Pauli exclusion principle. But Chandra, claimed Eddington, treated electrons as progressive waves because they are in motion inside a star, then transformed them into standing waves by applying a mathematical technique from relativity theory. Eddington claimed this was wrong: Chandra should have treated electrons like standing waves from the start in order to apply the Pauli principle correctly, but instead had combined special relativity with a version of quantum theory that did not agree with it – the 'unholy union' Eddington had referred to at the RAS. The type of wave, however, is totally irrelevant to the end result of any quantum calculation, rather like deciding whether to do calculations in miles or kilometres. Eddington's line of argument was entirely specious.

'Nobody ever dreamt of questioning Pauli's principle,' Rosenfeld wrote to Chandra. Chandra's use of relativity theory was also beyond question. 'These quite obvious arguments would seem to settle the question beyond any doubt. So I think you'd better cheer up and get on with your own work instead of losing one's time with fruitless arguments.'[7] This was the opinion not only of Rosenfeld, but of Bohr and Dirac too.

Chandra, however, was not prepared to listen to advice, even from such eminent sources. Soon after their confrontation, Eddington sent Chandra a manuscript with the same title as his lecture, 'On relativistic degeneracy'. In it he went into elaborate detail, setting out the arguments for his claim that Fowler's result for white dwarf stars, which did not take into account electrons moving at speeds comparable to that of light, and so did not include relativity theory, actually did.

Eddington accomplished this remarkable feat by playing fast and loose with relativity in ways that, although mathematically correct, have no meaning in physics. In other words, he brashly cooked up his own version of relativity theory and claimed it was the correct one. He even went so far as to insist, with his customary aura of authority, that Fowler himself should have realised this in 1926. Things were so confused at first that even Fowler was unsure of what Fowler himself had done.[8] From the remarks he made on 11 January, it was clear that Eddington had already been preparing this paper behind Chandra's back. When he visited his rooms, ostensibly to offer encouragement, his real motive had been to check on what Chandra was doing. It was sheer mean-spirited duplicity. Chandra was caught totally unawares.

Throughout early 1935, Eddington published a series of papers attacking Chandra. They were a hodgepodge of dubious claims and poorly argued assertions. He belittled Chandra's theory as a 'reductio ad absurdum', meaning that relativistic degeneracy might be an implication of the mathematics of Chandra's theory but did not actually exist in any real sense.[9] The only reason Chandra had deduced it, he argued, was because he was using the 'wrong' relativity – that is, not Eddington's version, which included his own mathematical formulation of Dirac's theory. Had he used the 'right' one, he would have ended up with Fowler's result – that white dwarfs remain stable regardless of their mass when they have exhausted all their fuel. What Chandra had done, in Eddington's view, was to defeat the 'original purpose of Fowler's investigation'.[10]

But what was Eddington himself really after? What was his real objective? Chandra agonised over this question. Perhaps everyone was missing something in Eddington's argument. Even Fowler had wavered for a moment, though he soon came round. He told Chandra, 'Don't be worried. I'm sure Eddington is wrong' and, noticing how miserable the young man looked, added in his usual genial way, 'Be cheerful!'[11] But he never confronted Eddington on Chandra's behalf.

Chandra continued to pester Rosenfeld: 'Honestly! Is it possible that there is anything' in Eddington's papers?[12] 'After having courageously read Eddington's paper twice I have nothing to change my previous statement: it is the wildest nonsense!'[13], Rosenfeld replied emphatically.

In fact, the real reason why Eddington was so opposed to Chandra's theory was that it entirely undermined his fundamental theory, which he had now been honing with obsessive care for seven years. Eddington's aim was to unify the general and special theories of relativity with quantum theory using a mathematics that could also produce the fine-structure constant. The crux of the fundamental theory was that there are no systems in nature in which the interactions between particles can be ignored. Therefore, electrons never operate in isolation. But now Chandra had produced a far-reaching, experimentally verifiable theory from studying an idealised quantum gas of *non-interacting* electrons. If Chandra's theory was valid, then Eddington's fundamental theory and his whole mathematical system of E-numbers would come tumbling down. Instead of being, as he thought, a unique way of deducing the physical world from its mathematical structure, it would be relegated to mathematical game-playing, as would his calculation of the number of electrons and protons in the universe, which connected so perfectly with its rate of expansion.

Determined to save his theory, Eddington suddenly and with great ferocity began including white dwarfs in his writings on it. He also began to make more and more fantastic claims about relativity theory, trying to rescue Fowler's more acceptable theory on the fate of the stars, while attacking Chandra's as mere mathematics which had nothing to do with the true nature of the heavens.[14] Like everyone else in the world of science, Chandra was aware of Eddington's determination to develop his fundamental theory and of the gravity and importance of the project. It was beginning to dawn on him that it was this that lay at the heart of Eddington's objections.

Rosenfeld declared Eddington's two papers of 1935 to be 'nonsense',[15] while Chandra wrote more temperately, 'we are quite unable to follow his arguments'.[16] Yet Eddington's authority and his uncanny skill at arriving at correct results from dubious premises meant that many astrophysicists were still not prepared to reject his arguments out of hand. Besides, Chandra's theory produced a result that was beyond anyone's imagining.

Stoner, of course, had already come up with the idea of a limiting mass for white dwarfs, though he had never pushed it through to its logical conclusion. Strangely, while Eddington chose to attack Chandra, he actually encouraged Stoner in his research. However, he kept tight control over Stoner's work, advising him to do no more than to explore what happened at the region where Fowler's theory broke down, thus making sure it did not threaten his own work.

Milne's way of dealing with Chandra's result had been to suggest that stars contain an absolutely incompressible core of some unknown material which ultimately prevents their complete collapse. Eddington's route was more direct. He simply tossed it out by declaring that it bore no relation to physical reality, then went on to create an unstoppable mathematical juggernaut to force everyone to accept the universal validity of Fowler's result. Fowler himself was not privy to this, and in fact, as Chandra recollected, he had assured Chandra that he supported the young man's conclusions. But he was not prepared to stand up in public and say so. To cap it all, Eddington hauled in Dirac's theory, rewritten in terms of his own E-numbers. By 1935, in response to Chandra's work, Eddington had tied up his fundamental theory so tightly that Chandra's result absolutely had to be incorrect – otherwise the whole structure would crumble. Eddington was playing for the highest of stakes. His fundamental theory was the culmination of his life's work, which he had always conceived of as the search for truth.

Rosenfeld was sympathetic to Chandra's predicament. Both he and Bohr, he wrote, 'realise your troubles and feel very sorry for you'.[17] He suggested forwarding Eddington's paper and Chandra's letters to Pauli, who by now was one of the leading figures in the world of physics. Pauli replied that Eddington was completely wrong and appeared to be driven by the desire to mould physics to fit his theory. But he would not go so far as to enter into the controversy and make a public statement supporting Chandra. Chandra had also been corresponding with his friend McCrea. Writing five days after the fateful meeting, McCrea was still undecided. He offered to write a '(polite) discussion of E.'s paper sentence by sentence' and send it to Eddington, but he never actually did. He added, 'No matter how [Eddington] may antagonise me on some points he never fails to entertain me, too.'[18] This was probably no consolation to Chandra.

A pattern was beginning to emerge. Senior figures, the likes of Bohr, Fowler, Dirac, Rosenfeld, Pauli and McCrea, were sympathetic. Chandra kept on running into colleagues who said, 'I'm sorry. I'm sorry.'[19] But they were not prepared to involve themselves any more than that. Chandra had helped Fowler prepare the second edition of his book on statistical mechanics, which Chandra had originally read as a teenager in India. He could not help noticing that Fowler had added a note pointing out that Eddington considered the relativistic degeneracy formula to be incorrect, without any further comment.[20]

As for Milne, on 13 January 1935 he sent off a letter to the editors of The Observatory. In view of Eddington's comments, he had recalculated some of his results but this time had ignored relativistic degeneracy, he wrote, adding wryly, 'whether it exists or not'. 'Sir Arthur Eddington's investigations may now confer on our work a justification to which it is only accidentally entitled,' he concluded.[21] To Chandra it must have seemed as if the world had turned upside down, with Milne now agreeing with Eddington.

For Milne, things had turned out unexpectedly well. Ever since Chandra first proposed his theory, Milne had done his best to bully him into not making it public. He tried several times to persuade him not to publish his early papers, until Chandra was driven to declare that he had 'lost all respect for Milne's objectivity'. Chandra knew very well that his theory demolished Milne's. In fact, he had imagined that Eddington would be overjoyed by it, in that it supported Eddington's standard model by demonstrating that in certain cases stars remain perfect gases and do not develop incompressible cores. Instead, to Chandra's and Milne's surprise, Eddington had rebuffed Chandra. Eddington's rebuttal of Chandra's theory actually provided backing for Milne's research. Thus Milne unexpectedly found himself on the same side as Eddington. It was too good an opportunity to miss. Despite his and Eddington's ancient antagonism and his friendship with Chandra, Milne seemed almost gleeful to find himself on the side of the enemy.

Chandra did not have the inner strength to ignore Eddington and continue with his own work. Although the heavyweights of the physics community were all on his side, they were unwilling to go head to head against the leading astrophysicist of the era – or not in public, anyway. The stakes were not high enough to make it worth them getting involved. Chandra would have to stick up for himself. He poured out his troubles to Babuji:

> My last paper on stellar structure involved me desperately in the rival jealousies of Eddington–Milne–Jeans. I am taking care to be scrupulously polite to all of them. Fortunately Fowler and Bohr are on my side. I cannot properly go into these matters in a letter. It is the continuation of the history of my differences in attitude in results with Milne which has been brewing for the last three years. The explosion hasn't yet occurred. The whole thing would have been smoothed over had it not been for an awful howler Eddington has started. He (A.S.E.) thinks that Pauli's principle is wrong! I do not know what he is up to. But the scientific politics in stellar structure in the triangular contest Eddington, Jeans, Milne strains me as I am in the middle and refuse to take anybody's side and E, M and J are all quarrelling with my work!!* [*I won't let them have any.] But Fowler supports me from the top![22]

Chandra was no longer a bystander caught between the warring factions. The controversy had made him a giant in a battle of giants. Babuji replied immediately. 'Your letter of 9th Feb has interested me very much,' he wrote. 'You are apparently standing at the median point, according to your diagram. You may let me know further developments.'[23] It was not all bad news, however. Chandra was now giving three lectures a week on 'Special problems in astrophysics'. He told Babuji proudly, 'By the way I am the first Indian to have given University lectures at Cambridge – I mean no other had had the *opportunity*.'[24] As far as he was concerned, it was a small step against the racism at Cambridge.

Towards the end of February, Milne wrote to Chandra reasserting his opposition: 'Your marshalling of authorities such as Bohr, Pauli, Fowler, Wilson, etc., very impressive as it is, leaves me cold.'[25] His comment was no doubt also coloured by his dislike for these scientists, who treated his own research with contempt, dismissing it as merely playing with complicated equations. For Chandra, Milne's antagonism was very hurtful. He often quoted these words of his and, in the 1970s, even had the letter transcribed and sent to those who were interested in the controversy with Eddington.[26]

In India, Chandra's Uncle Raman was also proving to be a foe. Chandra complained to his father: 'I have been told that C.V.R. [Raman] has said recently that I "am wasting my time in astrophysics and that there will be no astrophysicists miles near Bangalore". I wonder if anything could be done to stop him talking about me. He can be assured anyway that I will not get so much as scores of [miles] near Bangalore.'[27] By now Chandra was in Copenhagen, and spent March and April conferring with Rosenfeld on the Eddington situation. From there he wrote again to his father, saying that Cambridge had just granted him £50 to buy a Brunsviga 20 hand calculator, one of the best available. 'I am good enough for Cambridge and they are sympathetic to my work and do not regard me as "wasting my time",' he protested.[28]

Babuji was upset that Chandra was travelling so much, especially as India never seemed to be on his itinerary. 'DON'T postpone your marriage too long,' he wrote to him. 'I do not think you ought to become an "astrophysics" animal. Come over in xmas week of 1935 and get married and "let all intellectual ambition or work go a hang".'[29]

A month later, Chandra dropped a bombshell:

I had not written during the past weeks, partly as I was travelling and partly as I was thinking over my 'personal matters' . . . I realised there that my relations with Lalitha was purely illusory and that I really had not known her at all.

I had seen her in college but that was five years ago. I was simply deluding myself. So I have written to her breaking our understanding and this is final so far as this matter goes.[30]

Babuji replied that he was sorry to hear about Chandra's decision, but perhaps it was for the best. 'There are ever so many girls, who are reading in the BA class, much younger – whose inclination for family life are better.' Babuji was becoming concerned about his son's state of mind. He begged him to come home the following April for an extended visit. 'I do not like your way of letters since the incident,' he wrote. By 'incident' he meant the run-in with Eddington, of course, and Chandra's increasingly cynical tone in the letters that followed. 'The fact of it perhaps is that Miss D.L. has not sufficiently *known* and loved her father to be able to transfer that love to a husband in due course.' (Lalitha had only hazy memories of her father, Captain Doraiswamy, who had been away fighting throughout the Great War and died a year after returning.) He concluded with some fatherly advice:

If you stay in Madras for about 4 months you would be able to get to know some girls, then choose! The Indian way of love is far superior than the romantic attachments of the European: The wife you need is one who will be a comrade to you as well as see that your home interest will be attended to. That woman should be partly *potential* in her talent and not entirely *kinetic* is the best for married happiness.[31]

The last sentence was no doubt intended as a witty piece of physics punning.

The truth turned out to be quite different and rather mysterious. Two months after the dramatic exchange with his father, Chandra mentioned to Rosenfeld, in the middle of a technical discussion on relativistic degeneracy, that 'There is another "interesting" fact. Since I came back [returned to Cambridge from Copenhagen] I have got engaged with an Indian girl who was here. Lalitha by name . . . And I shall try to make her happy at least.'[32] What exactly was going on? While Chandra continued to write to Babuji right into the following summer, 1936, bewailing his broken romance, he was simultaneously telling his friend and confidant Rosenfeld that he had become engaged to a woman and – yet more mysterious – that she had been 'here', in Cambridge. Perhaps, despite Chandra's initial enthusiasm, he had developed cold feet. Sensing this, Lalitha took the first boat to England and turned up on his doorstep. Given the rigidity of Indian social mores, this must have required a great deal of courage and commitment on her part. Lalitha was a freethinker, quite different from most other Indian women. She did what she pleased. In later years she often said that Chandra was the love of her life. Perhaps she felt it was worth making a dramatic move to ensure that everything worked out between them.

Meanwhile, the dispute with Eddington simmered on. At a meeting of the RAS that May, Milne repeated the comment he had made in his letter to the editors of *The Observatory*: 'Sir Arthur Eddington is justified in his argument that the "relativistic" degeneracy formula leaves us with the old difficulty concerning white dwarfs, whereas the "non-relativistic" degeneracy formula does not lead to such a difficulty.'[33] Milne had clearly made a complete volte-face and was now with Eddington. For their own reasons, each needed to discredit Chandra's theory. Eddington rose for a minute and quipped, 'I have only to add that my degeneracy formula seems to have raised a hornet's nest about me. In my opinion, however, I have not yet been stung!'[34] Chandra decided that it was best to keep quiet.

All this was depressing him very much. 'Eddington is behaving in the most obscurantist fashion,' he wrote to his father. 'Though Fowler, Dirac, Bohr and many others agree with me, it is still very annoying to have such strained scientific relations with E. Indeed my differences with him (and also Milne) has to a considerable extent made me unhappy during the past months.'[35]

At the time, Chandra was discussing with the Hungarian-American mathematician John von Neumann how to extend his results on white dwarfs.[36] The thirty-two-year-old von Neumann was on leave from the Institute for Advanced Study in Princeton, and spent 1934 and 1935 in Cambridge. He often visited Chandra's rooms, and the two became close friends and confidants. They made a formidable pair in professional terms but a strange duo socially. Both were brilliant at numerical calculations and obsessed with detail. Von Neumann had the broader scientific base. Along with Edward Teller, he was among the leading lights of a galaxy of brilliant Hungarian émigrés. Fritz Houtermans, a German nuclear physicist who made important contributions to the study of what makes stars shine, joked that the Hungarian scientists 'were really visitors from Mars'.[37] Von Neumann had been one of the first professors at the Institute for Advanced Study, along with Einstein, for whom it had been founded. He had virtually invented the field which investigates how to arrive at the best possible decision from the information available – game theory – and also did ground-breaking work in computer science, economics, mathematics and physics. Despite the strong anti-Semitic climate in Germany in the late 1920s, his decision to leave in 1929 was purely for professional advancement, as he often said. But he was also a regular fixture at parties and loved nightlife. He had an endless repertoire of scatological jokes, which he told hilariously. Chandra recalled with some bitterness that, 'Von Neumann was one of the people who privately supported me against Eddington. Of course all these people who supported me never came out publicly. It was all private.'[38]

Chandra was becoming profoundly depressed. Milne wrote to him, saying:

> I was much distressed to hear from you that you have discontinued dining regularly in Hall at Trinity owing to the damping influence of Eddington. I think we must all struggle against this feeling of discouragement. We have a responsibility to our scientific posterity not to give up our efforts merely owing to the dominating influence of one man; Posterity will put us all in our proper, humble, places, and will judge us chiefly for our courage in stating the truth as we see it . . . I do beg you not to discontinue dining in Hall. The older fellows don't like it and it destroys the college spirit.[39]

That June, Chandra ventured to make a public mention of the

dispute. In the third instalment of his papers on the structure of stars, he mentioned in a discreet footnote that Eddington had questioned 'the validity of the relativistic equation of state for degenerate matter which', he pointed out, 'is still generally accepted'.[40] He was afraid that the *Monthly Notices of the Royal Astronomical Society* would refuse to publish it because he had dared to be critical of 'High Priest Eddington'. So he took care to mention Jeans's work on radiation theory right at the beginning. Gleefully he explained to Rosenfeld why he had included Jeans:

> I have been nice to Jeans. It was all politico! . . . The R.A.S., I knew, would have been reluctant to publish it (my paper), but I knew also that to save their faces they would send it to Jeans hoping to get a bad report! – then they can turn down the paper with no qualms! I was aware of this, and so I referred to Jeans very nicely. The trick worked! Jeans was emphatic about the publication! – It is all really sickening – these underhand methods, but what can one do?[41]

The dispute with Eddington had convinced Chandra that success in the world of science was not simply a matter of the brilliance of one's ideas – even more important, it seemed, was to play the game, and scheme along with everyone else. Or perhaps he was becoming paranoid, fearing enemies everywhere. Maybe the *Monthly Notices* would have published his paper with or without this subterfuge.

This 'politico', as he called it, forced Chandra to reconsider what the scientific process was and what had happened in his case. He mused in a letter to Babuji, 'By the way I do not agree with your statement that original thinkers *must* disagree. The whole quantum mechanics has been built by the beautiful conformity of opinion of some of the greatest thinkers of our time – Dirac, Heisenberg, Bohr, Pauli. In astrophysics, the differences are of a "political" nature.' It was a very important point, and not fully clear why this should be the case. 'Prejudices! Prejudices!', he went on:

> Eddington is simply stuck up! Take this piece of insolence. 'If

the worst comes to the worst we can believe your theory. You see I am looking at it from the point of view *not* of the stars but of Nature.' As if the two are different. 'Nature' simply means Eddington personified into a Lady!! Milne is more reasonable. He is a sport, and he is frank.[42]

Chandra's position was complicated by the fact that his allies and supporters were all abroad, in Germany and Copenhagen. The astrophysics community in England steadfastly refused to speak out on his behalf. But he was still determined to continue the fight.

That same month, Raman offered Chandra a job as a nuclear physicist. Chandra was furious. 'He [Raman] first says that he will not have "an astrophysicist miles near Bangalore" and now he wants to give me a clue that if I worked on nuclear physics, he might *tolerate* me – I suppose – some furlongs near his Institute,' he wrote to his father.[43]

Chandra had not yet responded to Eddington in print. The many colleagues he consulted had advised against it, though they had also acknowledged in private that they were on his side. Now, several months on, his confidence had returned. On 7 June, together with the Danish physicist Christian Møller, he submitted a paper to the *Monthly Notices*. Cleverly, they used the same mathematical representation of relativity theory that Eddington had used. But they used it in the accepted manner, putting in the solutions from Dirac's equation from his theory of the electron based on relativity. From this they were able to deduce Chandra's theory, including the upper limit for the mass of a white dwarf. In fact, they did the calculation twice, using the solutions from Dirac's equation for electrons as standing waves and then for electrons as progressive waves.[44] They presented their result without comment, except to add pointedly that 'we wish to state that we do not intend this note as a reply in any sense to Eddington's papers'.[45]

It was like a red rag to a bull. Eddington fired off a note to Chandra, once again attacking what he saw as Chandra's confusion between electrons as standing and as travelling waves, still refusing to concede that in the end it was totally irrelevant. That November he published a new paper in which he wrote that he had been led to 'amplify [his] attack on the relativistic degeneracy formula by showing why I am unable to accept Møller and Chandrasekhar's proof'.[46] They were, he wrote, violating Heisenberg's uncertainty principle – unthinkable in the world of physics. The two chose not to reply, at least not in print.[47] 'He is completely crazy,' Chandra wrote to Rosenfeld. 'I recently met H. N. Russell. He was frightfully enthusiastic. He finally whispered to me "Out there, we don't believe in E."!! . . . So finally I do feel a bit relieved. Met Milne recently. He really hates Eddington.'[48] So even that grand old man of astrophysics, Henry Norris Russell of Princeton University, was on Chandra's side, albeit only in private. Everyone was still fearful of offending Eddington. Rosenfeld replied with a jocular pun: 'The story of Eddington's degeneracy (if I may use such an ambiguous expression) takes the shape of the *Iliad*, with the various gods and heroes coming in.'[49]

As for Milne, he had now set off on a quixotic quest in search of a new theory of gravity to supersede Einstein's general theory of relativity – despite the fact that Einstein's theory had been confirmed twice by Eddington, in his eclipse expedition in 1919 and his work on the effect of gravity on the light from white dwarfs in 1924, and was totally unquestioned by anyone else.[50] His aim was to deduce a new law of gravity which would explain the shapes of galaxies. His book on the subject was published that year. As with his theory of stellar structure, he was full of bravado, declaring that his new cosmological theory 'destroys at one swoop much of the recent work of Einstein, Jeans and Eddington. How far it will ultimately reach I cannot tell, but I believe it to be the only really new idea in relativity since Einstein. And the joke is that Einstein (who has written to me about it) doesn't really see the point!'[51] Einstein dismissed it out of hand: 'Concerning Milne's ingenious reflections I can only say that I find their theoretical basis too narrow. From my point of view one cannot arrive, by way of theory, at any at least somewhat reliable results in the field of cosmology, if one makes no use of the principle of relativity.'[52] Some of Milne's ideas were certainly original. But, as Chandra put it, they would have been more fully appreciated had they been presented 'in a more modest framework'.[53]

Chandra wrote in exasperation to Rosenfeld:

Oh! Milne's book – it is a bitter disappointment. Can you imagine the following appearing in a scientific book? 'One can say if one pleases that we have found God in the universe – The physicist and the cosmogonist then need God only once to ensure creation. For the biologist, the world provides further opportunity for Divine planning . . . for a man, as more than a cosmologist, as more than a biologist, as possessing mind . . . an immortal soul . . . God is perhaps needed always.'!!! Hardy asked me the question 'Is Milne deteriorating?' . . . 'It looks as though the question is at least pertinent' that was my answer! – You see in Cambridge one learns to use bad language *but* politely![54]

In mid-July Chandra was in Paris, attending a meeting of the International Astronomical Union. Despite Eddington's public hostility, Chandra's reputation was soaring. 'At the moment I am thoroughly enjoying myself in the company of some "great" astronomers,' he wrote to his father. 'I have already had hearty discussions with H. N. Russell, H. Shapley, [Adrian] van Maanen, [Bertil] Lindblad.'[55] It was the first time that Chandra had ever flown, and he was thrilled. He defended his choice of transport against Babuji's dire warnings of how unsafe it was. Even Raman agreed that he should not take such risks.

The meeting was a splendid affair. Three hundred astronomers from thirty countries were there. Eddington too was present, and in fine form. The French President opened the proceedings, and there were sumptuous banquets and sightseeing tours. The gala dinner to celebrate 14 July, Bastille Day, was held on the bottom platform of the Eiffel Tower. Afterwards, guests climbed to the second platform to view the city of lights magically illuminated for France's national holiday. McCrea was one of those who attended. The Commission for Stellar Constitution, he wrote, 'met under conditions described by its president Sir Arthur Stanley Eddington (who knows about these things) as similar to those obtaining in the inside of a star!'[56] Perhaps the blazing heat Eddington was referring to was a reference not just to the summer weather, but to the fiery intensity of his remarks about Chandra, who was in the audience.

McCrea reported that 'Sir Arthur also reiterated his conviction of the validity of his recent conclusions concerning degenerate matter and their astrophysical significance.'[57] He did not mention Chandra's contribution. But Chandra himself remembered it with some pain. 'Eddington gave an hour's talk, criticising my work extensively and making it into a joke. I sent a note to Russell [who was presiding], telling him I would wish to reply. Russell sent back a note saying, "I prefer that you don't." And so I had no chance even to reply; and accept the pitiful glances of the audience.'[58] Once again, Eddington had humiliated him and he was not permitted to respond. Chandra soldiered on, perpetually frustrated in his yearning to engage in a verbal jousting match with Eddington. The great man always seemed to have the last word.

7

American Adventure

Chandra's star was rising; his brilliance was finally becoming recognised. Even before the momentous confrontation of 11 January, the director of the Harvard Observatory, Harlow Shapley, had noticed him and invited him to spend the following summer at Harvard. Chandra decided to accept, though he would continue to work at Cambridge for the remainder of his Fellowship at Trinity. He spent the last three weeks of September in Scotland, at a farmhouse in Kirkmichael in Perthshire, where he enjoyed 'long walks and the fresh air, the cool breeze and the open space – all violently(!)invigorating'.[1] There he absorbed himself in Dostoyevsky, Hardy and Balzac. 'Because of lack of "paper and pencil", I have been able to *think* more about stars, than I would have – as paper and pencil necessarily lands one in the intricacies of the calculation!' Full of new ideas, he was suddenly 'homesick for Cambridge'.[2] The world looked brighter.

Then, in October, Shapley offered him a lectureship in cosmic physics at Harvard for the next three months. He was rather apprehensive about taking up the appointment, but his friends assured him that it would be exciting.

On 29 November 1935 Chandra took the train from Cambridge to Liverpool. The following day he boarded the *White Star Britannica* and set off across the Atlantic on the eight-day journey from the Old World to the New. He spent the time absorbed in a volume of Ibsen's plays. It was a pleasant diversion from his usual preoccupation with the life of the stars.

The ship steamed into Boston harbour at midday on Sunday, 8 December. To Chandra's surprise and delight, Shapley himself had come to the dock to meet him. That evening he invited him to his home for tea. There Chandra was introduced to several distinguished astronomers. One was Gerard P. Kuiper. Kuiper had been observing white dwarfs and, to Chandra's excitement, told him he had 'obtained some remarkable confirmation of [Chandra's] last year's work on Highly Condensed Configuration'.[3] Then thirty-three, Kuiper had won a huge reputation for his work on binary star systems, particularly ones which included a white dwarf star. He went on to make a series of remarkable discoveries. In the late 1940s he discovered one of Uranus's moons, which was named Miranda, and the second moon of Neptune, Nereid. In 1951 he predicted the existence of a belt of chunks of ice, debris from the formation of the solar system, swinging through space just outside the orbit of Neptune. The first member of this icy swarm, called the Kuiper Belt, was actually discovered in 1992; it is about 100 miles in diameter. Since then hundreds have been found, and astronomers suspect that there may be tens of thousands more.

For the first time, to his great pleasure, Chandra had an office of his own. He enjoyed the social life of the astrophysics community, including what he referred to in a letter to Babuji as the 'bachelor-club(!)'.[4] He complained that the scientists spent too much time talking shop, in contrast to high table at Cambridge, where this was frowned upon. 'On the whole I prefer Cambridge – England . . . Here in spite of the extraordinary friendliness of the observatory people – I feel a stranger,' he wrote. Once again he was an outsider, as he had been in Cambridge and Copenhagen.[5] He often quoted Nehru's words, 'At home nowhere, a stranger everywhere.'

He was still haunted by the shadow of 11 January. In December he wrote to Balakrishnan, 'As for my work, it is going on but unfortunately, some controversies with Eddington and Milne during the past year has upset my enthusiasm a great deal . . . My controversy with Eddington has poisoned me and my peace quite considerably.'[6] But there were also opportunities for fun and some mild revenge. At Christmas Shapley arranged a review entitled 'The Follies of 1935'. Chandra was asked to make a 'scientific' contribution. 'I took my "revenge" on Eddington by telling some funny stories about him which were quite well appreciated,' he told Babuji. No doubt Eddington's story about the technological importance of women's zips was one. Chandra closed his letter in down-beat style: 'There is nothing particularly new except that 1936 begins tomorrow!'[7]

Cambridge remained on his mind, though he was always ambivalent about the place where he had experienced both triumph and despair. 'My visit to Harvard has upset my plans and quite a considerable work must await me at Cambridge on my return ... one always misses something, though at Cambridge everything is so beautiful that one misses only – I mean one misses the least number of things – one does not in any case miss Cambridge!'[8] Thus he expressed the confusion in his mind.

There was irritating news from India. Once again, Raman was making Chandra's life difficult. Despite sterling references from the great Cambridge mathematicians Hardy and Littlewood, Raman had rejected Chandra's old friend Chowla for a post. Chandra expressed his outrage to Babuji:

Oh! I can't stand Raman and his intrigues. He believes himself omniscient, though it would make no difference to his beliefs(!) what the physicists here and in Europe think of him and his students. Raman and his intrigues [are] "dirty double crossing" to use an Americanism ... My desire to settle finally in India and be of some service to Indian Science seems to dwindle day by day.[9]

In the New World, Chandra was going from success to success. In February 1936, Otto Struve, the director of the Yerkes Observatory, part of the University of Chicago, contacted him and invited him to give a lecture there. At the time, Yerkes was one of the most important observatories in the world, thanks entirely to Struve's dynamic leadership.

At thirty-eight, Struve was the most successful director Yerkes had ever had. Russian-born, of German descent, he was the fourth generation of a distinguished family of astronomers. In 1916, after two years at Kharkov University, he had joined the Russian army and fought the Turks in the Caucasus. After Russia and Germany made peace, he returned to university, only to be swept up in the revolution: he fought for the White Russians, so ended up on the losing side. Wounded and seriously ill, after many trials and tribulations, he finally managed to obtain a position at Yerkes using family connections. He completed his Ph.D. there in 1923. Along the way he established a reputation as an indefatigable researcher and went on to produce important papers in the field of stellar spectroscopy.

In 1932, Struve was appointed director. The observatory's standing had declined severely under his predecessor. Thanks to a no-nonsense managerial style that demanded a level of research excellence and productivity which he himself adhered to, he succeeded beyond his wildest dreams in re-establishing its pre-eminence. With his Prussian bearing and purposeful demeanour, the crystal-clear directives he issued and his vision of Yerkes as a great observatory, Struve was a formidable character.

He was also greatly impressed by Chandra, and immediately offered him a position as research associate. Harvard too had offered him a post, a three-year appointment as a member of its prestigious Society of Fellows, to commence in February 1937. This would enable him to pursue his research full time. Nevertheless, Struve's offer seemed the more attractive:

> It seems to me that your brilliant theoretical work could be made even more valuable to astronomers if it could be combined with practical investigations by observers. This would require close cooperation between yourself and the astronomers at the great observatories of America and might lead to new types of investigations of a non-routine character that would be especially valuable in the present state of astrophysics.[10]

Kuiper had already agreed to go to Yerkes, and played his part in convincing Chandra to join him there. Strömgren too had been hired. It was all most exciting. Chandra and Strömgren would be instrumental in leading the development of theoretical astrophysics in the United States, which up to then had been lagging behind Europe.

The problem was how to break the news to Babuji. Considering all the intrigues in India and the 'underhanded dealings going on in Indian scientific circles, it may after all be best that I spend some time in America,' Chandra wrote to him.[11] He also promised to return home for a visit first, before settling in America. As Chandra had feared, Babuji was deeply disappointed. He warned him of the shallowness of American culture and reminded him that settling there might ruin his chances of becoming a Fellow of the Royal Society (FRS), and thus of winning a top position in India. He concluded with stern optimism: 'I believe you will after all, return to India.'[12] He was still unshaken in his belief that Chandra would eventually come home.

Chandra went back to Cambridge in April. He had accepted the position at Yerkes, which he would take up on 1 January the following year. 'I had carefully considered the situation and Eddington fairly strongly advised me to accept [it],' he explained to Babuji. Perhaps the truth of the matter was that Eddington relished the prospect of Chandra leaving England. For Chandra's part, the time in America had renewed his confidence. 'FRS etc., are completely trivial,' he continued exuberantly. 'I may get it sometime within the next 5 or 10 years, but the chances are not renounced by my going to America. In any case I am quite indifferent to it.' As for Lalitha, 'I do not think that marriage for me is an important consideration.'[13]

Chandra was fully aware of the stark contrast between his welcome in America and his treatment in England. 'On the whole I am convinced that America has been the first in recognising me sufficiently to consider me worthy of an annual salary with a definitively senior position in a university,' he wrote to his father. 'In India they are blissfully unaware of my existence, and in England . . . there is some prejudice in giving Indians a definite appointment, tho' at Oxford, they have now appointed Radu Krishnan.'[14] It was indeed striking that the Cambridge heavyweights, Eddington and Fowler, had never considered offering

Chandra a position, while in the United States universities as prestigious as Harvard and Chicago had done so without hesitation. It is small wonder that, when given the opportunity, he chose to stay in America.

Chandra was also well aware that he was doing very little to raise the profile of science in India. Max Born spoke approvingly of Raman's efforts to promote science there. Chandra began to wonder whether he had been wrong in his criticisms of his uncle. A deeply divisive rivalry had arisen between the two stars of Indian science, Raman and Saha. Having talked to Saha, Chandra wrote to Babuji of 'his scheming, sneering attitude – an attitude which desires to destroy all things productive – an attitude which blurts out when in inviting me to come back to India he asks me to "join in the politics".'[15]

And why was Chandra stalling over his marriage? In India it was the custom for the eldest son to marry before his brothers. Chandra's hesitation made it difficult for Balakrishnan, for whom Babuji wanted to find a wife. He also wanted to find bridegrooms for his younger daughters. 'I think I shall have to postpone the girls' marriages,' he wrote in despair to Chandra. 'I do not have any suitable offers – in fact, you know I have to go begging in the present state of social affairs in India, and I do not feel much up to it.'[16]

Chandra kept at least part of his promise to Babuji: that July he sailed for India. He arrived in Bombay in the full heat of August. It had been six years since he left. As soon as he arrived, he wrote to Lalitha. According to Chandra's official biographer, she had not heard from him for a year, though she had seen his picture in the paper and knew he was in India.[17] But in reality she had of course secretly visited him a year earlier in Cambridge. So perhaps the two had kept up a surreptitious correspondence – or perhaps, in his old age, telling the story to his biographer, Chandra wanted to make things appear as respectable and above board as possible and therefore did not mention the secret visit. In any case, the two met in Madras. When Chandra saw her, whatever fears he might have had – that she might be a burden on him in some way, or perhaps hinder his work – melted away completely. They agreed that they had to marry, and as soon as possible. Together they walked along the Marina, discussing their future in America.

Chandra informed Babuji that there was no need for any of the traditional preliminaries, such as consulting an astrologer on the compatibility of their horoscopes. More important was to change his reservation on the ship back to England from a single to a double cabin. The marriage took place on 11 September. As a marriage of love, it was deeply unconventional for India. The wedding was very simple, quite different from the traditional elaborate ceremony which would have entailed vast expense for Babuji. Lalitha's unconventional family were most pleased. After a few happy days in Bangalore, the newlyweds sailed from Bombay on 13 October. Chandra had been in India for just two months.

On their way to the United States the couple visited Cambridge, where they spent a month. There Eddington and his sister Winifred invited them to tea. It was Lalitha's only encounter with the man who had caused her new husband such pain. He was all charm; she has nothing but warm memories of the meeting.

Eddington was fresh back from Harvard. He had been invited there as an honoured and popular guest on the occasion of the university's tercentenary. America's oldest and grandest university, Harvard was established in 1636 by John Harvard, a graduate of Cambridge. There were splendid celebrations, beginning with a spectacular fireworks display on the Charles River on 17 September. Half a million spectators lined the banks. The next morning, a distinguished audience of 17,000 took their seats in Harvard's inner sanctum, Harvard Yard, where the fledgling Continental Army had drilled in 1776 and George Washington had been headquartered. In a grand ceremony, sixty-two of the world's leading scholars were awarded honorary degrees, among them the psychologist Jean Piaget, the psychoanalyst Carl Jung, and Hardy and Eddington.[18] Harvard's dynamic young president, James Conant, read out Eddington's citation in ringing tones: 'Sir Arthur Stanley Eddington: Doctor of Science. A student of the cosmos who peers within the atom and surveys the expanding universe, an expounder to the multitude of the poetry of modern science.'[19] It was a fitting accolade for a lifetime's work.

Eddington lived up to his star billing. His subject for the Tercentenary Conference was white dwarf stars. He began his address with a witty barb directed at Ralph Fowler: 'My colleague

Fowler was in his youth a pure mathematician, and I am afraid he has never really recovered from this upbringing.'[20] His criticism of Fowler, he said, was that he had not pursued his result further using relativity – by which he meant, of course, Eddington's own brand of relativity. Had he done so, Eddington argued, Fowler would have realised that his non-relativistic result was actually relativistic.

Next Eddington turned his guns on Chandra. Chandra, he said,

> put the stars back in precisely the same difficulty from which Fowler had rescued them. The small stars could cool down all right and end their days as dark stars in a reasonable way. But above a certain mass . . . the star could never cool down, but must go on radiating and contracting until heaven knows what becomes of it. That did not worry Chandrasekhar; he seemed to like the stars to behave that way, and believes that is what really happens. But I felt the same objections as earlier to this stellar buffoonery; at least it was sufficient to rouse my suspicion that there must be something wrong with the physical formula used.[21]

He continued, using his favourite image: 'I was not surprised to find that in announcing these conclusions I had put my foot in a hornet's nest; and I have had the physicists buzzing about my ears, but I don't think I have been stung yet. Anyhow, for the purposes of this lecture, I will assume that I haven't dropped a brick.'[22] He went on to deliver a spellbinding lecture, giving his audience what they took to be the last word on the fate of the stars, conveyed with enormous authority, humour and panache. He had proved beyond all doubt that he deserved his reputation as the greatest astrophysicist of the day, recognised as such by all his peers.

To honour their esteemed guest, Eddington was invited to stay with Harvard's recently retired president, A. V. Lowell. He also went to see the Red Sox play at Fenway Park with his Cambridge colleague Hardy, who shared his love of baseball.[23]

That November, Eddington published a new book, *Relativity Theory of Electrons and Protons*. It was the first detailed exposition of his fundamental theory. He dismissed Chandra's work as a mere study in paradox: the ordinary degeneracy formula, he wrote, was first applied by R. H. Fowler [in 1926]. But for many years it has been discarded by astronomers in favour of a supposed relativistic degeneracy formula. The 'relativistic' degeneracy formula appears to be without foundation. The difference is important in the theory of the evolution of white dwarf stars; and it was the paradoxical results of the 'relativistic' formula, disclosed in an investigation by S. Chandrasekhar, which led me to examine its validity.[24]

Having thus disposed of Chandra and his theory, Eddington had put the heavens back in order, as far as he was concerned. Rosenfeld told Chandra that, 'I had a look through Eddington's new book. I find he is not only stupid, and irritatingly conceited, but most unfair especially toward you. But I suppose it is not worthwhile to come back on that subject!'[25]

By the end of 1936, the Chandrasekhars had settled into their daily routine at Yerkes. The observatory is beautifully situated on the shores of Lake Geneva, in the small township of Williams Bay, Wisconsin, 80 miles from Chicago, with lawns in front and rambling woodland at the back. It was the perfect site for an observatory, well away from the smoke, haze and fog of Chicago and also from the city's bright lights. The couple were to live in this idyllic setting for twenty-seven years.

They quickly bought a house there, only a few minutes' walk from the observatory. It was a big, old-fashioned wooden house with two floors, providing ample room for a couple without children, built at the end of the nineteenth century by Edward Emerson Barnard, a famous observational astronomer at Yerkes. When the Chandrasekhars lived there it was intimidatingly clean, 'very inhibiting to American graduate students whom the Chandrasekhars occasionally invited in for afternoon tea', with white rugs and plain white walls.[26]

Their days followed a regular pattern. Every morning they rose at seven. Chandra was in his office by nine. He usually walked home for lunch, stayed half an hour, and was back again for dinner at six. In the afternoon Lalitha would read, join him in his office or attend lectures. In the evening they went back to the observatory, where Lalitha read while Chandra worked for a couple more hours. Bedtime was usually 11 or 11.30. On Sundays they relaxed. 'There is not a great deal of social life here,' Chandra wrote to his father.[27] They spent time with the Strömgrens, the Kuipers or sometimes the Struves, drove around the countryside in Chandra's recently acquired second-hand Dodge, or went for picnics.

The two were eager to start a family. Early in 1944 Lalitha discovered that she was pregnant at long last. Delighted, she and Chandra began making preparations. But to their grief, she miscarried. They continued to be utterly devoted to each other, but remained childless.

Raman by now had decided that the time had come to repair his relationship with this nephew of his, now a major player on the astronomical scene. He wrote to Chandra in humorously conciliatory terms: 'you are right in deciding not to return to India except as a full professor. Your instinct for choosing Astrophysics for your special field was fundamentally right. I see papers on white dwarfs being solid neutrons, or double stars as cyclotrons, etc. etc. Apparently Astrophysics ≈ nuclear physics.'[28] The times had indeed changed. As Raman well knew, nuclear physics had begun to figure prominently in astrophysics, helping to solve the puzzle of where the Sun's energy came from.

In 1939 Chandra published a monograph entitled *An Introduction to the Study of Stellar Structure*. It was his farewell to white dwarfs. He had had enough of fighting with Eddington and having his ideas misconstrued by the astronomical community. According to Eddington's intimate friend Charles Trimble, the great man's only comment was a sardonic, 'How nice to have all the wrong things in one place.'[29]

Chandra's second book, *The Principles of Stellar Dynamics*, came out in 1943. Eddington was asked to review it. Predictably, he jumped at the opportunity to demolish Chandra once again. One of his criticisms of Chandra was his excessive use of mathematics. Eddington's review was cutting:

As a subject progresses the attractive simplicity of the early researches gives place to laborious elaboration. In the last three years, Dr. Chandrasekhar has been very active in the mathematical development of stellar dynamics. The trend of this work may be judged from the fact that one contribution alone contains more than 1800 numbered formulae. There is no denying that this heavy method of attack can be justified; but it leaves us with the depressing feeling that the subject which began thirty years ago as a joyous adventure has reached a stage of uninspiring ugliness.[30]

He had made a similar point in 1939, in referring to both Fowler's and Chandra's work: 'Almost as important as the elimination of actual error is the restoration of simplicity to a subject which had become highly and, as it now appears, unnecessarily complicated.'[31] It is true that Chandra used a lot of mathematical detail to buttress his argument. His goal was always to investigate the structure and dynamics of stars as systematically and as rigorously as possible. It is also true that some of Chandra's pages are over-burdened with mathematics. Nevertheless, his books marked significant advances. Every astrophysicist to this day has a copy of *An Introduction to the Study of Stellar Structure* on their desk, and, for all its ponderous mathematics, *The Principles of Stellar Dynamics* was an invaluable contribution to the subject. Chandra shrugged off the review as 'typically Eddington'.[32]

Eddington and Chandra were to meet once again – and, as it turned out, for the last time – at the Colloque International d'Astrophysique in Paris, in July 1939. The theme was novae and white dwarfs. By then, Chandra had been at Yerkes for more than two years and had gained in confidence. In group photographs he stands out, handsome and aristocratic, in an immaculate suit, utterly unlike the rumpled Cambridge dons. Despite his time in the United States, he still spoke with a refined Cambridge accent, tinged with his Indian lilt.

There had already been signs of trouble to come. Initially, Chandra was scheduled to speak on 'Novae, white dwarfs and planetary nebulae', while Eddington was supposed to address 'Theories of white dwarfs'.[33] Eddington informed Chandra in January that he assumed Chandra would stick to his assigned topic – Wolf–Rayet stars and novae – 'because I have no ideas at all' on these subjects.[34] Wolf–Rayet stars are extremely dense, hot and massive (greater than twenty-five times the mass of the Sun), in an advanced stage of evolution, ejecting very hot gases at high speeds. Chandra had conjectured that they could not pass directly to the white dwarf stage.[35] Eddington himself, he continued, would focus on planetary nebulae since the 'subject has been obscured by the prevalence of the "relativistic degeneracy" heresy'.[36] A planetary nebula is a colourful, glowing shell of gas ejected in the explosion of a star of relatively low mass, some four times the mass of the Sun. Astronomers already suspected that the burnt-out core left behind after the explosion was a white dwarf. In using the term 'heresy', Eddington was throwing down the gauntlet. He added a warning: 'If we stick to our original subjects, I do not think we shall overlap, as if you refer to details of white dwarf theory at all, I expect we shall completely contradict one another!'[37] It was a veiled threat.

The colloquium was sponsored by the Paris-based Fondation Singer-Polignac. Amos J. Shaler, a scientist from the Massachusetts Institute of Technology, was the Secretary-General in Astrophysics. Henri Mineur, an eminent astrophysicist at the Paris Observatory, Shapley (director of the Harvard Observatory) and Donald Menzel, also from the Harvard Observatory and a former student of Russell's, were on the planning committee.

It transpired that Mineur had deliberately put Eddington and Chandra together in the white dwarf session, without clearing it with anyone, expecting to spark a debate. Shaler then tried to shift Eddington out of the white dwarf session to deal with cosmological issues instead, but encountered fierce opposition from Eddington. He wrote to Chandra, 'Please do not think for a moment that you should abandon the white dwarfs, as a large number of the participants have voiced their desire to have you treat them rather than Sir Arthur. If you have any trouble with Sir Arthur, do not fail to "pass the buck" back to us, as it is not fair to have you shoulder any such burden.'[38] Here was ample evidence that Chandra's contemporaries now regarded him, not Eddington, as the authority on white dwarfs, from which Chandra could conclude that they also now accepted his discovery of the upper limit to a white dwarf's mass. No one, however, was yet prepared to accept the logical conclusion: that a burnt-out star far more massive than the Sun would collapse completely to an unimaginably tiny and dense point.

A few days later, another letter came from Shaler: 'I was very much afraid that Sir Arthur Eddington would put up a fight . . . I shall ask Dr. Mineur to do the dirty work right away. He started it anyway, by misinforming us in the first place, and he can damn well finish it.'[39] Having brought about the unpleasant confrontation in the first place, it was up to Mineur to take on the awkward task of insisting that Eddington switch sessions. It was an inauspicious beginning. But despite his apprehension about the forthcoming confrontation, Chandra was exuberant about the opportunity to travel in considerable luxury to Paris. 'My real object, however, is to get away from work – to be on vacation!' he wrote to Balakrishnan en route, adding that, 'I am leaving my wife behind in Yerkes.'[40] Their finances did not stretch to Lalitha joining him.

On his way to Paris, Chandra stopped off at Cambridge and took dinner at high table. He was seated with Dirac, Eddington and Maurice Pryce, a young Cambridge physicist who had just married the daughter of Max Born. Years later, Chandra could still remember what happened in vivid detail:

Pryce expressed surprise at seeing me and asked me whether I would join them in discussion with Eddington after Hall on the matter of relativistic degeneracy. After Hall, we adjourned to Pryce's room in Neville's Court. The discussion began with Pryce trying to tell Eddington his version of Eddington's arguments against relativistic degeneracy, so that Eddington could be satisfied that he, Pryce, understood Eddington's arguments. After Pryce had completed his narration, Eddington remarked that Pryce's account was entirely fair and accurate, and asked, 'What was the argument about?' [In other words, what was all the fuss about?] Pryce turned to Dirac and asked him, 'Did you agree with any of the things I have said?' Dirac said, 'No.' And Pryce added, 'I do not either.' Eddington became very angry – in fact, it was the only occasion when I saw him really angry. He got up from his chair, walked back and forth, and said, 'The matter is not for joking!' And he went on finding fault with Pryce's argument even though he had agreed with it a moment earlier. And for the next hour or so, it was Eddington's monologue. Next day, after Hall, Eddington came up to me and said that he was very disappointed that Dirac did not seem to understand the implications of his own relativity theory of the electron [i.e. the importance of Dirac's spinors, which played a central role in Eddington's fundamental theory]. I did not assent or dissent with Eddington's remark but asked instead, 'How much of your fundamental theory depends on your ideas on relativistic degeneracy?' He replied, 'Why, all of it!' And since I did not react to that remark, he asked me why I had asked that question. My response was, 'I am only sorry'; – not a polite remark to have made; but by that time I was really enraged with Eddington's supreme confidence in himself and in his own ideas.[41]

Thus Chandra forced Eddington to acknowledge that his fundamental theory depended entirely on his repudiation of relativistic degeneracy. If Chandra's theory was right, Eddington's whole intricately constructed theory – his life's work – would collapse.

Russell, the grand old man of American astronomy, presided over the Paris colloquium. At the conference luncheon, held at l'Hôtel de Ville, 'all of the Paris great were there . . . at high table', Chandra remembered, adding glumly, 'and I was way off in the corner somewhere'.[42] In his opening remarks Russell referred to Eddington's discovery that inside stars high-energy X-rays almost completely strip the atoms of their electrons, thereby increasing the volume available to squeeze in more atoms. It had, he said, gone some way towards solving the mystery of why white dwarfs were of such high density, and laid the groundwork for Fowler's results. He did not mention Chandra. He had led Chandra to believe that he was on his side, but he was not prepared to stand up for him in the presence of his friend Eddington. It was another slight for Chandra to add to his list.

Kuiper, the next to speak, introduced the subject of white dwarfs. Besides the three 'classical' white dwarfs, o^2 Eridani B,

Sirius B and van Maanen's Star, he had personally identified six-teen more.[43] As he reminded his audience, it was straightforward to calculate the mass of white dwarfs within a binary system and to deduce their radii from their luminosity and temperature.[44] Chandra's theory, he pointed out, was the only one that matched the observational evidence. Most white dwarfs are astoundingly small, little more than the size of the Earth, a hundredth the size of the Sun. Furthermore, every white dwarf that has ever been found has a mass less than the mass of the Sun, exactly as would be expected from Chandra's theory of relativistic degeneracy – from which it can be deduced that if a white dwarf is more than 1.4 times as massive as the Sun, it will collapse. Fowler's theory of non-relativistic degeneracy, on the other hand, predicts that the smaller the radius of a white dwarf, the bigger its mass, totally contradicting what is observed.[45] If Fowler was right, we would expect to see white dwarfs of a thousand or more times the mass of the Sun.

Then Chandra rose. Despite all the debate about who would speak on what, he ignored his assigned topic and devoted his entire talk to white dwarfs. He began with an audacious state-ment: 'It is now generally agreed that the equation of state appropriate for the discussion of the structure of white dwarfs is the degenerate form of the equation of state.'[46] In other words, his own theory of white dwarfs was the correct one. The perfect gas law from pre-quantum theory was definitely not appropriate for white dwarfs. And by not including the theory of relativity, Fowler had missed out on the appropriate equation of state, even though he had used quantum theory. In the discussion that followed, Eddington restricted himself to disarmingly simple questions.

Chandra's lecture had been on a Friday. Eddington's was the next morning. 'Our beliefs on Saturday must be different from our beliefs on Friday,' he said, as Chandra later recalled.[47] He began by summarising results he had published since 1935, with-out mentioning his fundamental theory. He stated firmly that the contraction of stars 'to a diameter of a few kilometres, until, according to the theory of relativity, gravitation becomes too great for the radiation to escape, is not a fatal difficulty, but it is never-theless surprising'.[48] Once again he refused to accept that a star might keep collapsing until its gravity became so great that not even light could escape. Fowler, he said, had resolved the question, while Edmund Stoner and Wilhelm Anderson had reopened it.[49] He himself had been led to re-examine everything, eventually rejecting the concept of an upper limit. Instead of crediting Chandra with the new equation of state, Eddington attributed it to Stoner and Anderson, who had simultaneously and independently discovered it but had never examined its ramifications nor placed it on a rigorous mathematical basis (as Chandra had). He added that it was impossible to devise an experimental test to verify whether Fowler's theory or the Stoner–Anderson theory (the same as Chandra's) was correct. In any case, it was unnecessary. Relativistic degeneracy was strictly a mathematical result with no astrophysical meaning – a *reductio ad absurdum*. However, further observational data would show whether modifications should be made to Fowler's theory, which predicted a mass for white dwarfs far higher than had been measured.

Kuiper was not going to let Eddington off the hook so easily. Were there 'any observational tests that would permit a choice between rival theories', he asked him.[50] Kuiper himself, of course, had just provided one. Eddington replied that it was incorrect to say that there were rival theories, and reiterated his claim that the Stoner–Anderson formula (i.e. Chandra's) existed only mathematically. Then Chandra leapt to the attack, arguing that he and Eddington differed 'in their fundamental premises'.[51]

Eddington suggested one area of compromise. Perhaps they could investigate some of the very small white dwarfs that Kuiper had found. If Fowler's theory were correct, they should be extremely massive, with extremely high densities and pressures. These 'extreme cases', as Eddington called them, might lead to the formation of neutrons and a rock-hard neutron core, thus preventing such massive stars from collapsing.[52]

According to the conference minutes, published in *The Observatory*, 'The meeting closed with a moderately outspoken duel between Sir Arthur Eddington and Dr. Chandrasekhar concerning the correct formula for degenerate matter under stellar conditions. Opinion was divided as to the moral victory.'[53] Chandra remembered the debate quite differently. As far as he was concerned, he was absolutely right. As he recalled, 'Eddington and I really talked to each other in strong language.'[54]

Just before departing, the two had a brief moment alone. Eddington attempted a reconciliation. 'I am sorry if I hurt you this morning. I hope you are not angry with what I said.' Chandra looked at him and asked, 'You haven't changed your mind, have you!' It was a statement, not a question. 'No,' Eddington retorted. Seething with frustration, Chandra looked him straight in the eye and demanded, 'What are you sorry about then?' and strode off. Eddington stood for a few moments, then shrugged his shoulders and went on his way.[55]†

Chandra sailed back to the United States on the *Normandie*. It was the ship's last voyage before being converted into a troop transport vessel. That September hostilities broke out. What had been known as the Great War was renamed 'World War I' as people began to see that a second world war was looming.

Chandra wrote to Babuji to report on the Paris conference. It had been, he said 'a relatively dull affair. I had a nerve racking controversy with Eddington who showed himself thoroughly opaque to reason. I had a trying time avoiding him, but he was equally after me – we were playing blind man's bluff!'[56] Babuji struggled to work out what had happened and what was going on in his son's mind. He had noticed a reference to Chandra's work in Eddington's *Relativity Theory of Protons and Electrons*: 'The relativistic degeneracy formula appears to be without any foundation.'[57] Babuji wrote back:

> Dr. Krishnan was telling me that in [the conference summary in *The Observatory*], your words were construed as an *insult* to Sir Arthur Stanley Eddington. I wonder why you have acted so. There must be always some genteel geniality in our speech ... (I am told that this was at the International Conference at Paris – where you said you played 'hide and seek' with Sir Arthur Stanley Eddington.)[58]

† The story of Chandra and Eddington is to a great extent also the story of Sirius B. I have brought the story of Sirius B up to date in Appendix A.

Dr Kariamanikkam Srinivasa Krishnan was one of India's most distinguished physicists, whose work with Raman had been critical to the discovery of the Raman effect. He had also made important contributions to the study of the magnetic properties of crystals.

Chandra had no choice but to reply. 'You have referred to my "insulting" Eddington,' he wrote:

> Actually the President of the Conference in Paris (H. N. Russell) referred to my restraint under extreme provocation . . . I have never been accused of using insulting language. Indeed, just today, I read an account of the Paris meeting, where my controversial discussion with Eddington is referred to as 'stimulating'. And in my 'dynamics' where I find and prove conclusively Eddington's work to be totally wrong, I refer to the circumstance in footnotes and obscure corners. If I had followed the traditional methods (set by Eddington among others) I should have made a real capital out of his errors – I should have obtained greater publicity in that way. Actually competent mathematicians [such as] von Neumann [and] Heisenberg have congratulated me on my dynamics [and] have actually accused me of just this restraint in exposing the Eddington fraud![59]

In 1942, Dirac, Rudolph Peierls and Pryce at last published a paper that defended Chandra and was highly critical of Eddington's views on relativity and quantum mechanics.[60] They pointed to Eddington's adroit use of mathematics to achieve his own ends. 'The issue is a little confused because Eddington's system of mechanics is in many respects completely different from quantum mechanics, and although Eddington's objection is to an alleged illogical practice in quantum mechanics he occasionally makes use of concepts that have no place there,' they wrote.[61] The 'alleged illogical practice' was Chandra's derivation of the relativistically degenerate equation of state, which Eddington claimed was incorrect because Chandra had represented electrons as standing waves.

One of Eddington's key criticisms of Chandra's theory was that Chandra assumed that electrons in a gas behave like free particles, not interacting with one another or with nuclei. He did this to simplify a difficult mathematical calculation. Dirac, Peierls and Pryce showed that allowing for electrons to interact with nuclei made no difference to Chandra's results for white dwarfs. Their arguments turned on certain basic concepts of relativity theory which Eddington did not agree with. In his determination to disprove Chandra's result on white dwarfs, Eddington had tried to force radical changes in the relativity and quantum theories.

Eddington replied by fudging the issue, maintaining that what was needed was a theory that incorporated relativity and took account of interactions among particles, instead of ignoring them as Chandra had done.[62] In the 1960s, scientists were finally able to calculate exactly the effect of interactions among the electrons in a relativistic degenerate gas. Far from disproving Chandra's theory, as Eddington had hoped it would, this simply led to a fine-tuning of his equation of state.[63] Dirac, Peierls and Pryce's paper was important and widely read. Yet in later years Chandra always chose to forget it. He continued to complain bitterly that the leading physicists had never come to his defence in his battle with Eddington.

Chandra was now well settled at Yerkes. But Raman continued to pursue him. From time to time Chandra was tempted to return to India, but he was always wary of the limited research positions available there and of the incessant internecine rivalries – he had had enough of that in Cambridge. When Raman offered Chandra the professorship which had previously been held by Saha at Allahabad University, Chandra turned it down. Struve wrote to Raman in Chandra's support: 'I thought that you might perhaps be interested in learning what we, his astronomical associates in the United States, think of him and of his place in our science.' He had, he said, invited Chandra to Yerkes because he was 'one of the most brilliant astrophysicists in the world. Chandrasekhar is in my opinion the only person who can successfully undertake the task of developing the theoretical side of American astrophysics. While he is here, no one could be a more effective "ambassador of good will" from India to the United States than Chandrasekhar.'[64] This was high praise indeed, especially coming from as demanding a taskmaster as Struve.

Foiled in his plan, Raman proposed instead to nominate Chandra for a Fellowship of the Royal Society. This was something he could do which would benefit both Chandra and Indian science. He asked Chandra to suggest five names from whom he could choose a seconder.[65] Chandra proposed Milne, Fowler, Eddington, Jeans and Sir Edmund Whittaker, an applied mathematician at Edinburgh University. Raman chose Milne, and requested supporting letters from Jeans, Fowler, Harry S. Plaskett (a Canadian astronomer who had known Chandra at Cambridge) and Eddington. Eddington wrote warmly of the fine work that Chandra had done in 'dynamical astronomy' (but making no reference to stellar structure) and of how they had been 'such good friends ever since [Chandra's] first coming to Cambridge from India'.[66] Chandra was elected a Fellow in March 1944.

Even though Raman had set the wheels in motion, in later years Chandra always downplayed his role. The reason seems to be a misunderstanding that took place in 1961 when Chandra was in India. Chandra never forgot a slight, even an imagined one. One day he walked into his uncle's office just as Raman was unwrapping Chandra's opus, *Hydrodynamic and Hydromagnetic Stability*. Raman remarked carelessly that 'The only book of this size I have seen before is a novel by Anthony Trollope – absolute trash.'[67] How had Chandra found the time to write such a tome, he asked. He himself, he said, much preferred research. In fact, in the late 1920s, Raman's arch-rival in research had taken time off to write a book; Raman, meanwhile, discovered the Raman effect and won a Nobel prize. Infuriated by all this, Chandra snapped that in that case he had already lost four Nobels. Raman was furious. Of course he had not intended to insult Chandra – it was Trollope's book he had called trash, not Chandra's. Ramaseshan, Chandra's cousin and a distinguished scientist, tried to convince Chandra of Raman's good intentions. Chandra's icy response was, 'You have the privilege to hold on to your views.'[68] Raman went so far as to write Chandra a note, reiterating his high opinion of his book. It was sent to a forwarding address in India but, according to Ramaseshan, Chandra never received it. Others think he did, but did not care. He had already made up his mind.

Raman made several more attempts to right matters between

them. He had already nominated Chandra for Fellowship of the Royal Society, and in 1948 he proposed putting him forward for a Nobel prize. Chandra replied courteously that he was most flattered. But, he said, astronomers were not eligible since there was no separate Nobel Prize for discoveries in astronomy, in which he included astrophysics, and in any case his interests in physics leant towards the 'less fashionable'.[69]

On Chandra's election to the Royal Society, Milne sent him a letter of congratulation. The contest had been close, he wrote. Chandra's name had not been as high as Milne had hoped. The aspect of Chandra's work that Milne had emphasised was his achievement in establishing theoretical astrophysics in America. Eddington too, said Milne, was impressed by this and felt that it merited Chandra's election. 'I thought you would be interested in this tribute to you from your and my ancient enemy,' he concluded.[70]

For Chandra, his move to America had been a new start in many ways. He had a new career at Yerkes, where he was universally respected and admired, and a new marriage. His election as a Fellow of the Royal Society – a signal honour for any British scientist and considerably more so for an Indian – gave him newfound confidence, as did the long-awaited public support of Dirac and others for his theory. He had every reason to put the feud with Eddington behind him.

8

An Era Ends

Eddington, Jeans, Milne and Fowler were never again to be major players in Chandra's life, though the reverberations of that early humiliation never faded away. By 1950 all four were dead. But even in death the dispute smouldered on.

Jeans died in 1946, at the age of sixty-nine. Milne wrote an obituary for the *Obituary Notices of the Royal Society*. He could not resist the opportunity to return to the burning controversies which had so marked their lives. Eddington, he noted, had never really come up with a convincing reply to Jeans's criticisms of his standard model, which resembled Milne's own misgivings. Eddington, arrogant to the end, simply 'made debating replies, and stood by his results . . . Like many of us at the time [Jeans] partly fell under the spell of Eddington's investigations.'[1] Perhaps that was the secret of Eddington's power – his sheer charisma. As for Jeans, Milne described him as someone who 'thought of the world as a mathematician thinks of the world; he tidied it up in his imagination'.[2] It was an apt description of all three. Chandra made a similar remark in his own obituary of Jeans, published in the journal *Science*. He also recalled that G. H. Hardy once told him that he had asked Eddington if he had ever bet on a horse. Eddington gave him one of his sardonic looks and replied, 'Just one. I bet on a horse named Jeans.' Hardy laughed and asked whether it had won. 'Eddington, with his characteristic smile, replied with the one word, "No."'[3]

The dispute consumed all of them. Chandra was convinced that Eddington 'effectively destroyed Milne'.[4] 'Milne's enormous originality', he recalled, was squandered on an obsessive desire to come up with a result 'which would contradict Eddington'.[5] Milne truly hated him. Their wrangling sent him into the blackest gloom. Tragically, his persistent unhappiness had a devastating effect on his wife, Margaret. Already in a delicate mental state, she fell into a deep depression and suffered bouts of mental illness which finally led to her suicide. Whenever Milne thought of those years, he 'could hardly control his tears', Chandra recalled.[6]

After Margaret's death, in 1938, Milne himself was overcome by great despair. The following year, hoping to cheer him up, Chandra invited his old friend to join him and Lalitha in Chicago. Together they travelled across the country to Austin, Texas, and up into the Davis Mountains, to attend the opening ceremonies for the McDonald Observatory, spectacularly located on a mountaintop plateau. There Chandra took a photograph of Milne, with his round boyish face and thick glasses, smiling broadly. The visit, it seemed, had had the intended effect.

Buoyed by his travels, Milne returned to England by steamer. On the voyage he met a woman named Beatrice Brevoort Renwick. Fascinating, beautiful and strong-willed, she came from one of New York's grandest families. Her great-uncle was the celebrated architect James Renwick, who designed St Patrick's Cathedral on Fifth Avenue and the Smithsonian Institution in Washington, DC. Milne and Beatrice fell head over heels in love and decided to marry. He was deliriously happy. His children were astounded that their usually reserved father was remarrying so quickly, and someone he had just met. Beatrice went back to the United States to sort out her affairs.

But by the time she left New York, at the end of May 1940, to re-join her fiancé, hostilities had already broken out. She took an Italian ship as Italy had not yet declared war. She landed in Genoa, then had to make her way overland to St Malo, on the coast of northern Brittany. The timing could not have been worse. Travelling through France, she found herself just one step ahead of the advancing German army. It was a terrifying journey. After many harrowing experiences, she and Milne were finally reunited in Oxford, and they married immediately. But Beatrice too suffered from depression: she never fully recovered from the trauma of her flight through France.

From 1939, Milne was back at work on munitions research, as he had been during World War I, once again working under Archibald Hill. 'To think of doing the same job for the same purpose 25 years later over again!' he wrote grimly to Chandra.[7] He updated his previous work, including theories of shell stability, calculations of the optimum distribution of anti-aircraft guns, and analysis of the results of experiments on armour penetration. His achievements remain valuable to this day.

Chandra sent food parcels to the Milnes, who were eking out a meagre existence in wartime London. In May 1944 Chandra was promoted to professor. Milne wrote to congratulate him and also asked after Lalitha, who was expecting a baby. That was the year of her miscarriage. The Chandrasekhars had been desperate to have children. 'My wife and I are deeply grieved to hear of the trouble and disappointment that befell your wife,' Milne told them, when he heard the sad news.[8]

Both Milne and Beatrice suffered greatly during the war. The long hours of war work and the Parkinsonian after-effects of the encephalitis he had contracted in 1924 exhausted Milne. For Beatrice, who had moved from high-society New York to war-torn London in the midst of the Blitz, the strain was even worse and she now had a one-year-old daughter to take care of. In the summer of 1944 a flying bomb destroyed their home. It was more than Beatrice could bear. She became deeply depressed, and took her own life in August 1945.

Colleagues were dying, too. Milne was called upon to write obituaries, first of Hardy and Jeans, then of his old friend Fowler. 'Fowler's death was a great shock and grief to me,' he told Chandra. 'He had looked ashen and very old of late.' Milne could not even attend the funeral because 'travelling is so appalling nowadays'.[9] He had great difficulty writing the obituary because his files had been lost when his house was destroyed. He had to send letters to colleagues asking for details of Fowler's personal life.

Typical, perhaps, of men of that time, Milne and Chandra's closeness seems to have been largely professional. Many years earlier, Milne had written to Chandra that he 'learned later that you had brought a bride from India'.[10] Friends though they were, Chandra had apparently never told Milne that he had married or even that he was considering it.

Sometimes Chandra wrote about Milne as if with a checklist in hand, alternating compliments and criticisms. The fact that after 11 January 1935 Milne had backed 'Eddington's conclusion (now considered false)',[11] and the letter he had written saying that Chandra's 'marshalling of authorities such as Bohr, Pauli, Fowler, Wilson, etc., very impressive as it is, leaves me cold',[12] were etched in his memory. Neither could he forget that Milne had tried to prevent him from publishing the paper he wrote in 1932, showing that stars with a radiation pressure of more than 10 per cent of their total pressure will collapse to an infinitely dense point. This was the paper that threatened to undermine Milne's theory, based as it was on the premise that stars will invariably develop an incompressible core to prevent them from ever collapsing into nothingness. It was this clash that first gave Chandra doubts about Milne.

Milne's obsessive anger after his showdown with Eddington in 1929, when Eddington had publicly demolished his arguments at the Royal Astronomical Society, had, in Chandra's opinion, been 'bound to have a detrimental affect on [his] scientific work'.[13] Some of his ex-students and colleagues disagreed. Thomas Cowling and Harry Plaskett, a contemporary of Milne, insisted that Milne and Eddington remained good friends and that Milne's lack of success in later years was caused not so much by resentment as by exhaustion and problems in his personal life.

Chandra's answer was that the fact that they remained 'personal friends [was] not contrary evidence. I can substantiate the role of "evil genius" which Eddington played in Milne's life by innumerable quotations from his letters to me and what he has said to me.'[14] Much the same applied to his own friendship with Eddington.

Students recalled Milne's passionate and involved lecture style. He would cover the blackboard with equations, a piece of chalk in each hand. 'When he lectured at Cambridge he finished with himself and all his students out of breath (I can believe it!),' wrote

Cowling, recalling a story that McCrea had told him.[15] Once, when Chandra was visiting Milne at Oxford, he had a strange premonition. He appeared as usual at Milne's study at the Radcliffe Library. 'I found that he had listed on the blackboard the names of the great mathematicians starting from Archimedes, through Newton, Hamilton, Lagrange, Laplace, Poincaré, and Einstein, and in the end his own name. I was shocked on seeing this and vaguely felt that Milne's scientific career must end in tragedy, as in fact it did.'[16]

In September 1950, Milne was on his way to an astronomical meeting in Dublin. He had told Chandra how much he was look-ing forward to it, and 'how, after years of depression and many personal tragedies, he was at last gaining serenity and some poise'.[17] He could not, however, pass up the opportunity to criti-cise Chandra's recent work on how light and heat travel through matter – radiative transfer. In a letter, he made the 'now oft-quoted statement that I have the habit of "choking the cat with the cream"', remembered Chandra.[18] Chandra was famous for swamping the books and papers he wrote with mathematical detail.

But before he could reach Dublin, Milne had a heart attack and died, at the age of fifty-four. Plaskett wrote in an obituary that Milne 'died as he had lived, undefeated'.[19] There were always tears in Chandra's eyes when he reminisced about the man who had been his first friend at Cambridge and who had suffered so much throughout his life. The photograph Chandra had taken of Milne in the carefree summer of 1939 was his favourite picture of his 'very great personal friend . . . the first and the best'.[20] The jutting veins on Milne's temples and neck show his passionate intensity of thought. But he also has laughter lines around his eyes. He retained his love of life, despite his terrible personal suffering.

And what of Sir Arthur Eddington, the 'evil genius' who drove Jeans out of academia, and had such an impact on the lives of Milne and Chandra? For many years after Chandra had departed for Yerkes, he continued to work on his fundamental theory. 'In fourteen years I have never had the smallest doubt that the direction which I took in 1928 was the one which leads to the unified relativity and quantum theory,' he declared with supreme confidence.[21] His lectures at the Royal Astronomical Society on his theory were utterly spellbinding. The president of the RAS, Harold Knox-Shaw, described them as a 'truly amazing performance', and even Jeans conceded that they were 'remarkable'.[22] Chandra was more caustic: 'Eddington became cocksure of his views on the [fundamental theory], on relativistic degeneracy, on the formation of black holes, and, indeed, on his whole approach to the "unification of quantum theory and relativity theory".'[23]

Eddington was consumed by the pursuit of his fundamental theory. He described it using Bottom's words from Shakespeare's A Midsummer Night's Dream as 'a most rare vision . . . It shall be called Bottom's dream, because it hath no bottom.'[24] In the summer of 1928, when he was beginning to shape his theory, he visited the monument to the seventeenth-century astronomer and mystic, Johannes Kepler, at Weil der Stadt in southern Germany. There he laid a wreath. The impromptu tribute was for a man who, like himself, had been a 'strange erratic genius, guided by a sense of mathematical form, an aesthetic instinct for the fitness of things'.[25]

The most direct way of verifying a scientific theory is to test its predictions through experiment, but Eddington was unable to produce any experimentally verifiable predictions from his fundamental theory. But even if a particular approach is unproven or unprovable, it can still inspire others to more fruitful lines of thought. Eddington's fundamental theory was of this sort. In 1937, Dirac, the personification of the coldly rational, austere, otherworldly physicist, surprised everyone by publishing a note on the significance of the dimensionless numbers that can be constructed from the fundamental constants of physics – such as the seven fundamental constants of Eddington's theory. The fine-structure constant, for example, has no dimensions, such as length and time, even though each of the fundamental constants that comprise it have dimensions. This had 'excited much interest in recent times', he wrote, adding that 'Eddington's arguments are not always rigorous' but nevertheless suggestive.[26] Even more out of character, Dirac had recently announced his wedding. It was at this time that he suddenly came up with these numerological speculations, prompting Bohr to remark, 'Look what happens to people when they get married!'[27]

Chandra, who at that time had just settled in Yerkes, was much taken with Dirac's paper. He decided to write up his own thoughts and send them to Dirac for his opinion. Later he paid homage to Eddington, 'the originator of this group of ideas'. Dirac and Eddington, he noted with respect, 'were playing for high stakes'.[28]

Dirac's paper had jogged Chandra's memory regarding 'certain "coincidences" which I had noticed some years ago, but which I have been hesitating to publish from the conviction that purely "dimensional arguments" will not lead one very far'.[29] Dirac's extension of Eddington's concepts inspired Chandra to take a fresh look at his own results, rewriting some in a way that pointed out their dependence on fundamental constants of nature. Developing the analogy with the fine-structure constant, he investigated how he could end up with the maximum mass for a stable white dwarf star by using a combination of the fundamental constants. Quantum considerations for the proper equation of state meant that he would have to use the Planck constant, as well as the speed of light and the mass of the proton. The universal gravitational constant also had to be included.[30] Dirac changed one or two sentences in Chandra's letter, and submitted it for publication.[31]

Chandra did not discuss fundamental constants again in any detail until he gave his Nobel prize lecture in 1983, where he argued that, just as the conditions that make atoms stable can be expressed in terms of fundamental constants, this should be the case for stars too.[32] In his foreword to a new printing of Eddington's The Internal Constitution of the Stars which appeared in 1988, he took Eddington to task on this point. According to Chandra, the main point of Eddington's famous discussion on how stars form – from the compression of interstellar gas, until eventually 'What "happens" is the stars' – was that if a star is indeed a gas which obeys the perfect gas law, an important assumption of the standard model, then it should be possible to express the ratio of the gas pressure and the radiation pressure of a star in terms of fundamental constants of nature, namely the Planck constant, the speed of light, the universal gravitational constant and the mass of the proton. This was surely the sort of thing Eddington would have loved. But strangely, this is a point 'on which Eddington is silent, Eddington does not isolate it – a surprising omission in view of his later preoccupations with material constants'.[33] For once, in his foreword, Chandra made no mention of the events of 11 January.

By 1937 Eddington had become increasingly isolated. He made eccentric comments, such as that 'one almost wishes that the Lorentz Transformation had never been invented; it is so continually applied without attention to its significance' – which is akin to saying that 'one almost wishes' that Einstein's relativity theory 'had never been invented'.[34] Throughout the early 1940s, the war years, Chandra maintained a cordial correspondence with Eddington. Chandra kept him up to date on his latest work in stellar dynamics. Eddington reported back that the atmosphere in Cambridge had changed: there were hardly any graduate students there, and the Observatory Club had stopped meeting. He tried to make light of the terrible effects of food rationing, joking that it was so bad that 'the regular dinners [at the RAS] have been dropped this session'.[35] Chandra sent Eddington parcels of rice. According to Lalitha, this was evidence that the two were close.

Life at Cambridge became worse and worse. In 1941 the town was bombed, causing some damage. Eddington wrote that the 'Christmas mail from America [to] here was lost' and that everyone at Cambridge 'missed a great many of the greetings from American astronomers.' 'My most severe deprivation is absence of the usual relaxations; I long to see a baseball match again!' he joked.[36] But he was still as hard at work as ever. He mentioned that his research on interstellar gas was going well. He remained cocksure: 'In the case of Einstein and Dirac people have thought it worth while to penetrate the obscurity. I believe they will understand me all right when they realise they have got to do so – and when it becomes the fashion "to explain Eddington".'[37]

During the summer and autumn of 1944, Eddington put everything aside to complete his book on his fundamental theory. Perhaps he had a premonition. He was now sixty-one and had begun to suffer bad stomach pains. He persevered stoically, not even informing 'his most intimate friend', Trimble, that he was unwell.[38]

Everyone thought he had overworked himself yet again. Doctors were busy coping with all the casualties from the war, and Eddington had to wait several weeks for an X-ray. It revealed a large tumour in his stomach. He had an emergency operation, but it was too late: the tumour was too far advanced. Eddington's last days were pitiful. He was put into a nursing home where there was very little care. 'My hands are so cold so do excuse the writing – we have to save fuel!'[39] Thus Winifred wrote to Shapley at Harvard, telling him of her brother's illness. He died on 22 November 1944, leaving his book on the fundamental theory unfinished. Russell, the grand old man of astrophysics, wrote in his obituary in the *Astrophysical Journal* that 'The death of Sir Arthur Eddington deprives astrophysics of its most distinguished representative.'[40]

After Eddington's death, Chandra and others turned out obituaries which were naturally full of praise, dealing only with the great man's achievements – which were indeed formidable. Nevertheless, some questions of a more personal nature remained to be answered. What was at the root of Eddington's bristly, antagonistic nature and willingness to cause pain? Was there more to it than the insensitivity of a certain class of Englishman at that time? What of Eddington the man? Was there any significance in the fact that his sister Winifred destroyed all his personal papers after his death? Such questions might seem irrelevant were it not for his treatment of Chandra, which threw everything into sharp focus.

On 11 January 1935, Eddington went beyond the lambasting he routinely meted out to Jeans and Milne. He turned the full force of his venom on a young man from a culture in which the verbal pyrotechnics practised in English academic and intellectual circles were completely alien. Yet Chandra's 'offence' was that he had solved a critical problem which Eddington himself had posed. Chandra was taken completely by surprise. After that, Eddington never relented. Until then he had been willing to consider anything that would advance the study of astrophysics, yet he went on to pervert the theory of relativity to support his view. Chandra's theory threatened the very basis of the fundamental theory that had become Eddington's obsession. Scientists have been known to back down when shown to be indubitably mistaken. So why did Eddington refuse to let go?

As Chandra put it, 'Eddington worked on the universe.'[41] His dream was to put everything together into one beautiful, harmonious package, a fundamental theory that would encompass everything, from atoms to stars. It was crucial that the number of electrons and protons in the universe he estimated from his fundamental theory should agree with the rate of expansion of the universe as measured by astronomers. But the numbers would match only if Chandra's theory of white dwarf stars was wrong and Fowler's correct. Eddington had fine-tuned his own version of Einstein's relativity theory to 'prove' that Fowler's theory, which did not agree with the standard interpretation of it, actually did, while Chandra's theory, or so he claimed, was merely an exercise in mathematics with no basis in reality. Chandra's results and his refusal to admit defeat threatened to scupper Eddington's beloved fundamental theory, taking with it everything he believed in and strove for. Something had to be done.

Like many Oxbridge dons of his day, Eddington lived a life in which personal relationships came a distant second to work. He devoted himself to his research with a single-minded passion, and over the years it provided him with friends, fame, travel and considerable wealth. His books were bestsellers; when he died he left £47,000, a fortune in those days. By the time he was knighted he had become a household name. Yet what the physicist W. H. Williams called his 'incommunicativeness', his stiff British diffidence, condemned him to a life of isolation. After his mother died, his only family was his sister, Winifred. Then there was his friend Trimble, the one person with whom he could relax. We have no idea how far their friendship went, for the very good reason that in those days intimate male friendships were not just forbidden but illegal. If he was homosexual, as many of his colleagues were, like them he would have had to lead a life of concealment, a life on the edge. Perhaps when Chandra inadvertently undermined his fundamental theory it also threatened to topple his fragile psychological balance.

This was one tile in the mosaic of Eddington's brutal treatment of Chandra. But there was a healthy dose of colonialism involved as well. It was expressed with panache and subtlety; but it was there all the same. It was an essential element in the British attitude towards Indians. As the poet of imperialism, Rudyard Kipling, wrote, 'East is East and West is West and ne'er the twain shall meet.' Herbert Compton, an old India hand, wrote less poetically in 1904:

> There is no assimilation between black and white. They are, and always will remain, races foreign to one another in sentiment, sympathies, feelings, and habits. Between you and a native friend there is a great gulf which no intimacy can bridge – the gulf of case and custom. Amalgamation is utterly impossible in any but the most superficial sense, and affinity out of the question.[42]

Gilbert Slater spent 1915 to 1921 at the University of Madras as its first Professor of Indian Economics. In his book about his experiences there, he describes the Madras Club:

> Some time about the end of last century a move was made to establish a club in which Europeans and Indians could mix on equal terms, and which, it was hoped, they would join in equal numbers. The club-house was built, and the club started, with every influential backing, as 'The Cosmopolitan'. But in the evening, after the day's work is over, men need to relax in the company of their own kind . . . While I was in Madras the Cosmopolitan was flourishing, but its membership was exclusively Indian. On one occasion I received and accepted an invitation from a member to dine with him there, but he gave me an ordinary European dinner, and I met none of the other members.[43]

Chandra recalled a story that Raman told him about a visit he made to Cambridge in the 1920s. Rutherford was his host there. As they walked around, Raman noticed that all the young men seemed to be playing tennis. 'Do they ever do any work?' he asked Rutherford. Rutherford replied with a loud guffaw, 'My dear Professor Raman, we don't want bookworms. We want governors for our empire.'[44] Raman, taken aback, could do no more than laugh politely. Chandra recalled that 'most of the people in academic circles were very conservative in Cambridge'.[45] Dirac, Fowler and Milne were not sympathetic to the cause of India's independence, and he avoided the subject with them. But Chandra was always careful to mention the interest that British scientists took in Indian science. Important figures such as Eddington and Rutherford often visited India. Chandra pointed out that 'Ramanujan would have died unknown, but for Hardy.'[46] Left unsaid is that Eddington was far from a Hardy for Chandra. Moreover, Ramanujan never wanted to stay in England, whereas Chandra did. Chandra dreamed of becoming Lucasian Professor of Mathematics at Cambridge, following in the footsteps of Newton and Dirac.[47]

Littlewood describes an unpleasant episode when the Trinity dons were discussing the issue of electing Ramanujan to a Fellowship there in 1919. He had already been elected a Fellow of the Royal Society, so election to Trinity should have been a foregone conclusion. Yet there was a great deal of opposition. One voter said bluntly that 'he wasn't going to have a black man as Fellow'.[48] Nevertheless, Ramanujan's application was successful simply because 'You can't reject an F.R.S.'[49] Fourteen years later, in 1933, such crude and overt racism had vanished. Chandra was elected a Fellow, despite Fowler's pessimism. Yet no British university offered him a position, even though there were openings available. Nor did any of Chandra's colleagues at Trinity think of recommending him as a Fellow of the Royal Society. Chandra wrote to his father that he was the first Indian ever to teach a course at Cambridge, adding that there was 'some prejudice giving Indians a definite appointment' there.[50]

Chandra also experienced racism in Chicago, and in the Deep South when he worked at the Aberdeen Proving Grounds during World War II. In New York, the Barbizon Plaza, a top hotel, once refused the Chandrasekhars accommodation.[51] Unpleasant as this was, in some ways overt racism was easier to deal with. At least you knew where you stood.

When Chandra wrote of Eddington as working on the universe, he meant the universe of the mind as well as the physical universe. Eddington had a dream that goes back to the very beginning of Western science, in ancient Greece: of finding one theory that would explain all of nature. He thought that he had discovered this through pure mathematics.

Eddington described his books about the fundamental theory as 'workshops', messy and cluttered with scrappy ideas waiting to be plucked out and developed – such as nature's curious choice of the number 137. His other books were 'showrooms', where tested products were displayed.[52] Yet even in the showrooms there were untidy corners, which the scientist might choose to ignore for the moment. In one of these corners was the question of white dwarfs. That was where Chandra got his start, tidying up Eddington's showroom.

Things might have been different had Eddington engaged in more rational debate. There are always Byzantine machinations when issues as fundamental as the nature of physical reality are under question. The question was debated fiercely in 1905, when Einstein presented his special theory of relativity, and even more strenuously between 1926 and 1933, when a viable interpretation of quantum mechanics was being hammered out. It continues today. But what occurred on 11 January 1935 was singular in the annals of modern science.

Chandra's discovery might well have transformed and accelerated developments in both physics and astrophysics in the 1930s. Instead, Eddington's heavy-handed intervention lent weighty support to the conservative astrophysics community, who steadfastly refused even to consider the idea that stars might collapse to nothing. As a result, Chandra's work was almost forgotten.

It was left to physicists to take up the baton. They thrived on challenges. They were happy to take chances, to construct apparently outrageous models of real phenomena, replete with nitty-gritty detail, which most astrophysicists preferred to avoid. Getting to grips with the fundamental issue of what makes stars shine demanded an understanding of nuclear reactions, as Eddington, for one, had foreseen. Physicists were on the look-out for new applications for the burgeoning field of nuclear physics. The heavens furnished an untouched arena in which to put their theories through their paces. The quest to understand the universe was about to pass out of the hands of astrophysicists and into the domain of the physicists.

PART II

How Stars Shine and How They Die

In 1932, when Chandra was still an eager young graduate student, physicists had discovered the neutron and gone neutron crazy. This particle, electrically neutral and with a mass about the same as the proton's, was detected at the beginning of the year by James Chadwick, a young colleague of Rutherford's, at the Cavendish Laboratory in Cambridge. Physicists immediately seized upon it as the panacea for all previous confusion about what made up the nucleus, and began to formulate promising theories of how the nucleus is held together and how it decays. Nuclear physics was born. But neither Chandra (who was at the Cavendish at the time) nor most other experienced astrophysicists spotted that it might have any significance for astrophysics. The Cavendish was the undisputed centre for nuclear physics in England, but the schism between physicists and astrophysicists remained as wide as ever.

In fact, this remarkable advance in scientific knowledge would eventually have an enormous impact on astrophysics. The extraordinary discovery of the neutron also helped open people's minds to the possibility that the demise of a star might be far more spectacular than anyone had so far dared imagine. Milne, for one, took an immediate interest in the debate on the formation of neutrons and how it might relate to the structure of stars. Perhaps, he suggested, electrons and protons might form neutrons at the high temperature and density inside stars; the energy liberated in this process might play an important part in how a star cools down. He did not, however, follow up on the subject.[1] Instead, it would be physicists who took the lead in applying new scientific developments to the study of the stars. It was the beginning of a long path of exploration that would lead eventually to the rediscovery of Chandra's theory and the realisation that he had been right all along.

The discovery of the neutron launched nuclear physics – i.e. the study of the nucleus – and made it one of the most exciting areas of enquiry. Previously, scientists had believed that the nucleus was made up of protons and just enough electrons to give it a net positive charge, so as to balance the negative charge of the electrons revolving around it. The result was an electrically neutral atom. A conspicuous drawback to this model was that, according to Heisenberg's uncertainty principle, electrons confined in the nucleus would have more than enough energy to burst out.

But there were also problems with the new model of a nucleus made up of neutrons and protons. What force could bind the neutral neutrons and positive protons together? Any such force required a considerable stretch of the imagination, because only particles with opposite charges were supposed to attract one another. Fortunately, though, one did not need to understand the nature of this force to begin speculating on how the newly discovered neutron might relate to the fate of the stars. Its very existence was enough to spark the imagination of the maverick Swiss-born physicist Fritz Zwicky.

One of the photographs commemorating Albert Einstein's visit to the California Institute of Technology (Caltech) in January 1931 shows an impish man, grinning joyously from the front row, two seats away from the president of Caltech, Robert A. Millikan, next to whom sits Einstein. Zwicky was one of the least popular members of staff and had probably grabbed the chair for himself. Born in 1898, Zwicky completed his Ph.D. on crystal structure at Einstein's old university, the Federal Institute of Technology in Zurich.[2] He remained an ardent and patriotic Swiss all his life, returning regularly to vote in Swiss elections.

Millikan, a Nobel laureate in physics, brought Zwicky to

Caltech in 1925 to initiate a research programme in crystallography. Then fifty-seven, Millikan was famous as a talent spotter. Between 1910 and 1921 he built up the physics department at the University of Chicago. In 1921, George Ellery Hale, one of the greatest fund-raisers in the history of astronomy, invited Millikan to perform his magic at a small institute in Pasadena, Los Angeles, which had recently been renamed the California Institute of Technology. In the years that followed, Millikan transformed Caltech into a Mecca of science.

Zwicky – a self-proclaimed genius – had a reputation for being difficult as well as for spinning fantastic schemes. But Millikan was willing to take a chance on him. He put the word out that he was Zwicky's protector. Zwicky certainly needed one: almost immediately he got off on the wrong foot, viciously criticising colleagues in the physics department and dismissing most of them as 'spherical bastards': bastards no matter which way you viewed them. Millikan suggested that Zwicky should switch to astronomy. Zwicky did so, and was soon antagonising his new colleagues as much as he had his old. But his abilities as an astronomer could not be faulted, especially his observations on the temporarily brilliant stars that astronomers had dubbed novae. Chandra attended a lecture he gave on this very topic in Cambridge in the autumn of 1930.[3]

In 1931 Zwicky teamed up with Walter Baade, a new arrival from the Hamburg Observatory. Five years Zwicky's senior, Baade was an acclaimed observational astronomer. Baade and Zwicky admired each other's brilliance and were often seen in the Caltech corridors chatting away in German. Short, with a sharp, angular face and a beak-like nose, Baade described himself as having a voice like a barking dog. Talkative, stimulating and ebullient, he had a certain quizzical look that conveyed the message, 'Tell me more, and with greater precision and clarity.' His style was the diametric opposite of Zwicky's. Cracks in their friendship began to appear when Zwicky accused Baade of taking all the credit for their joint work.[4] Baade formally distanced himself from Zwicky in 1936 after an embarrassing incident with Cecilia Payne. She had sent them a paper she was planning to publish in which she questioned some of their results. Zwicky fired back a note, heavily criticising Payne's paper and calling her a fool, signed with both their names.[5] Their final estrangement came during World War II, in a scene which showed Zwicky's hair-trigger personality all too well. During the war Baade found himself classified as an enemy alien. One day, totally out of the blue, Zwicky accused Baade of being a Nazi and threatened to kill him. 'They were a dangerous pair to put in the same room,' recalled a colleague some years later.[6]

But in the 1930s Baade and Zwicky made a superb team. With a great leap of imagination, Zwicky realised that the newly discovered neutron might offer a way of understanding the stars. He conferred with Baade. The two of them decided to present the idea at a meeting of the American Physical Society to be held at Stanford University in December 1933. In a famous abstract for the *Physical Review*, they suggested a new way that stars might end their lives.

At the time they were trying to understand why novae suddenly flared up and became a hundred thousand times brighter, followed by a gradual decline towards their former brightness which sometimes lasted for months. Flare-ups had also been observed in stars so distant that they had to be enormously brighter than novae to be seen from Earth. Astronomers called them 'giant novae', 'exceptional novae' and 'chief novae'; in their paper, Baade and Zwicky referred to them as 'supernovae', a term they coined in 1931.[7] A supernova can be ten thousand times brighter than a nova, as bright as all 200 billion stars in our Galaxy put together. The amount of light that supernovae emit is incredible. One which appeared in AD 1054 was visible even during the day and lasted for several weeks. Astronomers in China and Japan saw it, as did the Anasazi tribe of the south-western United States. Observations made at Mount Wilson Observatory in 1921 and refined by Baade, among others, finally identified the cause. Astronomers observing the Crab Nebula, a huge cloud of luminous gas in the constellation Taurus, discovered that it was expanding, and worked out its rate of expansion. Winding back the clock, they calculated that the expansion had begun nine centuries ago – the Crab Nebula is actually the wreckage of a massive star that 'went supernova' in AD 1054.[8]

In a nova, the amount of mass lost from the exploding star is about a thousandth of 1 per cent. In a supernova the star can lose 90 per cent of its mass, blown off in an explosion of unimaginable violence. The Crab Nebula is 40 thousand trillion miles away, 60 trillion miles wide and 7500 times as luminous as the Sun, bright enough to be visible with a small telescope. Fragments of the exploded star are still rushing outward at more than 1000 miles per second. From the measured distance between the Earth and the Crab Nebula, we know that the star exploded in about 5500 BC, some 6500 years earlier than the Chinese observations. This is because light from the exploding star, travelling at 186,000 miles per second, took 6500 years to reach the eyes of the Chinese observers.[9]

As late as the 1960s, astronomers were content to assume that most stars turned into white dwarfs when they died, while the rest simply blew themselves to smithereens. Baade and Zwicky, however, dared to speculate otherwise.[10] Zwicky's brainwave was to focus on what might be left behind after a supernova explosion. He realised that to find the answer he had to bring physics into the picture. On Zwicky's instigation, with great verve and imagination, he and Baade proposed that in some cases, rather than simply blowing itself up entirely, a supernova explosion might leave a highly compact core, which they dubbed a neutron star. Perhaps 'supernovae represent[ed] the transitions from ordinary stars into *neutron stars*'.[11] By 'neutron stars' they meant stars made entirely of neutrons. They qualified their proposal with the phrase 'with all due reserve', most likely at the insistence of Baade. But Zwicky was sure they were right. 'Nobody else would have dared to say that at the time. I thought it was pure fantasy. How could such a thing be?' recalled Hans Bethe, a physicist at the University of Tübingen who had become fascinated with nuclear physics after the discovery of the neutron.[12] What made the idea so daring was that Baade and Zwicky were attempting to understand some of the largest known objects, stars, in terms of a tiny part of the nucleus of the atom.

Between 1934 and 1939, Baade and Zwicky used data derived from observations of supernovae to hone their neutron star hypothesis. Using state of the art equipment at the Mount Wilson

Observatory, they made extensive measurements of the brightness of supernovae and produced wide-angle photographic surveys
covering thousands of galaxies in a single exposure, taken at intervals of weeks apart. They then compared them, looking for any
bright supernova which had suddenly flared up.[13] Zwicky was sure
that the 'tremendous *rate* of generation of energy in supernovae
would require an explanation along new lines'.[14] Indeed it did. His
conclusion was that a supernova is the result of a massive star
exploding with such violence that its core collapses into a highly
condensed star a mere 12 miles across, the size of Manhattan, yet
of nuclear density – a colossal 100 trillion grams per cubic centimetre. This is 100 million times greater than the density inside
a white dwarf. On Earth a teaspoonful of white dwarf matter
would weigh more than 6 tons. The same tiny amount of neutron
star matter would weigh a billion tons, most probably enough to
take it plunging through the Earth. Over the next twenty years,
Zwicky and Baade continued to seek out more observational evidence for supernovae.

Chandra was among those fired by Zwicky and Baade's discovery. As early as 1935, he mentioned the 'super-nova phenomenon'
in a paper published in the *Monthly Notices of the Royal
Astronomical Society*.[15] He took a longer look in 1939, when he
spoke on the 'origin of the supernova phenomenon' at the
Colloque International d'Astrophysique in Paris, where he clashed
so memorably with Eddington.[16] There he tied it in with his own
work, suggesting that when a star that had exhausted its fuel was
left with a mass more than 1.4 times the mass of the Sun (i.e. the
Chandrasekhar limit, as it had now become known), the outer
layers of the star would collapse under the pressure of gravity, and
there would be an enormous release of gravitational energy. This
would blast the star's outer layers into interstellar space, while the
electrons and protons in the core would be squeezed together to
form an incredibly dense neutron core. Eddington too, as we have
seen, backed the scenario of a neutron star.

In Russia, Lev Landau had been speculating about highly condensed star cores since 1932, in terms that made more sense to
physicists than to astrophysicists. He was fond of quipping that he
had entered physics too late: 'all the nice girls are already married,

and all the nice problems are already solved'.[17] Born in 1908, Landau was a child prodigy who could 'scarcely remember not being able to differentiate and integrate'.[18] He entered Baku University at the age of fourteen, then transferred to the more prestigious University of Leningrad two years later. After completing his Ph.D. at nineteen, he made a Grand Tour of the great European centres of theoretical physics. His brilliance was clear to everyone – even the hypercritical Pauli, whom he visited in 1929. Landau ended his tour at Niels Bohr's institute in 1930 and immediately began research on the foundations of quantum theory. Tall and lanky, with bushy black hair piled into an eccentric quiff, Landau was a formidable physicist whose interests spanned the whole spectrum of theoretical physics. He more than held his own wherever he went. In Copenhagen he refused to be ground down by Bohr's legendary relentless arguing. Once, Bohr was lambasting Landau after Landau had presented a lecture. Landau ambled nonchalantly over to an unoccupied front bench, where he stretched out and gazed quietly at the ceiling as Bohr stood over him arguing intensely, his face red and his veins bulging.

In 1931 Landau returned to the Soviet Union. A fervent Marxist and patriot, he was determined to bring the latest in theoretical physics to his country. He was immensely successful. Under his leadership the physics department at Kharkov Gorky State University in the Ukraine became the best in the Soviet Union; in 1937 he became head of the theoretical division at the Institute for Physical Problems in Moscow. Landau also brought astrophysics to the Soviet Union. He suggested a radically new way of approaching the subject – to construct stellar models using the 'methods of theoretical physics'. As far as physicists such as Landau were concerned, astrophysicists like Milne, who made 'assumptions only for the sake of mathematical convenience', were merely playing with unrealistic mathematical models of stars.[19] He then made a brief and elegant calculation which enabled him to construct a model of a star made entirely of cold degenerate matter – highly compressed, rock-hard matter that is no longer radiating light or heat. The model yielded the result that a star of this kind which was greater than 1.5 times the mass of the Sun would be unstable and liable to collapse indefinitely to an infinitely dense point. He had rediscovered Chandra's upper limit to the mass of a white dwarf star (although his exact figure was slightly different), but through the path of theoretical physics.

On reaching this conclusion, Landau immediately declared that 'in reality' stars could not possibly 'exhibit any such ridiculous tendencies'.[20] As much as Milne and the rest of the astrophysics community, Landau was convinced that a star simply had to stop collapsing at some point. He concluded that at that point it 'must consist of a core of highly condensed matter, surrounded by matter in an ordinary state'.[21] Along with almost everyone else, he accepted the view that as a star burnt up the last of its fuel, it would somehow shed sufficient matter to end up with a mass below the upper limit and thus avoid disappearing into nothingness. As a physicist, Landau did not read astrophysical journals and so was unaware of Chandra's earlier work. For years to come, it was Landau's paper of 1932 rather than Chandra's article, written in 1931, that was credited with first identifying the limiting mass for white dwarfs, even though Landau himself did not initially believe his own discovery.

Six years later, Landau spun a variation on his earlier result. He proposed that, somehow, deep inside a star where the pressure is extremely high, a core could develop, very much like a white dwarf. If the mass of the core were to exceed the upper limit, this core would become unstable and collapse to nuclear density – the density at which neutrons and protons are packed into a nucleus, 10^{14} grams per cubic centimetre. If the Earth were compressed to this density, it would be only 1000 feet across, instead of 7926 miles, its actual diameter. The electrons and nuclei in the star would be squeezed so tightly together that they would form a neutron star. Like electrons, neutrons obey the Pauli exclusion principle and generate a degeneracy pressure so that they can counteract any further gravitational compression. Landau had worked out the *maximum* mass for a stable white dwarf. But at this point, he realised, the question that needed to be asked was, what was the *minimum* mass needed to form a neutron core? Combining Newton's theory of gravity, as applied to planets, with quantum statistics, he estimated that it must be around one-thousandth the mass of the Sun.

Landau went on to develop a new theory of how stars shine, quite different from Eddington's, which claimed that it was through subatomic processes such as fusion. He suggested that when gas particles from the shell surrounding the neutron core fall into the core, their energy of motion turns into heat, which we observe as starlight.

Landau was driven to this theoretical flight of fancy by more than just a desire to push back the frontiers of knowledge. His personal survival was at stake. Since 1931, Josef Stalin's regime had become more and more brutal, and by 1937 life in the Soviet Union was becoming intolerable. The forced collectivisation of agriculture led to the deaths of some 7 million people, and there were purges of the country's leading political, intellectual and military figures. As one of the country's highest-profile scientists, Landau found himself under scrutiny. The best way to keep out of the dreaded political prisons, almost invariably a one-way journey, was to keep as visible as possible. Landau hoped that his paper on neutron cores would do the trick.

Landau sent his paper to the journal *Nature* and to a leading Russian physics periodical, and also sent an English translation to Bohr. Bohr was an honorary member of the Soviet Academy of Sciences and highly respected in Soviet scientific circles. On hearing that he was impressed with the paper, Landau arranged for the leading Soviet newspaper, *Izvestia*, to invite Bohr to publish a comment there.[22] But despite his efforts, Landau was arrested, on 28 April 1938. He was taken off to Butyrka, one of the worst prisons run by the NKVD, the forerunner of the KGB. As always, the charges were trumped up and absurd. Although Jewish and a Marxist, Landau was accused of spying for Nazi Germany. According to NKVD files, he had been overheard criticising the way in which the state organised scientific research, and also Stalin's reign of terror. As far as the authorities were concerned, this was anti-Soviet activity.

The Russian physicist Pyotr Kapitza, a discoverer of superconductivity, a friend of Bohr and Rutherford and creator of the Institute for Physical Problems in Moscow, where Landau had worked, wrote to Stalin and to the People's Commissar, Vyacheslav Molotov, on Landau's behalf. He pleaded that only

Landau had the capacity to unravel the puzzle of superconductivity, which would greatly enhance the stature of Soviet physics. Taking a chance, he even threatened to halt his own scientific research if Landau was not released. It worked. Landau was freed a year later. He had suffered terribly and emerged a broken man. He would have been 'unable to live for even another half year', he recalled in an interview in 1964 for the Soviet newspaper *Pravda*. He re-established himself as a scientist, but he never again put himself in a position that could be interpreted, even vaguely, as politically offensive.[23]

Across the world, two refugees from Communism, George Gamow and Edward Teller, were leading figures on the staff of George Washington University in Washington, DC. Every spring, beginning in 1936, they organised a conference on theoretical physics at the Department of Terrestrial Magnetism at Washington's Carnegie Institution. The conference in March 1938 was on how stars shine. It marked a turning point, both in astrophysics and in nuclear physics.

Gamow and Teller made a potent combination. Gamow was from Russia, Teller from Hungary. Through the force of their personalities and the acclaimed originality of their research, they were able to bring together key figures in astrophysics and physics who were at the cutting edge of the burgeoning new field of nuclear physics. Bethe, who was to make the first breakthrough in cracking the mystery of stellar energy, declared that the meetings were the 'most stimulating that I have ever attended'.[24]

Chandra, naturally, was among the twenty scientists who were invited to the 1938 meeting. By then he was happily settled at Yerkes. Bengt Strömgren, Chandra's old friend from Copenhagen and colleague at Yerkes, was also present. He was an astrophysicist who could speak the same language as physicists, and Gamow regarded him as the 'ace' of the meeting.[25] Gamow as coordinator sparked many lively exchanges. The discussions were informal. There were no recorded minutes, and participants were not required to submit papers. In such a relaxed atmosphere, no one felt that they would lose face in any interchange. It was different in every way from the meetings at the RAS. Einstein and Russell were also invited, but neither attended. The subject matter was outside Einstein's range of interests. Russell's absence, however, was more surprising because he had long been interested in the question of stellar energy. Perhaps speculations by fast-talking physicists such as Gamow were not to Russell's patrician tastes. One can imagine how he might have reacted to a 'Gamowian' statement such as, 'A physicist doing stars feels happy as long as he dose not nead tuch the astronomical tables.'[26] It was now physicists, not astrophysicists, who dominated the field.

Gamow was one of physics' most colourful characters. As Léon Rosenfeld said, 'How could anyone who has ever met Gamow forget his first meeting with him – a Slav giant, fair haired and speaking a very picturesque German; in fact he was picturesque in everything, even in his physics.'[27] No one ever forgot Gamow's visit to Niels Bohr's institute in Copenhagen. He was a chief instigator of the high jinx that became a trademark there, with table-tennis tournaments (which Heisenberg usually won), movie-going (American Westerns were preferred) and mock theatrical portrayals of Bohr, Heisenberg and Pauli, usually written, directed and illustrated by Gamow.

Born in 1904 in Odessa, Gamow always wanted to be a scientist. He spent a year at Novorossyshky University in Odessa in 1922, then moved to the better equipped University of Leningrad. There, Landau was among his friends. They formed a study group and rapidly got up to speed with modern physics. By 1926 Gamow had mastered general relativity and quantum physics. Looking for fresh areas to research, in 1928 he set off for Max Born's institute in Göttingen. There he decided to focus on nuclear physics. He set to work to investigate the nuclear processes that make stars shine.

It turned out to be a magical summer. Studying the literature on the subject, Gamow discovered an intriguing article which Rutherford had published the previous year, describing the extraordinary result of striking uranium atoms with alpha particles (the nuclei of helium atoms).[28] Physicists knew that uranium nuclei could emit an alpha particle in a form of radioactive decay. Rutherford found that in some cases alpha particles with twice the energy of ones that had been emitted by the uranium nucleus could not penetrate it. The puzzle was how the low-energy alpha particles managed to escape from the uranium nucleus if particles with a higher energy could not get in. The uranium nucleus has ninety-two protons, and their combined positive charge acts as a strong barrier which the alpha particle would have to penetrate in order to escape. Gamow decided to attack this problem by applying quantum mechanics to the nucleus – something which no one had previously thought of doing.

According to classical Newtonian physics, particles inside the nucleus behave like billiard balls inside a glass fruit bowl. In order to escape they have to generate enough speed to roll up and over the side of the bowl. But quantum theory states that particles can also behave like waves. An alpha particle could therefore pass through the barrier erected by the massive positive charge of the uranium nucleus, in much the same way as a light wave passes through the glass of the bowl or from water to air. Gamow dubbed this process 'tunnelling'.[29] Alpha particles confined within a nucleus can tunnel their way out, rather like prisoners laboriously digging a tunnel to escape. Quantum mechanics suggests, however, that alpha particles hitting the nucleus from the outside are more likely to bounce off. Gamow was the first physicist to formulate a theory of nuclear phenomena; his astounding discovery was also to have an impact on the verification of Chandra's theory.

The logical next step was for Gamow to apply his tunnelling theory to the interior of stars, to try to establish how stars shine. But this required lengthy calculations, work he detested. Fortunately he soon found a way around this problem.

During that incredibly creative summer, he had met another maverick physicist, Fritz Houtermans. Of Dutch and Austrian parentage, Houtermans was a Communist as well as half-Jewish, a lethal mixture in the 1930s. Having grown increasingly uneasy at living in what was then Nazi Germany, he emigrated to the Soviet Union in 1933, and in 1937, at the height of the Stalinist purges, he was arrested as a German spy. He went on to spend two harrowing years in a series of NKVD prisons, including Butyrka, where Landau was interned. In 1939, when the Hitler–Stalin pact was made, he was repatriated, barely alive.

But he was immediately arrested by the Gestapo on suspicion of being a Russian spy, and further mistreated. The Nobel laureate Max von Laue, a close friend of Einstein's and one of the very few German scientists willing to stand up to Hitler, arranged for Houtermans' release. He spent the war working in a private research laboratory.[30]

When Gamow met him in 1928, Houtermans had just received a Ph.D. in experimental physics at Göttingen. He convinced Gamow that he was Viennese at heart and that bohemian spirits such as they could do their best thinking in cafés. During their conversations, Houtermans confided that what he really wanted was to be a theoretical physicist. Gamow could not believe his luck. The two worked together in their favourite café, their papers and coffee cups littering the table.

Houtermans introduced Gamow to Robert d'Escourt Atkinson, a Welshman who had just completed his Ph.D. at the same laboratory as Houtermans in Göttingen. Like Houtermans, he too dreamt of being a theoretical physicist. Gamow suggested that the two of them look into the problem he had isolated, of applying his tunnelling theory to the interior of stars. Astrophysicists had not yet realised that stars are made up mostly of hydrogen, so Atkinson and Houtermans assumed that they contained a mixture of elements, including lithium, beryllium, boron, carbon, calcium, nitrogen and oxygen. At the high temperatures inside a star, the nuclei of these elements have been stripped of all their electrons. The heat is so extreme that they are able to overcome the repulsive force caused by their mutual positive charges and move close enough to tunnel into one another, causing nuclear reactions that release the enormous energy that powers the stars. They concluded that the nuclei with the fewest protons are most likely to tunnel into each other; only a few at a time burn up (i.e. convert into energy) completely, which means that the nuclear reactions proceed slowly enough to enable a star like the Sun to shine for billions of years.

The day after they completed their paper on thermonuclear reactions in stars, Houtermans 'went for a walk with a pretty girl. As soon as it grew dark the stars came out, one after another, in all their splendour. "Don't they shine beautifully?"

cried my companion. But I simply stuck my chest out and said proudly: "I've known since yesterday why it is they shine."[31] Two years later, they married.

When he left Göttingen, Gamow visited Niels Bohr's institute, and returned to Leningrad in 1931. He found the mood in Russia increasingly intolerable and resolved to leave as soon as possible. But international travel for Soviet citizens was severely restricted. In 1933 he was invited to the prestigious Solvay Conference, in Brussels, which that year happened to be on nuclear physics. He organised an invitation for himself and his wife. They never returned. After the conference, Bohr invited Gamow back to Copenhagen, where he met Teller.

Physically, the blond Slavic giant and the dark intense Hungarian with his prosthetic foot were complete opposites. Teller had had a bright future as a professor of physics in Germany, but that ended in 1933 with Hitler's rise to power. This was all the more devastating for him, as he had witnessed the excesses of communism and fascism in Hungary. Teller went straight to Bohr's institute at Copenhagen. He and Gamow struck up a close friendship, both being birds of passage. After a year, Teller went to England, where he met Chandra, while Gamow moved on to George Washington University. In 1935 Gamow offered Teller a position there. They worked together closely and made several important discoveries in nuclear physics. Teller was greatly affected by the ease with which Gamow came up with new ideas and liked to believe he could do the same.

Physicists had now homed in on the crucial problem: what was the chain of nuclear reactions that enabled stars to shine for billions of years? By 1932 it had been pretty well established that stars contain a high percentage of hydrogen, as Strömgren had realised. Everyone already suspected, as Eddington had proposed, that stars shine by proton–proton fusion – burning hydrogen to make helium. Francis Aston had shown this was possible in the laboratory. This inspired Atkinson to rethink his original work with Houtermans. He looked into several chains of nuclear reactions that might power stars like the Sun, but they were all fired by nuclei which, under further scrutiny, turned out to be unstable and would thus result in stars burning their fuel too quickly.

Another question was whether fusion could be initiated by thermonuclear reactions. This was a term originally coined by Gamow for the unimaginably violent reaction that occurs when the intense heat and high density inside stars permit two protons to overcome their mutual electrical repulsion, smash into each other and become glued, or fused, together.

Gamow, Atkinson and Houtermans were the first to verify theoretically during that highly creative 1928 summer at Göttingen that nuclear fusion could power the stars. Ten years later, Gamow's doctoral student, Charles Critchfield, investigated further. Before the all-important 1938 conference at the Carnegie Institution in Washington on how stars shine, Critchfield sent Hans Bethe a manuscript of his paper. Bethe made many comments and corrections, and it became known as the Bethe–Critchfield paper.[32]

Chandra recalled the impression Bethe made on everyone. Well over six feet tall and of massive build, he had every aspect of nuclear physics at his disposal: he grasped the essence of a problem immediately and 'just bull-dozed in'.[33] Born in 1906 in Strasbourg, Bethe studied under Sommerfeld, Pauli and Fermi. He was the acknowledged expert on nuclear physics and a natural for the Washington conference. His three massively detailed articles in the *Reviews of Modern Physics* were known as 'Bethe's Bible'. He loved getting to grips with problems. He did not thrive in chaotic situations in which scientists have to create something 'out of very little'.[34] Like Chandra, Bethe preferred to 'take a subject in which the foundations have already been laid and then try to exploit them'.[35]

At the 1938 meeting in Washington, every important figure whose work touched on how stars shine was there. The astrophysicists were the first to speak. Chandra expounded on white dwarfs, and Strömgren on the hydrogen content of stars. Everyone agreed that fusion processes were what powered stars and caused them to shine, but 'people were really at a loss as to what to do and what reactions to consider', Bethe recalled. He was struck 'by the total ignorance which pervaded everybody at the meeting'.[36] The astrophysicists were suddenly apprehensive, realising that the 'thermonuclear reactions in the stars were [being studied] by physicists – [Gamow] and Bethe, among them – because astronomers didn't know about nuclear physics. They were sitting on their astronomical things,' as Gamow recalled in his colourful way.[37]

Bethe was hooked on the problem of how stars shine. For the entire month after the conference he obsessed over it and finally solved it, at least for stars of about the same mass as the Sun's. He worked at it fifteen hours a day, seven days a week, poring over his 'Bible' and papers by Gamow and Teller.[38] But one puzzle remained to be solved. What were the nuclear reactions that powered stars more massive and more luminous than the Sun, such as Sirius A? After numerous attempts, Bethe discovered the chain of nuclear reactions that occurred at the much higher internal temperatures of stars such as these, burning hydrogen and transforming it into helium. The key was to find elements that would react at the temperatures believed to exist inside such stars – about 10 million degrees kelvin – and would burn at a rate that agreed with the lifetime that very luminous stars such as these were believed to have, around a million years.[39] Eddington guessed this when he wrote that, in the early lifetime of a star, there is a 'small scale rehearsal of the great development which appears to set in at' about 10 million degrees kelvin. Chandra liked to point out that this was a case where astrophysicists had something to teach physicists – the temperatures and conditions for the proper nuclear reactions.[40]

Bethe published all this in his epoch-making paper of 1939.[41] But there were still big questions that had not been answered. What might happen to a star once it had burnt up its hydrogen? What sort of cataclysmic event might have generated the heaviest elements, such as uranium? In November 1938, Gamow summed up his thoughts on the issues that had been raised at the Washington meeting. The question of the neutron core, he concluded, was of merely academic interest.[42] But unlike Gamow and the others, Bethe was prepared to speculate. He agreed that stars about four times as massive as the Sun would be able to burn up enough fuel to get below Chandra's upper limit and expire as white dwarfs. But for stars more massive still, he thought otherwise. On the basis of his new theory, he proposed that by the time such a heavyweight had completely burnt up its fuel, it would be made up of heavier stable elements and would only stop collapsing 'when a neutron core is formed'. But, he concluded, 'these questions obviously require much further investigation'.[43]

During the summer of 1938, Bethe met up with J. Robert Oppenheimer and his group of brilliant young graduate and post-doctoral students at the University of California in Berkeley. They promptly decided to turn their full attention to the fate of the stars. Oppenheimer was born on New York City's Riverside Drive in 1904, into great affluence and culture. Two events in his early life uncannily presaged his future. In 1921, as a wealthy young American, he made the usual Grand Tour of Europe, during which he visited the Joachimstal mines in northern Czechoslovakia. Among the many ores mined there was a dense, black, sticky material called pitchblende. In 1789, Martin Heinrich Klaproth, the first professor of chemistry at the University of Berlin, extracted a greyish metal from this ore. The English astronomer William Herschel had recently discovered the planet Uranus, so Klaproth named his newly isolated metal 'uranium'. It was used mainly for colouring ceramics until Marie and Pierre Curie discovered it contained the radioactive elements radium and polonium. As late as the 1940s, the Joachimstal mines were still Europe's only source of uranium, whose use by then went far beyond decorative purposes.

To toughen up his six-foot, gangly son, Oppenheimer's father sent him out west to a dude ranch in the Jemez Mountains northwest of Santa Fe in New Mexico, to camp and ride horses. It was there that, in 1922, the eighteen-year-old first saw the Los Alamos mesa, a flat-topped rock formation which took its name from the cottonwood trees that grew there. Both uranium and Los Alamos were to loom large in his future career.

A man of enormous complexity, Oppenheimer hid a deep insecurity behind a charismatic yet arrogant exterior. 'Robert could make people feel they were fools. He made me [feel that], but I didn't mind. [Others] did,' recalled Bethe.[44] After completing his Ph.D. with Max Born at Göttingen in 1927, Oppenheimer set out to learn the new quantum theory at first hand from the masters in Europe. Born, Heisenberg and Pauli were much impressed with him, as were many others. Back in the United States, Oppenheimer rejected a job at Harvard and went instead to Berkeley 'because it was a desert', with no school of theoretical physics there. 'I just thought it would be nice to start something,' he recalled.[45]

Berkeley was the ideal location for him. Caltech, where superb theorists such as Richard Chase Tolman and the world's best cosmic-ray group led by Robert Millikan were working, was not far away. Carl Anderson, who had discovered the positron, was a member of Millikan's group.[46] Cosmic-ray research was at the frontier of physics. In the days before particle accelerators, it was the only way to study high-velocity particles, a subject on which Oppenheimer had become an expert. Jointly appointed to Caltech and Berkeley, Oppenheimer could make use of the sagacious advice and criticism of the established physicists at Caltech while starting his own school of theoretical physics at Berkeley. Like Landau in Russia, Oppenheimer was determined to thrust his country to the forefront of physics research. He succeeded, creating the greatest school of theoretical physics ever seen in the United States.

Everyone agreed that Oppenheimer's lectures were difficult to follow. But they conveyed the beauty of the subject and the excitement of its development. He liked to cultivate the look of the tortured European intellectual with a cigarette or pipe hanging from his mouth, and spoke in a velvety whisper with long dramatic pauses punctuating perfectly formed sentences. The word went out that Oppenheimer was the person to study with. A collection of brilliant young men found their way to him. Most were to leave their mark on post-war physics, having earned their research spurs with the Manhattan Project.

Oppenheimer had invited Bethe to address his students in the summer of 1938, following the Washington meeting. But, looking over the sea of hungry eyes, Bethe decided not to draw their attention to the problem of what makes massive stars shine and what happens when they grow dark. Oppenheimer's group were like sharks. They pounced on whatever titbits he had to offer, and Bethe wanted to hang on to his hard-won discoveries.[47] Seeking to cast a little light on Bethe's teasingly enigmatic remarks about the fate of massive stars, Oppenheimer thought of applying general relativity. He approached the general relativity guru at Caltech, Richard Chase Tolman, whose book on statistical physics Chandra had devoured as a teenager. Tolman, a Massachusetts-born Quaker, was then fifty-seven. He had made significant contributions to Einstein's general theory of relativity and was regarded as one of the world's leading cosmologists. When Einstein visited Caltech in 1931, it was because Tolman was there. He was highly respected for the depth and breadth of his intellectual interests as well as for his wisdom and urbanity. Oppenheimer frequently visited Tolman's home. There was an added attraction: Oppenheimer was said to be having an affair with Tolman's wife, Ruth.[48]

Tolman was intrigued by Landau's paper, which suggested (as had Chandra's earlier paper) that a white dwarf star might implode if it exceeded a certain mass. Later he would make important contributions to the Manhattan Project: it was Tolman who was the first to suggest implosion as a way of squeezing nuclear fuel so it would blow up.[49] He now suggested to Oppenheimer and Robert Serber, a young collaborator of Oppenheimer's, that they should read that paper and also Landau's more recent article, written earlier that year, 1938, which looked into the minimum mass needed for a neutron core to form inside a star. For Tolman, the concept of a neutron star was breathtaking. It was a classic case of a conundrum that cried out for the general theory of relativity. At neutron star densities – ten million times denser than a white dwarf – relativity predicts an incredible warping of space around the star which cannot be described using Newton's theory of gravity.

Landau's 1938 paper seemed full of mistakes. So Oppenheimer and Serber set to work. They saw immediately that Landau had used Newton's theory of gravity instead of the general theory of relativity to estimate the lowest possible mass at which the neutron core could become rock-hard and resist being crushed by gravity. 'This figure appears to be wrong,' they wrote bluntly.[50] Had they known that Landau was languishing in an NKVD prison at that very moment, they might have been more circumspect. In their research they focused on the part played by nuclear forces.

Inside the nucleus, neutrons and protons are pressed incredibly closely together, a ten-thousandth of a billionth of a centimetre (just 10^{-13} centimetres) apart. At this degree of closeness the nuclear force is enormous, much stronger than the pull of gravity. This meant that the neutron core could contain far more matter than Landau had estimated.

On the basis of the little that was known about nuclear forces at the time, Oppenheimer and Serber estimated that the minimum mass for a stable neutron core was about a tenth of the mass of the Sun, a hundred times bigger than Landau had thought. But if that was correct, it would lead to a complete breakdown of Eddington's model for stars like the Sun. If the Sun had a neutron core of 10 per cent of its mass, it should give rise to effects that could be observed by astronomers, effects completely different from those that we would expect to find if the Sun was, as previously assumed, a perfect gas. However, no such effects had been seen.

But what if the neutron core was a star in itself? Was there a *maximum* mass for a neutron star, as one would expect there to be, given that neutrons function much as electrons do and could generate a degeneracy pressure in the same way? And what would happen if this mass were exceeded? Together with two of his graduate students, George Volkoff and Hartland Snyder, Oppenheimer looked into these questions. Combining general relativity with the embryonic nuclear physics, they began to lay the modern theoretical framework for understanding the fate of collapsed stars. Their work showed how massive stars could perish as either white dwarfs or neutron stars – or could collapse completely.

Oppenheimer and Volkoff's landmark paper, 'On massive neutron cores', was published in the *Physical Review* early in 1939.[51] With great verve and some elegantly minimalist calculations, they addressed the question that the papers by Landau and by Oppenheimer and Serber had failed to consider: what is the maximum mass for a stable neutron star? Working with only limited knowledge of developments in astrophysics, as provided by Tolman, they calculated that the maximum mass for such a star would be between a tenth and seven-tenths the mass of the Sun. A star like this would be massively dense yet a minuscule 12 miles across. They stopped at this, putting aside the key question of what would happen if the maximum mass were exceeded.

There was, of course, one astrophysicist at Caltech who had been a physicist and could have filled them in more completely on developments in astrophysics: Zwicky. But he was generally avoided. After all, who needed to be called a 'spherical bastard'? The sophisticated and cultured Oppenheimer kept his distance, and Zwicky returned the compliment. Four months after Oppenheimer and Volkoff's paper was published, he submitted an article to the *Physical Review*. In this he attacked the problem of neutron stars head on, linking them to supernovae.[52] But he made no reference to his physicist neighbours.

Chandra recalled that during 1934 and 1935, he and von Neumann were jointly exploring topics similar to those which Oppenheimer and Volkoff had considered.[53] The work on neutron stars must have been more von Neumann's than Chandra's. Chandra preferred to consider ideal systems rather than delve into the nitty-gritty of nuclear physics, and in any case in the 1930s there was very little nuclear physics to consider. Nor was he eager to ponder the effects of general relativity, a subject he considered a graveyard for physicists, having seen what it did to Eddington and Milne. Eddington's perverse version of relativity, spun specifically to prove Chandra's theory wrong, and Milne's attempt to formulate a cosmological theory to supersede relativity, did great harm to both their careers. 'I did not want to go into that region; I just wanted to settle on safe grounds,' Chandra recalled.[54] He refused to reformulate his conclusions along lines more familiar to physicists, and continued to publish papers in the classical style of astrophysics. Most physicists ignored them.

Chandra was, of course, up to date with all the latest developments in nuclear physics. He read texts on the subject, taught a course on it at Yerkes and was on hand for discussions at the Washington meeting. Bethe referred to conversations with Chandra, during and after the Washington meeting, in the paper he published in 1939. But Oppenheimer and his group did not take Chandra seriously: they thought he was not a 'real' physicist.[55] In their paper on neutron stars, Oppenheimer and Volkoff praised Eddington for opening up the field of modern astrophysics, and reviewed Landau's research, but relegated Chandra's to a footnote.

Chandra had been in the United States since the end of 1936. Yet even after Oppenheimer's papers on astrophysics began to appear in 1938, he was not curious to find out more. He preferred to remain ensconced in the remote rural beauty of Williams Bay, as far as possible from Trinity College, in domestic peace and tranquillity. Like the great scientists he so admired – Einstein, Newton, Poincaré – Chandra preferred to deal in grand mathematical theories, leaving the details to others. The trouble was that in the study of stars it was the details that were all-important. In his otherwise exhaustive monograph *An Introduction to the Study of Stellar Structure*, written in 1938, the concluding chapter on why stars shine is curiously incomplete and unsatisfactory. Chandra was fully aware of Bethe's theory, that it was the reactions of carbon with protons that makes stars shine, but he chose not to mention it. Instead, he discussed other theories which he already knew were wrong. He had, it seemed, absolutely no interest in collaborating with Oppenheimer's group. He simply abandoned his work on collapsed stars and let Oppenheimer steal a march on him.

No one yet had come up with an answer to the great question posed by Chandra's theory: what happens to a massive white dwarf star which exceeds the upper limit Chandra had discovered? Was it actually possible that such a star might shrink to a hugely dense and unimaginably tiny point? It was Oppenheimer who finally addressed the problem. He examined the fourth way in which a star could expire: by collapsing and collapsing until it sucks the very space around it into its giant maw. This was an attack on the very frontier of astrophysics, and Oppenheimer organised it as carefully as he would later plan the Manhattan Project. He broke the problem down into carefully defined units, then selected the right people to deal with each. With Serber and Volkoff, Oppenheimer had calculated the minimum mass necessary to form a stable neutron core, and the maximum mass at which a neutron star could remain stable. The logical next step was to work out what would happen if this maximum mass was exceeded.

Oppenheimer assigned this task to Hartland Snyder, who was adept at the difficult mathematics of general relativity.

Oppenheimer and Snyder set out to use the general theory of relativity to explore collapsing stars. They studied stars so massive that, even after they have burnt up all their nuclear fuel, they are unable to blow off enough material to shrink below the maximum mass limit and thus cannot form a stable neutron core. Tolman was on hand to provide advice on general relativity. To simplify their calculations, Oppenheimer and Snyder limited themselves to a spherical cloud of gas. They traced the collapse of the cloud and explored what would happen when it shrank beyond what they called its gravitational radius. This is now known as the Schwarzschild radius, after the German astronomer Karl Schwarzschild.

Schwarzschild was a highly theoretical physicist and a practical astrophysicist. When World War I broke out in 1914, he gave up the directorship of the Potsdam Observatory and, at the age of forty-one, volunteered for the German army. Late in 1915, while he was calculating artillery trajectories on the Russian front, he happened to read Einstein's published papers on general relativity. Einstein formulated his general theory of relativity using elegant mathematical hypotheses relating gravity to the geometry of space and time. But the equations governing the theory were so difficult to solve that Einstein assumed that extreme approximations would be the best that anyone would manage. Remarkably, Schwarzschild worked out the first exact solution almost immediately, using the theory to investigate the effect of a spherical object on space and time around it. Einstein was astonished that his equations could be solved so quickly and simply. He wrote to Schwarzschild that the work 'appears to me splendid'.[56] Schwarzschild would have made an ideal co-worker for Einstein. Sadly, in March 1916 he returned to Berlin seriously ill, with a rare skin disease contracted in the trenches. He died soon afterwards. Eddington fondly remembered meeting Schwarzschild at a scientific meeting in Hamburg in 1913. He participated in a motor race with 'Schwarzschild and five mad Englishmen', he wrote to his mother.[57] Eddington won, of course.

Schwarzschild was concerned only with how a spherical object affected space and time around it. He did not discuss the fact that his solution gave the result of infinity extremely close to the centre of the object. Later, the place at which the solution becomes infinite became known as the Schwarzschild radius. The Schwarzschild radius defines a region whose gravitational pull is so strong that nothing can escape, not even light; in other words, it takes an infinite amount of time for light to escape. Schwarzschild's solution becomes infinite once an object has dwindled enough to be bounded by its Schwarzschild radius. Every object has a Schwarzschild radius. If a star were to shrink until it was smaller than its Schwarzschild radius, its gravity would become so powerful that it would distort space around it so that nothing could escape. The same applies to the Sun, the Earth, you, me or a can of baked beans. The Sun's Schwarzschild radius is about 2 miles, as opposed to its actual radius of 432,000 miles. Human beings have a minute Schwarzschild radius – about the size of a proton. The Schwarzschild radius of a proton, in its turn, is unimaginably tiny. At the time, scientists considered all this pure science fiction. After all, they thought, in reality nothing could shrink to such minuscule proportions.

In 1926, Eddington had imagined what might happen to a star that shrank so much that it distorted space around it. It would, he wrote, 'produce so much curvature of [space] that space would close up around the star, leaving us outside (i.e. nowhere)'.[58] He was describing what would happen to a star that shrank below its Schwarzschild radius: it would collapse for an infinity of time, becoming infinitely small and infinitely dense – in other words, a singularity.

Unlike their predecessors, Oppenheimer and Snyder took the idea of the Schwarzschild radius seriously and set about investigating whether it might apply to real stars. Their astonishing discovery was that under certain conditions a massive star could actually implode until it was smaller than its Schwarzschild radius, pulling space around it and disappearing from view. Scientists first used the term 'Schwarzschild singularity' to refer to the Schwarzschild radius, because it takes an infinite amount of time for light to get out from inside it. But this turned out to be a misnomer. While the Schwarzschild radius defines the region, it is the imploding star that is the singularity, because it ends up infinitely small and infinitely dense.

Mysterious and unfathomable though this solution was, it was exactly what Chandra had described when he discovered that stars above a certain mass would do just that. His finding had finally been substantiated.

Oppenheimer and Snyder's work threw up an intriguing puzzle which the two were at a loss to explain. The viewpoints of an observer moving alongside a star which has shrunk almost to the size of its Schwarzschild radius, and another observer viewing it from a distance, seemed irreconcilable. The observer alongside sees matter sweeping inwards faster and faster as it is tugged by the increasingly strong gravitational field near the horizon defined by its Schwarzschild radius. But when the star falls over the horizon, no more light can emerge from it. The distant observer, on the other hand, will report that the collapsing star 'freezes' when it reaches the size of its Schwarzschild radius. Because of the vice-like grip of the shrinking star's gravity, light takes longer and longer to escape, and eventually the star appears not to be moving any more, frozen in space and time. The 'star thus tends to close itself off from any communication with a distant observer', they wrote.[59] They noted that the speed of the collapse might slow once they were able to take proper account of how hugely compressed matter behaved. But they could not see how to reconcile the two apparently conflicting views.

Ironically, in 1939, at the same time that Oppenheimer and Snyder were engaged in their ground-breaking work, Einstein himself was attempting to prove that 'the "Schwarzschild singularity" does not exist in physical reality [because] matter cannot be concentrated arbitrarily'.[60] No object, he asserted, and certainly not a star, could ever be compressed to a size smaller than its Schwarzschild radius. Most uncharacteristically for Einstein, there were errors in his results. Like most others, he refused even to contemplate the idea that stars might collapse completely.

Meanwhile, dark clouds were gathering over Europe. Nazism and Communism were about to change the course of physics and astrophysics and the lives of its foremost figures in ways that could never have been imagined. Among those who fled Hitler and emigrated to the United States were Bethe, Bohr, Einstein,

Fermi and the eminent astrophysicist Martin Schwarzschild, son of Karl. Stalin's 'gifts' to the United States included Gamow and Teller. For the first time, science was to be a major contributor to warfare. Oppenheimer was to find his vocation as the administrator of the biggest scientific enterprise of the twentieth century: the Manhattan Project, centred at Los Alamos. In 1942, he began full-scale recruiting for the project. Its purpose was to produce a weapon of such massive destructive power that it could obliterate an entire city – the atomic bomb. In his late thirties, Oppenheimer was one of the older scientists on the project. The average age was twenty-four.

The Nobel laureate Luis Alvarez worked closely with Oppenheimer during the Manhattan Project days. He believed that if Oppenheimer had lived into the 1970s, when neutron stars were an established fact and the search for black holes was well under way, he would have received a Nobel prize 'for his contributions to astrophysics'.[61]

Chandra was not yet a US citizen in 1941 and initially had problems being cleared for defence work. His old friend from Cambridge, John von Neumann, was instrumental in resolving them. Chandra became a civilian consultant at the Army Ordnance Department's Ballistic Laboratory at the Aberdeen Proving Grounds in Maryland. There he worked on ballistics problems, some similar to those on which Milne was working in England. But shock waves and radiative transfer (the way in which radiation moves through matter) were what really interested him.[62] In 1942 he gave a colloquium on the topic at the University of Chicago. The huge audience included Eugene Wigner (another brilliant Hungarian émigré physicist and Dirac's brother-in-law) and Fermi. They and many others were about to disappear to the south-west, where they could be reached at P.O. Box 1663, Santa Fe, New Mexico. Two years later, Chandra began receiving letters from that address from someone he had not seen since the Washington meeting in 1938: Hans Bethe, now Head of the Theoretical Division at Los Alamos.

Bethe's first letter was dated 20 March 1944. 'Johnny von Neumann has written to you on our behalf asking whether you would be interested in joining our project,' he wrote:

We are in great need of your help and we believe that you would be the best man to ask to take charge of certain calculations which have some loose connection with the work you have been doing at Aberdeen. We have no other person here who understands this type of problem with the exception of Johnny, who is here only a fraction of the time. You know that many of your friends are here . . .[63]

The bureaucratic red tape that had plagued Chandra when he arrived in London in 1930 also entangled him at Yerkes in 1944. It was not until September 1944 that he received clearance to go to Los Alamos. By then, the Allied forces had fought their way off the Normandy beaches and were pouring across France towards Germany; there was talk that the war might be over by Christmas. Chandra was reluctant to move Lalitha away from their home. He was also deeply disturbed by the racial abuse he had suffered in Maryland: 'even at the Aberdeen Proving Grounds I used to encounter racial prejudice in many forms – in restaurants and places like that – and I was slightly afraid of driving down south', he recalled.[64] In the end he decided to decline Bethe's offer.

Chandra was well aware of what was going on at P.O. Box 1663. He had kept abreast of developments in nuclear physics and was aware of the 'disappearance' of top researchers in the field. Bethe repeated his offer. He agreed that the war in Europe seemed to be over but wrote, in September 1944, that the 'work here is likely to last at least until the end of the Pacific war and possibly longer'.[65] The first atomic bomb fell on Hiroshima on 6 August 1945, the second on Nagasaki three days later. In an interview Chandra agreed that the first atomic bomb was unavoidable, but objected that the 'second did not seem to me necessary at all'.[66] And he went further. At first he denied that there had been any racist overtones in the decision to drop an atomic bomb on Japan. 'But it did occur to me', he added, 'that if the war in Germany had not been over, the bomb probably would not have been dropped on Germany.' There was an awkward pause, broken by the interviewer, who said, 'Well, back to your astrophysical work.'[67]

When Chandra finally made his debut at Los Alamos in the 1950s, it was to work on the next generation of nuclear weapons – with explosive characteristics remarkably similar to those of supernovae.

10

Supernovae in the Heavens and on Earth

Looking back on the turbulent times in Cambridge after 11 January 1935, Chandra wrote that he was astonished that he was 'never completely crushed by these stalwarts. I decided that there was no good my fighting all the time, that I am right and that the others were all wrong. I would write a book. I would state my views. And I would leave the subject.'[1] And that was exactly what he did. It took him less than a year to complete his first book on white dwarfs, *An Introduction to the Study of Stellar Structure*, though he got, he said, no 'special exhilaration' from writing it.[2] The spark was gone, beaten out of him by Eddington's onslaught.

'This idea of permanently leaving a subject after writing a book did not occur in a natural way,' he recalled. He 'hated to relinquish the subject' but 'had no choice'. There was so much more to investigate about white dwarfs, such as what might happen if they rotated or pulsated. But he had had enough trouble: 'Adhering to my point of view would only make more of a Don Quixote of me than I already appeared to be among the astronomical community.'[3] He finally realised that he was just tilting at windmills. Eddington would never change his mind. So he gave up speculating on white dwarfs. Later he published *The Principles of Stellar Dynamics*, an elegant, highly sophisticated mathematical study dealing with the equilibrium of galaxies and solar systems. But he carefully avoided considering the structure of white dwarfs.[4]

Then, in 1939, Gamow launched into a particularly intriguing area of speculation. Gamow's idea was to use Bethe's new nuclear physics, which analysed how stars shine, to explore the turning point in the lives of stars, when they have burnt all their fuel and begin to get old.[5] He developed an evolutionary scenario which assumed that robust stars, those on the main sequence of the HR diagram, stayed there, becoming brighter and moving up the main sequence until they had burnt up all their hydrogen. Then, he postulated, they moved left – off the main sequence – contracting and becoming dimmer, and eventually expiring as white dwarfs, except for those that exceeded Chandra's maximum mass, which would blow off mass in a supernova explosion and end up as neutron stars. Like Bethe, he concluded that, as heavier stars like red giants were not on the main sequence, they were obviously not fuelled by hydrogen. He suggested that they were young stars which shone due to gravity compressing their gas particles. Eventually they would start to burn hydrogen, and would move onto the main sequence, like stars with a smaller mass, but would go on to explode into white dwarf fragments.

Chandra had a keen eye for lurking stability problems; his forte was identifying the exact point at which stars are liable to collapse. He spotted a flaw in Gamow's argument. The challenge was irresistible, and he decided to turn his attention to white dwarfs one last time.

In the billions of years during which a star is in its prime of life, it burns the hydrogen in its core and converts it into helium, leaving behind a core of helium 'ash'. The hydrogen at the centre burns first, because that is where it is at its most concentrated and the internal temperature is the greatest; this phase is called hydrogen-core burning. When the nuclei of two hydrogen atoms (protons) fuse, the fusion process produces deuterium, a variant of hydrogen with a single electron and therefore the same chemical constitution as hydrogen, but with a neutron as well as a proton in its nucleus. (Variants such as these which have the same number of protons in their nuclei but differ only in the weight are called isotopes.) The deuterium nuclei (deuterons) then fuse with the protons (hydrogen nuclei), producing helium. Thus the original hydrogen core gradually turns to helium.

Gamow had made the cavalier assumption that a star burns up *all* its hydrogen; but Chandra realised that this meant that it must accumulate an ever-growing helium core as it moved in its evolutionary path across the HR diagram. This made him uneasy. He decided to look into the problem with a Brazilian postdoctoral student, Mario Schönberg, who was better informed than Chandra about the nuclear physics of stars.

Chandra and Schönberg investigated whether this emerging helium core could remain stable, without collapsing or exploding, throughout the process of burning hydrogen. They came up with the surprising result that the helium core reached the maximum mass it could attain without collapsing when a mere 10 per cent of the hydrogen fuel had been consumed.[6] This is now known as the Chandrasekhar–Schönberg limit. But what happened next?

If the star's total mass was below Chandra's upper limit, it would expire as a white dwarf. If not, then – shades of Milne – it would have to eject enough mass to get below the upper limit, then pass through a series of degenerate cores, finally reaching a completely degenerate, rock-hard state. One way to do this would be for the star to blow up – to become a supernova. At this point they did not consider the possibility that the star might collapse to nothingness. They assumed that somehow it had to become a white dwarf.

Chandra and Schönberg's work showed that stars did not spend as long on the main sequence as everyone had thought.[7] The next question was to identify the intermediate stages before they began their demise, either as white dwarfs or as neutron stars. To investigate this, astrophysicists began to study stars in star clusters using the HR diagram.

In some regions of interstellar space, called 'nebulae', there are enough gas and dust particles to begin to swirl together and form stars. In these stellar nurseries, most of the stars have been born at roughly the same time. As Strömgren discovered, a star's position on the HR diagram is determined by its mass and hydrogen content, and a star's appearance changes as it burns up hydrogen. Therefore, as the stars have different masses, we can use their positions on the HR diagram to work out their ages and evolution. By the late 1940s, studies of star clusters revealed a 'bridge' between the strip where giants are found and the main sequence. On the basis of theoretical discoveries such as Chandra's and Schönberg's, astrophysicists were able to construct models of stars that suggested that after the Chandrasekhar–Schönberg limit has been reached, when 10 per cent of a star's hydrogen has been burnt up, a new evolutionary process begins to occur.[8] (See Figure 1(b) in the astronomical picture section for a modern HR diagram.)

Once the star has reached the Chandrasekhar–Schönberg limit, its helium core begins to cool off. As radiation pressure decreases, gravity begins to dominate in the tug-of-war between the two opposing forces. This causes the helium core to contract, which in turn causes it to heat up again, this time to such a degree that the shell of the original hydrogen core that surrounds it starts to burn. This creates more helium, which falls onto the core, causing it to contract yet more. As a result of this internal heating, the vast outer layer of the star expands and the star becomes a red giant. Far from being young stars, as Gamow suggested, red giants are actually stars in the final phase of life. They are ten to a hundred times larger than the Sun, about 100 million miles across (the radius of the Earth's orbit around the Sun) and a hundred to a thousand times more luminous.

The Sun has enough fuel to burn its hydrogen core for another 10 billion years. It has already been doing so for about 5 billion years; in another 5 billion years it will become a red giant, vaporising the inner planets of the solar system, including Earth, and burning off the atmospheres of the rest. In red giants that have developed from stars no more than eight times the mass of the Sun, stellar winds begin to blow away the outer layers, which have cooled off and are loosely held, so that the mass decreases. Meanwhile, the helium core continues to contract until eventually its temperature rises to 100 million degrees kelvin, and it reignites. This is followed by successive paroxysms of ignition and reignition, creating a core of carbon and oxygen and blowing off a huge amount of hot, glowing gases that produce a brilliantly colourful display which we see as a planetary nebula.

The remaining carbon–oxygen core, covered with a thin surface layer of hydrogen and helium, is a white dwarf star. Electron degeneracy pressure creates a rock-hard core, without enough mass to generate the gravity needed to make it collapse any further, so the burning caused by the core collapsing stops. It has burnt up nearly all its fuel, making it dim, but has undergone extreme con- traction under the inward crush of gravity, making it hot. Its mass is less than Chandra's limit, which saves it from the fate of eternal collapse. The white dwarf cools down, the carbon and oxygen crystallise, and this mysterious object, which has opened the doors to all the wonders of the cosmos, expires – a diamond in the sky.[9]

During the war, Chandra recalled, he could only discuss astro- physics with students, as everyone else was working full time on defence. Most of the students were women who could not be drafted and were free to continue their research. In 1944 he began to study radiative transfer. This was the happiest time of his life: 'I'm always nostalgic about that period . . . Before that, there was unhappiness connected with the Eddington controversy, and a certain diffidence whether I could make the grade in science.'[10] He enjoyed rediscovering and improving on the results of the British school of high mathematical physics, which had included such venerable nineteenth-century figures as Lord Rayleigh and George Stokes, who were heroes to Chandra and had discovered many equations and effects that bear their names. By 1948 Chandra felt that he had reformulated the theory of radiative transfer into a 'coherent mathematical picture with elegance and beauty'.[11] He decided to bring his research to a close with another of his mag- nificent tomes packed with equations, wrapping up the subject as it then stood and stating the problems that still needed to be solved, in a definitive volume that would be referred to for years to come. The book was *Radiative Transfer*.

For Chandra, this was a satisfying and rewarding period in his teaching as well. In contrast to the white cottons of southern India, at Cambridge his wardrobe had become extremely Anglicised. He carried this style over to Yerkes. In winter he wore a dark suit, in summer a light grey one (the shade came to be called 'Chandra grey'). He finished it off with a white shirt and conservative tie. Students remember that he always spoke in complete sentences, like a formal Englishman. But what they remember best was his blackboard technique. With his long delicate fingers he handled the chalk like a paintbrush, writing out equations in beautiful script. His lectures were so tightly organised and logical that his students later used their class notes as sources for their own teaching and research. He worked through the solutions of equations in great detail, rarely making a mistake. The only criticism anyone ever made was that he emphasised mathematical methods rather than the basic physics in the equations. Colleagues joked that whatever the official title of Chandra's course, it should have been called 'Mathematical methods of . . . '.[12] Chandra's principal selection criterion for Ph.D. students was their skill in mathematics. They all considered him a superb teacher.

In 1945, Struve asked Chandra to be the associate managing editor of the *Astrophysical Journal*. In 1952 he took over as managing editor. He insisted on clarity of presentation. 'A bad sentence cannot be corrected; it should never have been written,' he often said.[13] As part of this job he began travelling to Chicago, first for one day a week, then two, for consultations with the University of Chicago Press, the journal's publisher. He usually took several students with him in his car, and the trips became the occasion for stimulating discussions. 'We will meet at 6.00 a.m., plus or minus fifteen minutes,' he would tell them, adding playfully, 'I can be plus or minus. You should be plus.'[14] Usually he spent Thursdays and Fridays on campus.

By the late 1940s, Chandra had an international reputation. Around this time he began to initiate serious contact with the physicists at the main campus in Chicago. The astronomy department there had organised a series of courses for students who hoped to research their theses at Yerkes. Struve asked Chandra to teach the first course, entitled 'Topics in the theory of stellar atmospheres (Astronomy 301)', thus giving students their first chance to see this mysterious figure who stayed out in the wilds of Wisconsin. The course has become legendary thanks to an apocryphal story dreamt up by the President of the University, John T. Wilson, when he was introducing Chandra before Chandra gave the 1975 Ryerson Lecture. Chandra liked the story so much that he never set the record straight.

Twice a week during the term of 1948 and 1949, Chandra made the arduous three-hour drive from Yerkes to the Chicago campus to teach a course on astrophysics to just two students, Tsung-dao Lee and Chen-ning Yang (so the story went). Yang was an instructor in the physics department (he had just completed his Ph.D. under Teller), Lee a graduate student of Fermi's working on the internal structure of white dwarfs. It was Fermi who had suggested that Lee should consult with Chandra. It was the only physics class in history in which every student involved went on to win a Nobel prize. Both Lee and Yang won theirs in 1957, for the overthrow of parity, discovering that nature can in fact distinguish between left and right.[15] The professor would have to wait another twenty-six years.

The real story was ferreted out by Donald E. Osterbrock, an eminent astrophysicist turned historian of the subject and a former student of Chandra's. There were actually several students registered for the course. Lee and Yang sat in, as did Fermi and other members of the physics department. As the term went on, the numbers dwindled. Lee and Yang stuck it out, along with six registered students, Osterbrock among them. One staff member who lasted almost to the end was Marcel Schein, a Belgian-born cosmic-ray physicist. He looked very much like a well-known professional wrestler of the time called 'The Angel', so Chandra's students referred to him as 'The Belgian Angel'. He always sat in the front row and invariably went to sleep with his head hanging over the back of his chair, snoring loudly. Chandra would stand right over him while he lectured, with a look of disgust, but he never woke him up. There was one class, however, where Chandra really did teach only Lee and Yang. That winter the area was hit by a severe blizzard that all but shut it down. True to form, and against everyone's advice to stay in Williams Bay, Chandra heroically made the trip by train and foot into Chicago, then hiked to the university. Only Lee and Yang showed up for that class, having been told by Fermi that the professor was on his way.[16]

For Chandra, the post-war years were ones of mixed happiness. Like an author in search of a plot, he was at a loss for what to do after completing his work on radiative transfer. He did not recover his stride until 1952, when he settled on his next topic, hydrodynamic and hydromagnetic stabilities. This is the study of flow: in terrestrial terms the transition from the normal flow of a fluid to highly turbulent, chaotic flow, such as in whirlpools; and in astrophysics the conditions for generating magnetic fields that maintain their shape. The mathematics is extremely complex. It was probably his contacts with physicists on campus that set Chandra on this course. By now he had moved far away from his early focus on white dwarfs. Besides his editorial duties in Chicago, he attended colloquia in the physics department, which was packed with brilliant physicists, and also at the Institute for Nuclear Studies, headed by the charismatic Enrico Fermi.

During the war, Fermi had become curious about astrophysics and wanted to learn more. After hearing Chandra's lectures, he suggested that they collaborate on a problem. Fermi's method for learning something new was to talk to the experts, rather than delve deeply into the relevant literature, and then combine his new knowledge with his own deep understanding of physics. To Chandra, Fermi seemed 'like a master musician who, when presented with a new piece of music, would play on sight with great conviction'.[17] The difference in styles between Chandra and Fermi went beyond their approach to physics problems. In his immaculate suit and tie, Chandra was formal and aloof, while Fermi was outgoing, with sleeves rolled up and no tie, and often lunched on hot dogs with his students in a local diner. Short and stocky, he was a smiling, popular man with an Italian's ease of manner.

Fermi made it all look easy. He had a natural grasp of the big picture: he could work out almost anything from basic equations. Research scientists have bookcases filled with advanced texts. Fermi had only one book in his office – a table of integrals (a handbook for mathematical calculations). In 1953 Chandra and Fermi produced two papers on the magnetic field in the arms of spiral galaxies. It was a brilliant collaboration. Chandra's forte was patiently to seek appropriate and elegant mathematical methods for solving problems. Fermi was just the opposite: he had no interest in elegance in mathematical presentation. Chandra was entranced with Fermi's command of physics and his 'use of whatever means got him the answer'.[18] Their first paper was written primarily by Fermi and contained few equations. The second, written by Chandra, was mostly equations and very few words.

Little by little, however, the atmosphere at Yerkes began to go sour. From the very beginning, Chandra harboured a particular resentment against Struve. In later years he was to look back on his time at Yerkes with the conviction that he had been unfairly treated. He had discovered that Kuiper and Strömgren, who had arrived at the same time, had been hired as assistant professors while he had entered as a mere research associate. They had been promoted and received salary rises and tenure the following year, while Struve appointed Chandra assistant professor with no increase in salary. It was six years before he was promoted to associate professor, and he was not made a full professor until 1944.

It turned out that Struve had a very good reason for recruiting Chandra at a lower level. Building Yerkes into a first-rate astronomy department, he was recruiting brilliant young men regardless of nationality. But in introducing Chandra into the department, Struve was contending with entrenched racial prejudice. Chicago was effectively segregated, a state of affairs supported by many powerful staff members, notably Henry G. Gale, the dean of physical sciences, and Gilbert Ames Bliss, a famous mathematician and head of the mathematics department. Both were strongly opposed to the appointment of someone who was 'Indian and black'.[19] Astronomy was linked to the mathematics department, which made Struve's job all the harder. So Struve bypassed Gale and Bliss and went straight to Robert Maynard Hutchins, the president of the university, who supported Struve's strategy of hiring young talent. This infuriated Gale. He made Struve promise never to allow Chandra on the Chicago campus. Struve ignored him. Gale retired in 1940, but by then other professors had made a point of encouraging Chandra to spend more time in Chicago. Kuiper and Strömgren were much better known to American astronomers than Chandra, and there was nothing to hinder their progress. It was natural for Struve to hire them as assistant professors. In Chandra's case, Struve had to work around Gale and Bliss. It caused less upheaval to appoint him at a more junior level.

Russell and Shapley had warned Chandra about the racial tensions in Chicago and advised him to accept the Harvard position. Shapley feared that Chandra's radical political views would be unacceptable in the rather conservative political atmosphere of

Chicago. Knowing about Chandra's visit to the Soviet Union in 1934, he even warned Struve that Chandra was a Communist. Many people interpreted Chandra's strong reactions to the racism Indians encountered in England as radical politics. But Hutchins, a man of extremely liberal political views, made it clear to Struve that he was interested only in whether Chandra was the best person for the position. Tall and handsome, Hutchins was a born leader, with great charisma. He had become president in 1929, at the age of just thirty. Young though he was, he had already proved himself a dynamic administrator as dean of the law school at Yale. He continued to serve as president until 1945, and was then chancellor until his resignation in 1951. He was directly responsible for the university's meteoric rise.

It was not until the 1960s that Chandra learned of Hutchins' courageous stance. At the time there was a violent reaction against desegregation. In a public lecture, Hutchins claimed that even the hallowed halls of high academia had had their share of racist incidents. He cited two instances of racial prejudice which had occurred at the University of Chicago while he was president. One was in the medical school. The other was when the head of a certain science department opposed the appointment of a theoretical astronomer because he was Indian and black. The speech made waves across the nation. Chandra and Hutchins maintained close contact over the years. When Hutchins announced his retirement, Chandra dashed him off a note saying that his resignation was a 'disaster for the University'.[20] Hutchins replied with warmth, 'My leaving is not a disaster for the University. You wait and you'll see. I am very proud that I helped bring you here.'[21] At his retirement ball, students flocked around him to shake his hand. Such popularity was unusual for a university president, then and now.

In 1946 Chandra had been offered a professorship at Princeton to succeed Russell, at double his Chicago salary, and he accepted with alacrity. Hutchins convinced him to change his mind. While it would be a great honour to succeed someone like Russell, he argued, would it not be more advantageous to initiate his own research programme instead of continuing someone else's? To illustrate his point, he asked Chandra to name Lord Kelvin's successor at Glasgow University. Chandra could not. Hutchins then offered Chandra a distinguished service professorship at a salary that matched Princeton's offer. Chandra accepted the honour and the sorely needed salary boost. Six years later, he was promoted again, to Morton D. Hull Distinguished Service Professor.

In 1951, Hutchins invited Strömgren, who had returned to Copenhagen before the war, to take over as director of Yerkes. Strömgren believed curriculum reform to be long overdue. For well over a decade, Chandra had taught the theoretical courses himself. He would usually teach a course for three years running, after which his enthusiasm would wane. He would then write a book on the subject, and move on to a new research topic and institute new courses. Rather than producing well-rounded research students, this teaching method turned out students who could work with Chandra on whatever his current research project happened to be. It was not a satisfactory system. So Strömgren appointed a committee, made up mainly of younger members of staff, to redraft the curriculum. Chandra saw them as a cabal. In his absence they called a vote and threw out his curriculum. As far as he was concerned it was yet another byzantine plot against him, akin to Eddington's in 1935.

The following year, to everyone's relief, Fermi invited Chandra to leave Yerkes and join the physics department in Chicago. After that, Chandra rarely taught at Yerkes. His research interests had begun to diverge sharply from astrophysics. At first he stayed on the campus when he visited Chicago. Then, in 1959, the Chandrasekhars rented a pied-à-terre near the campus. After being an unofficial member for twelve years, Chandra's appointment to the physics department was confirmed in 1964.[22] That year he and Lalitha bought an apartment in the city. Chandra looked upon his experiences at Yerkes as another humiliation, but at least now he was associating with people such as Fermi.

Fermi was also eager for Chandra to work at the Institute for Nuclear Studies, which was linked to but separate from the physics department. The Institute was perpetually buzzing with excitement. Staff there held joint appointments, initially in chemistry and physics. Soon the circle widened to include astrophysics, astronomy, geophysics and mathematics, making it the greatest concentration of research talent in the United States. The core figures from Los Alamos had all relocated to Chicago. Besides Fermi, who had received his Nobel prize in 1938, there were three other past or future Nobel prizewinners: Maria Goeppert-Mayer, Harold C. Urey and Willard F. Libby.[23] After Fermi's death in 1954, the Institute for Nuclear Studies was renamed the Enrico Fermi Institute. By then, Chandra was a member of this hybrid organisation.

Roger Hildebrand, then a young staff member, remembered the fifties as Chicago's golden years. Throughout the Institute, there was an 'enthusiasm, a love of science, no snobbery; the whole spectrum of physics was discussed, whatever was interesting'. The most popular meetings were the Thursday afternoon seminars, for staff only, run by the German-born physicist Gregor Wentzel, who had arrived at the university in 1948 at the age of fifty. Wentzel's credentials were impeccable. After completing his Ph.D. under Sommerfeld, he had worked with Heisenberg and Pauli, producing important results in elementary particle physics. He and Chandra became so close that Chandra even let him smoke his huge cigars in his presence. Each of the luminaries had their own special chair in the Institute's colloquium room. If anybody, God forbid, sat in one by accident, the 'owner' would stand in front of them, uncomfortably close, and stare. It was a daunting experience for a young staff member. When there was a lull in the discussion, Wentzel might turn to one of the junior lecturers, seated at the back of the room, and say, 'Well, Roger, what are you thinking about these days?' Hildebrand recalled that 'you'd damn well better have something good to say or you could be demolished!'.[24]

As it happened, Chandra's new area of interest, hydrodynamic and hydromagnetic stabilities, in addition to his reputation in the field of radiative transfer, made him more attractive than ever to the new powers that had emerged at Los Alamos, such as his old friend Edward Teller, who had left Chicago in 1949 to return to defence work. Their project was to build a bomb so powerful that it would make the atomic bombs dropped on Japan look like mere firecrackers. The new weapon was a hydrogen bomb, and its creators would depend heavily on the physics of stars for its development.

A detonating atomic bomb is like an exploding star; a hydrogen bomb even more so. In each case radiation has to find a way out. If it escapes too easily, the star or bomb will burn up too quickly; if it is too constrained, the explosion will be premature. When scientists design a nuclear bomb the aim is to hold in the radiation just long enough so that nearly all the nuclear fuel is cooked, then to let it escape as rapidly as possible. Thus the study of what makes radiation flow in a smooth, stable way, and the resistance of the medium through which it flows, is critical to both bombs and stars. Similarly, hydrodynamic and hydromagnetic stabilities, essentially the study of flow, are relevant to exploring the behaviour of shock waves and how to use them to compress matter – a vital area of knowledge in the making of bombs. Chandra's expertise in all these fields was essential.

Back in 1934, Fermi, then in Rome, had thought of bombarding non-radioactive elements with neutrons to see if he could produce radioactive elements. Charged particles are repelled by similarly charged ones, and therefore have difficulty penetrating a nucleus. Until the discovery of the neutron, alpha particles (the nuclei of helium atoms) were the projectiles used in collision experiments such as these. The disadvantage was that alpha particles, which consist of two protons and two neutrons, have a positive charge. A heavy nucleus, however, is made up of many protons and therefore has a huge positive charge which the alpha particle has to overcome in order to penetrate it. But the neutron has no electric charge. This made it the perfect 'bullet' with which to bombard atoms and smash right into the nucleus.

A logical extension of Fermi's experiments was to explore what happened when neutrons struck a naturally radioactive substance such as uranium. To his amazement, he found that this produced elements heavier than uranium, the heaviest element known to occur naturally on Earth. These new elements were highly radioactive and broke up into stable, lighter elements over periods of time ranging from minutes to millions of years; they are therefore no longer naturally present on Earth.[25] In 1938 Fermi received the Nobel Prize for Physics for his idea of using neutrons to produce radioactive elements and for discovering elements heavier than uranium. Uniquely in twentieth-century physics, he did outstanding research in both theory and experiment.

Then, in 1938 – as the world was on the brink of war – Fermi's experiments led to an earth-shattering discovery. Re-examining his data, some German scientists realised that he had not simply created a new element, as he had thought; he had actually split the nucleus of the uranium atom.[26] Once a neutron has smashed its way into the uranium nucleus, there is pandemonium as the nucleus's 146 neutrons and 92 protons try to eject the intruder. The whole nucleus begins to vibrate as if it were a drop of water. First it stretches into a dumbbell shape, pinched in the middle, then it splits, like an amoeba reproducing. This is nuclear fission, and it happens in a tiny fraction of a second. When the nucleus splits, it divides into smaller nuclei of two other elements, barium and krypton. The total mass of the new particles, however, is slightly less than that of the original neutron plus the uranium atom. This 'missing mass' is converted into energy as predicted by Einstein's equation $E = mc^2$. But the minuscule amount of missing mass multiplied by the huge quantity c^2, the square of the velocity of light (186,000×186,000), results in the release of a colossal amount of energy. The two new nuclei, barium and krypton, are radioactive and release massive radioactivity. The fission also produces at least two neutrons which can then smash into other uranium nuclei, leading to an uncontrolled chain reaction and thus creating a nuclear explosion. For this to happen there has to be a critical mass of uranium, of about 10 kilograms; if there is less than this there will be no explosion, only a fizzle. The nuclei also have to be packed tightly enough together to provide a sufficiently large number of targets for each neutron.

When Bohr heard that the nucleus had been split, he immediately grasped the horrific potential of the enormous amount of energy generated. Perhaps he remembered the haunted face of Alfred Nobel, who discovered dynamite – which he hoped would be used for peaceful purposes. As had happened with dynamite, someone was bound very soon to think up a lethal use for fission.

In September 1941, Fermi was in his office at Columbia University, New York, mulling over the possibility of an atomic bomb (a fission bomb). With him was Teller. Teller believed that the fission bomb would be limited in power, as the vital critical mass placed an upper limit on just how much fissionable material could be packed in. The more powerful the device, the quicker it would blow itself apart. In fact, the maximum explosive power of an atomic bomb is around 1 megaton, the equivalent of a million tons of TNT. A mere teaspoonful of TNT can cause vast damage. Fermi suggested that a fission bomb might be used to create the temperature needed to start the process of fusing hydrogen atoms to make a hydrogen bomb (a fusion bomb). Such a bomb would be modelled on a star, for it is hydrogen atoms fusing that produces the energy that makes a star shine. While fission blows apart, fusion glues together: two nuclei are fused to make a heavier one. Unlike a fission bomb, there is no critical mass needed and no upper limit on the power a fusion bomb can have. Like stoking a fire, it burns ever more brightly as more fuel is added. It has unlimited explosive power.

At first Teller was incredulous. A year later, when the atomic bomb project was well under way, Teller returned to Fermi's suggestion. Initial calculations indicated that an atomic bomb might generate a temperature of 10 million degrees kelvin. From his work on stars, Teller knew that this was high enough to ignite hydrogen. Compared with a fission bomb, the fuel for a thermonuclear bomb (fusion bomb) cost a pittance. Deuterium could be extracted from water for a few pence per gram. In contrast, extracting the isotope of uranium that can most easily be induced to undergo fission, and which accounts for less than 1 per cent of the element, was immensely expensive.[27] The fusion bomb was immediately named the 'super'. 'It was a terrible thing' was Bethe's first reaction.[28] For Teller, the super became an obsession. 'Something changed in him after he joined the Los Alamos Project,' recalled his old friend Gamow.[29]

From the moment that Teller arrived at Los Alamos in April 1943, there were problems. To start with, Teller was offended when Oppenheimer appointed Bethe, not him, as head of the theoretical division. Teller's group was relegated to creating a fission bomb in which the fissile uranium was squeezed together in a process called implosion, and told not to worry about the super. Tolman, the expert on the general theory of relativity at Caltech, had suggested the previous year that an amount of a fissile substance just below the critical mass could be made to explode if it was compressed sufficiently for the neutrons in it to set off a chain reaction. To create an implosion – blowing things together rather than apart – scientists put chemical explosives around the inside of a hollowed-out sphere containing a sphere of fissionable material at the centre, and explode them simultaneously. The result is a series of concentric shock waves moving inwards, which crush the fissionable material until it is dense enough to spark a chain reaction. But Teller's heart was in the super and not in the fission bomb. He sulked, and did not put his heart into his work. Bethe became increasingly annoyed. Oppenheimer did not want to lose someone as innovative as Teller. In 1944 he finally allowed Teller to leave Bethe's theoretical division and work on the super with his group.

For the atomic bomb, scientists could perform table-top experiments to determine the feasibility of a particular design. But the hydrogen bomb needed an atomic bomb to ignite it, so this was impossible. All that Teller and his group could do was to make simulations based on mathematical models to work out the ideal configuration of atomic bomb and fusion material that would create the temperature that would set off a thermonuclear reaction. But the calculations were so horrendously complicated that they were beyond the capacity of the mechanical calculating machines provided by IBM during the war. Then, in 1945, the ENIAC (Electronic Numerical Integrator And Computer), the world's fastest electronic computing machine, which had just been built at the University of Pennsylvania, went on line. It could handle these calculations. But it was soon outmoded, and in 1949 was replaced by the MANIAC (Mathematical Analyzer, Numerical Integrator, And Computer). Besides the one at Los Alamos, another was built at the Institute for Advanced Study in Princeton, New Jersey, at the instigation of John Archibald Wheeler, a physicist there. A veteran of the Manhattan Project and a key consultant at Los Alamos, Wheeler was disappointed with the progress at Los Alamos on the super. He wrote to the head of the physics department at Princeton that 'thermonuclear research involves strong Princeton specialties – nuclear and atomic theory, ideas from astrophysics, and hydrodynamics'.[30] Hydrodynamics was what Chandra was working on over at Chicago. So Princeton became the second centre for the hydrogen-bomb project.

Despite his extreme right-wing anti-Communist views and salt-of-the-earth demeanour, Wheeler had one of the most way-out minds in physics. He had spent most of the war in Hanford, Washington, studying basic problems of reactor physics essential for the production of plutonium, an alternative fissile material to uranium.[31] During the 1930s, Wheeler and Bohr had worked together closely on the new field of nuclear physics. When Bohr arrived in New York City in January 1939, Wheeler met him. Bohr immediately filled him in on the latest development – nuclear fission. The two set about trying to understand the full significance and workings of this ground-breaking new phenomenon. Wheeler agreed with Teller that it was vital to develop the next phase of nuclear weaponry, the super. A primary obstacle to its development was Oppenheimer, who was reluctant to sanction a bomb even more powerful than the atomic bomb. As chairman of the General Advisory Committee (GAC) set up to advise the newly created Atomic Energy Commission (AEC), he had a great deal of say in the matter. Teller was growing increasingly antagonistic towards him.

Teller wanted to go immediately for a high-megaton bomb that would dwarf the atomic bombs that had been dropped on Hiroshima and Nagasaki. The Hiroshima bomb, dubbed Little Boy, had been equivalent to 20,000 tons of TNT. The ENIAC had been built for the specific purpose of testing Teller's 'classical super', which consisted of a fission bomb attached to one end of a pipe filled with liquid deuterium. His theory was that the heat from the exploding fission bomb would travel down the pipe, triggering fusion reactions in the deuterium. He estimated that the classical super packed around 10 megatons in explosive power. Another of Teller's designs was a hemispherical bomb made of concentric shells of deuterium and fissionable material. He called it the Alarm Clock, because it would wake up the world. But Teller's preliminary calculations showed that it would not be very powerful, so he did not take it any further.

Then, on 29 August 1949, the Soviet Union exploded Joe 1, pretty much an exact copy of the US bomb dropped on Nagasaki, called Fat Man, which used the implosion mechanism. The design came courtesy of Klaus Fuchs, a thirty-eight-year-old German émigré scientist who had arrived in Los Alamos in 1942. There he held a position so central that he was privy to the most highly classified information – which he passed to the Soviets.[32] Ironically, Fuchs had joined the implosion group as Teller's replacement. President Harry S. Truman made the news public on 23 September with the dramatically terse statement, announced in banner headlines across the USA, that 'We have evidence that within recent weeks an atomic explosion occurred in the USSR.' The United States' monopoly on nuclear weapons was over. There was near panic in the country. What should be the response of the United States? The official debate was carried out in the highest secrecy. Part of the scientific community agreed with the AEC that the Soviets had realised that the next step was the super and were determined to get there first. Suddenly everything had changed. Sorting out the design problems for the super was now top priority.

Most of the GAC, chaired by Oppenheimer, cautioned against entering a potentially disastrous arms race. But powerful and vocal scientists such as Teller supported the super. In alliance with key congressmen and senators, they carried the day. What tipped the scales was that Fuchs had also been privy to Teller's designs for the classical super and the Alarm Clock. On 31 January 1950, Truman announced to the world that the United States had embarked on a project to build a hydrogen bomb. The nuclear arms race was on. Teller asked Oppenheimer to work on the super, but Oppenheimer bluntly refused. He had concluded that any advances the Soviets could make based on Fuchs's information would be 'marvellous indeed' because simulations indicated that Teller's classical super was unworkable.[33] The bomb designers had reached an impasse, which Oppenheimer interpreted to mean that the super was simply not feasible.

Then, in December 1950, the mathematician Stanislaw Ulam had a flash of inspiration. Ulam had been working on the super since 1943, when he was assigned to Teller's renegade group in Los

Alamos. Born in 1909 in Lemburg, Poland (now Lvov, Ukraine), Ulam was a mathematical prodigy. He learned his trade in the remarkable school of pre-war Polish mathematics, and completed his Ph.D. under the world-renowned Stefan Banach. Dark, urbane and always impeccably groomed and dressed, Ulam radiated an air of carefree confidence. At Los Alamos he soon became known as Stan. In 1934, sensing the approaching menace of Nazism, Ulam had decided to start making contacts elsewhere. Travelling around Europe, he stopped off in Cambridge, where he met and impressed Eddington and Chandra, among others. His take on their relationship was that Chandra collaborated with Eddington, 'for whom he had mixed feelings of admiration and rivalry'.[34] Their paths were to cross again much later in Los Alamos when Chandra was a consultant on the theory of turbulence and other hydrodynamical problems.[35]

Staring out of the window of his living room in Los Alamos, Ulam suddenly realised that the key was implosion, the very same implosion that Chandra had predicted might happen in the case of massive white dwarfs.[36] He suggested that a fission bomb could be used to trigger a hydrodynamic shock wave. The shock wave would be carried by a flow of neutrons. To shape it properly, the fission bomb (or primary) should be physically separated from the fusion fuel (or secondary). Ulam described his idea to Teller late in January 1951. Almost immediately, Teller saw that Ulam had hit upon something quite new. A fission bomb creates not only a flood of neutrons, but also high-energy radiation (i.e. X-rays). Teller was well aware that it would be very difficult to focus the neutron shock wave so as to implode the fusion material. The radiation, however, could be focused; moreover, it would reach the fusion material before the neutrons because light travels faster than any particle with mass. The solution was therefore to use radiation to induce implosion.

What emerged from their collaboration became known as the Teller–Ulam configuration, the basic design for all fusion bombs. The details are still classified, but the general idea is that the fission bomb (the primary) generates X-rays which are somehow used to implode the fusion material (the secondary) and begin a thermonuclear reaction (fusion of hydrogen). The result is a massive explosion, very similar to what happens to a star squeezed by gravity. The design was all too feasible, the signal for everyone to move ahead quickly, for the Soviets were probably not far behind. Even Oppenheimer, who had been the most outspoken of the scientists opposed to the production of hydrogen bombs, conceded that the Teller–Ulam configuration was 'technically sweet'.[37]

The bomb was completed in 1952. It was tested on 1 November, on the island of Elugelab in the Eniwetok Atoll, 3000 miles west of Hawaii. Referred to as 'Mike', it had a yield of 10 megatons, bigger than anyone had expected. The fireball alone was over three miles across, big enough to destroy New York City, fry most of its inhabitants and blast the rest to bits. In a billionth of a second the Mike fireball created every element in the universe and artificial ones as well. Elugelab disappeared completely. But one problem remained. Mike was not a deliverable weapon because no means had yet been found to solidify the fusion components, which were packed in a huge refrigerator. The entire bomb weighed 65 tons. Two years later, the hydrogen bomb was 'weaponised' – that is, redesigned so that it could be transported by aeroplane. The new weaponised bomb, called 'Bravo', was detonated in March 1954. The explosion was more than twice as big as expected – 15 megatons, creating a fireball four miles across.

In the Soviet Union, meanwhile, physicists had been working round the clock to develop an atomic bomb. 'Leave [the physicists] in peace. We can always shoot them later,' Josef Stalin was quoted as saying to the feared head of the secret police, Lavrenti Beria, director of the gulags and the instigator of purges that killed hundreds of thousands of Soviet citizens.[38] At the time, Soviet physicists were under attack for being ideologically impure. Stalin understood, however, that scientists would be indispensable in any future conflict with the West. He demanded immediate results, which is why the Soviets' initial atomic bomb was a copy of the plutonium bomb, Fat Man, whose design had been handed over by Fuchs. Failure was not an option, especially as Beria was overseeing the bomb project personally. The Soviets then tried to produce a hydrogen bomb modelled on Teller's super. The work was carried out at a top secret site referred to as 'The Installation', in reality the town of Sarov, 240 miles east of Moscow. It was also known as Arzamas-16 or 'Los Arzamas'.[39]

The two most important Soviet scientists involved in the effort were Andrei Sakharov and Yakov Zel'dovich. Like Gamow and Teller, they were physical opposites. Tall, thin, invariably dressed in black and terribly introverted, Sakharov's studies for his Ph.D. were dramatically interrupted by World War II, which he spent as a factory worker. In 1947, at twenty-seven, he finally completed his Ph.D., in nuclear physics. He was immediately assigned to the team led by Zel'dovich, which was looking into the possibility of developing nuclear weapons.[40]

Zel'dovich was almost the exact opposite. A powerfully built man with a rapid stride, a swarthy complexion and close-cropped hair on his bullet head, he carried himself with chest stuck out and head held high. He was a very confident and assertive character, and gave the sense of being much taller than he actually was. Zel'dovich was a human dynamo who thrived on physical challenges. In the middle of discussing astrophysics in his office at home, he would suddenly suggest to colleagues that some exercise was in order. Gymnasiums in Moscow were not easy to find, so Zel'dovich devised exercises such as playing catch with a medicine ball in the stairwell of his housing complex. Bold and brash, he was the life and soul of any party, always on the dance floor with any women he could find.

In 1929, at the age of fifteen, Zel'dovich graduated from high school and began studying for a career as a laboratory technician. In 1931 he went on a school visit to the famous Physical-Technical Institute in Leningrad. The professors were struck by the young man's penetrating questions and invited him to work there in his free time. Over the next ten years, while still a student, he lectured at the Institute on the latest developments in quantum physics in such depth and breadth that he was invited to become a member of the chemical physics department. Three years after that initial visit he completed his Ph.D. It was an amazing career path for someone who was effectively self-taught. Zel'dovich's work on shock waves, gas dynamics and their application to explosions were and still are highly regarded.

At the time of the German invasion of the Soviet Union in

June 1941, Zel'dovich was studying chain reactions in uranium. Along with his fellow scientists, he moved back to working on conventional explosives. He helped invent the famous Katyusha missiles, legendary for terrifying the Germans. In 1943, the Soviet government decided to embark on its own nuclear weapons programme and drafted Zel'dovich into a small elite team. Working at The Installation, Zel'dovich headed one of several competing design groups. Part of the research was delegated to universities, such as Moscow State University, where Landau was working. At first, Sakharov's team was delegated to check and refine the calculations produced by Zel'dovich's. They resented this subsidiary role. 'Our job is to kiss Zel'dovich's ass,' one of Sakharov's colleagues put it – in private, of course.[41]

While the first Soviet atomic bomb was developed largely thanks to Fuchs, the first successful hydrogen bomb, similar to Teller's Alarm Clock, was developed by Sakharov. It was exploded on 12 August 1953 at the testing ground at Semipalatinsk, in Siberia. It yielded 0.4 megatons. That made it a mere ten times bigger than the atomic bombs used in World War II. The Soviet scientists immersed themselves deeply in the search for design innovations that would produce a higher-yield bomb to rival Mike.[42]

During those years, Zel'dovich branched out into particle physics and astrophysics. Working with him, Sakharov read all he could on gas dynamics and astrophysics. They had all realised that 'the physics of stars and the physics of a nuclear explosion have much in common'.[43] Sakharov was familiar with the papers on collapsed stars written by the Soviet Union's top theorist – Landau – in 1932, when he rediscovered Chandra's upper limit for the mass of a stable white dwarf, and in 1938 when he looked into the minimum mass necessary for a neutron core to form within a star. Zel'dovich had first come to appreciate the role of explosions in astrophysics in 1934, after hearing Chandra's lecture in Moscow on his upper limit and on the possibility that stars might collapse or explode. Wheeler, too, recalled that scientists working on implosion began to look towards the 'astronomer's techniques'.[44] The groups at Los Alamos and 'Los Arzamas' agreed that 'the high temperatures generated in thermonuclear reactions led to the development of the physics of high pressures and high temperatures', very much like reactions in stars.[45]

In the spring of 1954, Sakharov and Zel'dovich finally cracked the Teller–Ulam configuration. With the help of the Soviets' own high-speed computers, developments followed quickly. On 22 November 1955 the two tested a bomb capable of being delivered. It had a yield of 1.6 megatons. They went on to produce larger bombs, and on 30 October 1961 detonated a 50-megaton bomb, dubbed the Czar bomb, which to this day remains the largest nuclear weapon ever exploded.[46]

The temperaments of Sakharov and Zel'dovich mirrored those of Ulam and Teller. Ulam and Sakharov were both interested in computational aspects of physics; Teller and Zel'dovich were more intuitive and tried to avoid detailed calculations. 'Zel'dovich could see the solution of many problems without long calculations', recalled his former student and collaborator Sergey Blinnikov.[47] Landau suffered a tragic car accident in 1962 which seriously impaired his mental ability, and from then on Zel'dovich was the undisputed leader of Soviet theoretical physics. He wrote dozens of books and papers. When Stephen Hawking, one of the most important figures in astrophysics today, met Zel'dovich, he was amazed to discover that he was just one man. He had assumed that the name was a pseudonym for a group of authors who had worked collectively – how else to explain such phenomenal productivity? In recognition of his achievements in the Russian nuclear programme, Zel'dovich was made a laureate of Lenin and Hero of Socialist Labour (the highest civilian award) three times, an extraordinary honour.

The scientists who took part in wartime research never forgot the lessons they learnt. Take, for example, Robert F. Christy. His contributions to the design of the bomb dropped on Nagasaki were so great that it was often referred to as the 'Christy gadget'.[48] During a sabbatical from Caltech in 1960 at Princeton, Christy decided to 'learn something about the stars'. After all, he reasoned, the 'mathematical approach was very similar to what we had been working on in Los Alamos during the war. And I thought it was interesting, in a way, that the theory used to make atomic implosion bombs was the same theory I could apply to certain kinds of variable stars. It's interesting to see how things relate to each other.' He had always been impressed with Eddington's work and decided to look further into the mechanism of Cepheid variable stars. He was delighted when the RAS awarded him the Eddington Medal for his results. 'Eddington himself I always considered a physicist', he wrote, but in fact 'he's a great astrophysicist'.[49]

Wheeler, too, was fully aware of the intimate connection between bombs and stars. Early in the 1950s, at Princeton, armed with his state-of-the-art MANIAC computer, he set out to resolve what he considered to be nothing less than *the* fundamental problem of physics – 'the fate of great masses of matter'.[50] He began by exploring what happens to very cold matter. By 'cold matter' he meant the burnt-out core of a star in which the nuclei have given up the last of their energy. His method of studying this cold matter was to formulate an equation describing its state that took account of densities ranging from very low to very high, applying, whenever necessary, the principles of general relativity and what was then known about nuclear forces. Wheeler had worked on the atomic and hydrogen bombs and done nuclear physics research with Bohr in the late 1930s. This gave him the ability to deal with the way matter behaves over broad ranges of temperature and pressure. Two graduate students, B. Kent Harrison and Masami Wakano, worked with him. They input data into their equation of state for just about every known star, working through several per hour. MANIAC allowed them to do something that would have been a hopelessly laborious process before the digital computer was invented – to explore in detail the evolution of a large number of stars. Chandra had made laborious calculations on a mechanical calculator, as had Serber and Volkoff in their work with Oppenheimer.

The MANIAC computer enabled Wheeler and his team to come up with some truly astonishing results. It was clear from their super-equation for the state of cold matter that, whereas a star with a mass less than the Chandrasekhar limit would collapse to a white dwarf, a burnt-out relic with a larger mass would collapse non-stop and become a neutron star. There was no alternative, no intermediate stage.

Harrison, Wakano and Wheeler provided the first detailed glimpse of the fate of a star more than eight times the mass of the Sun. Their results shed new light on the life of stars. What they found is that the high gravity of such a massive star makes its interior hotter than at the centre of the Sun, so it takes less time – about 10 million years – for the hydrogen to burn up. Once the hydrogen fuel is spent, the star begins to die. First the helium core cools and begins to collapse. Then the crush of gravity raises the temperature again, to the point where helium begins to burn. After another million years, the helium too is spent, leaving a carbon core surrounded by layers of hydrogen and helium ash. This in turn starts to cool and then collapse under the gravitational pressure of the mass around it. Again the temperature rises and the core reignites, and, after six hundred years, burns up, leaving a neon core. The same process repeats again and again, more and more rapidly; the heavier elements burn more quickly because they are more stable and so the same amount of fuel will generate less energy. Neon burns off in a year, oxygen in six months, and silicon in a day. A star that started out eighteen times the mass of the Sun ends up as an onion-like structure made up of layers of different elements.[51]

Silicon is the last element in the core to burn. What is left then is iron, the element with the most stable nucleus. Iron will undergo fusion or fission only if energy is supplied; it cannot generate any energy by itself. The formation of an iron core signals the beginning of the end for a massive star. At this point the temperature of the iron core is a million degrees kelvin, and the pressure is 10 million grams per cubic centimetre. The onion layer of silicon ash is 4000 miles across, twice the size of the Moon and 50 million times as massive. Inside it is the iron core, 1000 miles across. This is a tiny fraction of the volume of the whole star, which by now has bloated up to 20 million miles across, exerting an incredible pressure on the core.

The interior of the star is so hot that all the atoms have been stripped of their electrons. The core now consists of iron nuclei and electrons moving close to the speed of light – relativistic electrons. But the immense pressure of the 20-million-mile, onion-like structure above it squeezes the mixture of iron nuclei and relativistic electrons closer and closer together. Eventually, the electrons smash their way into the iron nuclei and combine with the nuclear protons there to produce neutrons and neutrinos. This creates a host of nuclei of elements heavier than iron that contain more neutrons than usual – 'neutron-rich' nuclei. Many things now begin to happen. The total effect is to reduce the number of electrons – lessening the degeneracy pressure which caused the iron core to be hard as rock – as well as drive the mass of the core above the Chandrasekhar limit. The core becomes unstable under the enormous inward pull of gravity and collapses to a neutron star.[52]

Harrison, Wakano and Wheeler's analysis also suggested something else – that it was not out of the question that a star with a large enough mass might carry on collapsing and collapsing for ever, until it was 'nowhere', as Eddington would have put it. Was there anything that could prevent the core of a star from collapsing right through the white dwarf and neutron star stages into nothingness? This is what Eddington and Milne had sought in their day and what almost everyone else continued to hope for right into the 1960s. Wheeler and his team could not find any such process. So what did the core then collapse to? Wheeler suspected that at such extremely high temperatures and pressures the behaviour of matter might be governed by a whole new set of physical laws. The answer, he suggested, might be found at the 'untamed frontier between elementary particle physics and general relativity', in the realm of a quantum theory of gravity.[53] But in fact, the definitive solution would come from an unexpected quarter.

11

How the Unthinkable Became Thinkable

Wheeler's analysis of the fate of stars as they burn up their fuel seemed to reveal that a star far larger than the Sun might ultimately collapse to nothing – a notion so outrageous that he simply refused to believe it himself. The complicated equations he and his team had processed through the MANIAC computer indicated that a very massive star would shrink until it was unimaginably tiny and unimaginably dense. Its gravity would be so strong that it would pull space around itself like a shroud and be swallowed up by it. But to Wheeler, this was simply unacceptable. Quite apart from the absurdity of it, there was no observational evidence whatsoever to support it.

At the Solvay Conference on physics in 1958, Wheeler argued that something had to be missing. There had to be a way for stars to shed mass so as to get below the maximum possible mass for a neutron star, which he and his team had calculated to be about twice the mass of the Sun. It was like a rerun of the sessions at the Royal Astronomical Society in the early 1930s, with Wheeler insisting – armed with no evidence, just his own preconceived notion of the way things had to be – that something had to put a brake on a star's collapse.

Oppenheimer took on the role that Chandra had played in those earlier debates. He for one was prepared to consider the possibility that Wheeler's results could be taken at face value. He had an answer that cut through Wheeler's byzantine arguments:

'Would not the simplest assumption about the fate of a star of more than the critical mass be this, that it undergoes continued gravitational contraction and ultimately cuts itself off more and more from the rest of the universe?'[1] But his was virtually a lone voice – added to which he carried little weight in the astrophysics community. Oppenheimer had written a few papers based on purely theoretical concepts. Wheeler, in contrast, was running a huge research programme which included applying nuclear physics to the study of realistic stars. At this stage, for all their theoretical work, no one had actually observed any way in which a star might die, other than as a white dwarf. But a discovery would soon be made that would finally convince Wheeler – and bring Chandra and his theory back to the centre of the debate.

It all came about when a man named Stirling Colgate had a startling insight. At the time he was America's premier diagnostician of thermonuclear weapons (hydrogen bombs) at the Livermore National Laboratory in California. He realised that flashes of light from supernovae might set off the detection devices in American satellites spying on the Soviet Union and start World War III, even though supernovae were some hundred thousand trillion miles away and the explosion had actually taken place more than a hundred thousand years before. Lean and rugged and with a perpetual suntan, Colgate has a raffish, devil-may-care style that belies his age (now touching eighty) and his on-going serious dedication to scientific research. A man of extremely varied interests, his expertise spans experimental and theoretical physics, and he has a feel for machinery that is right out of Robert Pirsig's *Zen and the Art of Motorcycle Maintenance*.

Colgate's introduction to nuclear weapons came dramatically in the place where it would all happen, long before it did. He was a student at the Los Alamos Ranch School when, in December 1942, a delegation arrived. It included an impressive number of military officers together with two civilians, one in a pork-pie hat, the other in a fedora. Immediately after the visit there was a shocking announcement: the school was to close forthwith. Colgate and the two other seniors would graduate immediately, ahead of schedule. Colgate and his precocious friends quickly worked out what was going on. The two civilians, plain 'Mr Jones and Mr Smith', were in fact scientists. They were actually Oppenheimer and Ernest O. Lawrence, a nuclear physicist from Berkeley and a 1939 Nobel laureate. Oppenheimer personally delivered the commencement address to the three-man graduating class, after which he had several buildings bulldozed to make room for laboratories and offices. Oppenheimer, Colgate remembers, did not seem at all uncomfortable with any of this, though Lawrence was a little abashed. From that moment Colgate developed a serious mistrust of Oppenheimer which he maintained throughout his later dealings with him, though he also sympathised with the regret and guilt Oppenheimer felt as a result of his role as the leader of the Manhattan Project, which had wreaked destruction by creating the atomic bomb.

Colgate is a scion of one of America's wealthiest families, though he has always gone out of his way to be his own man. Two years after that fateful meeting in Los Alamos, he enlisted in the merchant marine, which was, he recalled, 'perfect for me'.[2] He had to establish himself among the older experienced seamen, for whom being a whiz-kid in science meant little; what counted was being able to get the job done. On 6 August 1945, the captain summoned the crew to the mess hall and told them that America had dropped an atomic bomb on Hiroshima, adding that he would 'appreciate it if Mr. Colgate would tell us what it means'. Colgate knew the answer – though, he points out, at that time his conjectures were classified. Most classified of all was his explanation that each fission reaction would release two or more neutrons, producing an uncontrolled chain reaction and leading to an apocalyptic explosion.

When he was discharged in 1946, Colgate returned to Cornell University. He had already spent a year there studying electrical engineering. 'Physics', he decided, was 'the thing for me.' Edwin Salpeter, a professor there, remembers how impressed everyone was with this extraordinary young man. Colgate completed a B.S. and a Ph.D. in nuclear physics, then took a position as a postdoctoral fellow at Berkeley.

In 1952 he moved to Livermore National Laboratory, recently created by Teller with the encouragement of the United States Air Force to provide competition for Los Alamos, which Teller felt was lagging behind in weapons research. Teller made it amply clear that Livermore's sole mission was to research and develop a hydrogen bomb. Aware of Colgate's reputation, Teller suggested that he take on the important task of making diagnostic measurements for the forthcoming tests. 'Well, of course,' he replied. It was a natural progression for him. 'I was always enamoured with explosives, and eventually I graduated to dynamite and then nuclear bombs,' he explains nonchalantly.

The diagnostic work was divided into two parts. One was to analyse what the bomb yielded by studying the radioactive products of the explosion, scooped from the atmosphere by purpose-built aircraft. Colgate's responsibility was 'fast diagnostics' – measuring the range of energy of the neutrons and high-energy gamma rays bursting out of the explosion. Scientists thus hoped to deduce how well the bomb's mechanisms functioned and the precise rates of successive events in the progression from fission to fusion during the Teller–Ulam process, before all the components had been vaporised.

New diagnostic instrumentation had to be designed as quickly as possible. The test of the first hydrogen bomb, Mike, was already scheduled to take place that same year, 1952, on the island of Eniwetok. Colgate's work required him to shuttle back and forth between Livermore and Los Alamos. He would fly to El Paso in Texas, take another plane to Albuquerque in New Mexico, then drive from there to Los Alamos. On one of these trips he decided that some R&R was in order. He crossed the border into Mexico and strolled into a bar in Juarez, a short distance from El Paso. There he got into conversation with an ex-Marine who happened to be working as a graduate student in astrophysics. Colgate's new pal was from the Deep South, and this was the 1950s, long before political correctness had become de rigueur. In a long slow drawl, the ex-Marine declared that he worked for one of the blackest people you ever would see. He had been brought up to believe such people could not be too bright, but this guy was just the smartest person he had ever met. 'And, if you can believe it,' he concluded, 'he has an absolutely crazy name, too: Chandrasekhar!' Colgate had come across Chandra in the context of his work on radiative transfer, but had never realised that Chandra had any connection with astrophysics.

Back at the bar, Colgate recalls, scientific bonding led to innumerable rounds of drinks. The next morning the pair were so hung-over that Colgate still can't remember how they got from Juarez to El Paso. They were almost thrown off the El Paso to Albuquerque flight. From there, Colgate's new friend went on to Chicago. Colgate arrived at Los Alamos in a terrible state. As he tells it, the first person he ran into there was Chandra, working as a consultant on the upcoming Mike test. Without thinking, he walked straight up to him and blurted out the whole story. Suddenly he felt 'sad and terribly embarrassed'. But Chandra just roared with laughter. Unfortunately, charming though the tale is, the only person who can remember Chandra being at Los Alamos at the time is Colgate himself, so perhaps the tale has become embellished over the years. But it does show the impact that Chandra had on people and the kind of myths that grew up around him. It might even be true. This was how Colgate remembers being introduced to Chandra. Their paths were to cross again in the world of astrophysics a decade later.

Since Colgate did not belong to the 'glamorous' category of bomb design, and his work was top secret, he is not usually mentioned in books on the development of the hydrogen bomb. Nevertheless, two years later, at the test of Bravo, the first deliverable thermonuclear bomb, the twenty-nine-year-old scientist was in charge of a staff of thousands. The Bravo test was an immense success, in terms of both its yield and the diagnostic methods used to assess it. Colgate had won his spurs. Teller gave him carte blanche to choose his next research project, and he opted to look into thermonuclear fusion and plasma physics, both peaceful uses of nuclear energy. But one can never escape one's past. In 1959, on the advice of the Livermore and Los Alamos National Laboratories, the State Department recruited him as the scientific consultant on nuclear test ban negotiations in Geneva. Both sides needed to 'agree on a detection system so we don't all lob nukes at one another'. Colgate proposed accomplishing this with satellites designed to detect nuclear testing.

Then he remembered some research he had done in 1956 with his colleague Montgomery Johnson. At the time, the US

Government had suggested adding another battlefield to air, land and sea: making outer space a fourth dimension for warfare. Colgate and Johnson had been recruited to investigate the blow-off of radiation and other debris from a hydrogen bomb exploded in space. They had conducted simulations suggesting that the result would be 'monster' amounts of X-rays and gamma rays, just as if it were a supernova.[3] A hydrogen bomb, it seemed, was remarkably similar to a supernova. But the problem was that if they sent up a satellite to detect nuclear testing, a supernova might set off the detection device, thus accidentally precipitating a cataclysmic war. 'It was just natural to bring up supernovae in Geneva if we wanted to get a good spy satellite,' recalls Colgate. 'The Russians obliged and they too made a big thing about gamma-ray bursts' – highly energetic radiation from outer space, which could come from a supernova. Colgate was a very junior member at this august gathering of senior Soviet scientists. He felt inadequate in his knowledge of supernovae, but, 'as I learned in later years, so were they'.

By 1959, research on X-ray and gamma-ray emissions from space had become quite sophisticated. X-rays cannot penetrate the Earth's atmosphere, so experiments were carried out using equipment sent up on rockets, at first on board captured German V-2 missiles. Then in 1961, satellites detected gamma rays emitted by the 50-megaton Czar bomb which the Soviets had just exploded in violation of the Soviet–American moratorium on nuclear weapons. The Americans were apoplectic. Clearly the Soviets had agreed to the moratorium as a ruse, to buy time to prepare for their next series of tests while the United States was honouring it. Livermore and Los Alamos went into overtime. Understanding the supernova phenomenon became top priority.

Teller urged Colgate to follow up on this line of enquiry, but in fact Colgate had already started to move over to astrophysics. As soon as he realised the connections between supernovae and nuclear weapons, he went to see William ('Willy') Fowler, Caltech's expert on nuclear reactions in stars. They struck a deal. For a couple of days each week, Colgate gave lectures on explosions – his area of expertise – at Caltech, and in exchange boned up on astrophysics. Both the staff and graduate students much enjoyed shooting down his budding ideas on supernovae. After hours with his head down, Colgate would spend his afternoons and evenings at nearby Venice Beach.

Colgate carried out his initial research on supernovae with his colleague Montgomery Johnson. Their work was based on the classic 1957 paper by the husband-and-wife team of Margaret and Geoffrey Burbidge at the University of California, San Diego, together with Fowler and the Cambridge astrophysicist Fred Hoyle, a quartet known in the trade as B^2FH. The Burbidges, Fowler and Hoyle argued persuasively that as stars evolve they produce heavier and heavier elements. When a star much more massive than the Sun has completed its evolution, there is an iron core right at the centre with a density of 10 million grams per cubic centimetre. The natural next question for astrophysicists to ask was, what makes a super-heavy star such as this go supernova, and what is left behind? Everyone agreed that the hypothesis that Zwicky had come up with in the 1930s was almost certainly right: there had to be something like a neutron star left behind after a cataclysmic supernova explosion.

An ageing star is a huge ball – layer upon layer of burnt-out nuclear ash surrounding an unburnable iron core, which is what is left over after all the silicon has burned up. At this point there can no longer be any nuclear reactions that will give the star energy, so it begins to cool off. Crushed by the enormous weight of the onion-like structure above it, the core begins to implode. By now it is a mixture of neutron-rich nuclei, electrons and protons at an unimaginably high temperature, about 5 billion degrees kelvin, caused by the extreme compression. At that temperature the high-energy radiation in the star (X-rays and gamma rays) is powerful enough to smash the neutron-rich nuclei into helium nuclei (alpha particles), protons and neutrons. It is the fusion of these light nuclei that produces hydrogen bombs and makes stars shine for billions of years.

Suddenly, in the blink of an eye, all this is undone. All these reactions absorb energy rather than provide it, so the only way the star can get any more energy is for it to shrink under the massive inward shove of its own gravity. As a result, the core begins to heat up again. The high-energy radiation then turns on the alpha particles, ripping them apart into their constituent protons and neutrons. More and more electrons collide with the protons in the heavier nuclei that remain intact and turn themselves into neutrons and neutrinos.[4] This depletes the supply of electrons, and thus weakens the electron degeneracy pressure, which can no longer support the core and keep it rock-hard. Without electron degeneracy pressure the entire star collapses – but not simultaneously.

At a speed of 36,000 miles per second, the iron core and its shell of silicon ash shrivels in a fraction of a second from a ball more or less the size of the Earth to an incredibly dense sphere a mere 12 miles across. But this core contains only a third of the total mass of the star. The collapse happens so quickly that the onion layers of carbon, oxygen, neon, helium and hydrogen are left behind – a shell with no centre. As the core is shrinking at lightning speed, nuclear ash from the onion-like structure above it is falling after it. The core reaches its limit of compression, then stiffens and rebounds like a coiled spring. This 'core bounce' creates a shock wave which undulates outwards at 6000 miles per second, ploughing through the ash still plummeting in from the onion layers above the core. When scientists create a model of a supernova explosion, one problem is to make sure that the shock wave does not stall – that it is not stopped dead by the mass of material plunging in. If it rolls on, it creates an explosion that produces the exact yield of chemical elements one sees in an actual supernova explosion.

Inside, the collapsed core is like a cubist painting, an Alice in Wonderland scenario with no distinction between inner and outer. Everything merges into everything else, with exotic elementary particles, perhaps even free quarks, springing up, as well as huge numbers of neutrons and some protons and electrons. This is a neutron star.[5]

For a neutron star to be left after the star has imploded, the core has to have a strong enough repulsive force to counterbalance the attraction between neutrons and protons (which naturally bond to create nuclei, drawn together by the nuclear force). Otherwise the attraction between them, together with the force of the star's own gravity, will be enough to make the star carry on collapsing, dwindling ever smaller and smaller until it disappears into nothingness. The star also has to blow off enough mass to get below the maximum mass for a stable neutron star. Oppenheimer and Volkoff calculated this mass to be seven-tenths the mass of the Sun. (With developments in nuclear physics enabling them to formulate realistic models of supernovae, astrophysicists currently calculate the mass above which a neutron star becomes unstable to be two or three times the mass of the Sun.) If the burnt-out core of a star constitutes more than this maximum mass, the options of shrivelling to a neutron star or ending up as a white dwarf are both closed to it, for the maximum mass for a stable neutron star is greater than the Chandrasekhar limit for a white dwarf (1.4 times the mass of the Sun).

Colgate and Johnson's research was a first attempt to understand the mechanism of a supernova explosion. In terms of practical application, what they were trying to find out was how to blow up a star. They assumed that the shock wave from the core bounce smashes into nuclear ash plummeting inwards due to the inward tug of gravity. They postulated that the shock wave turns this matter around and heats it up, and that this is what causes the supernova explosion. But this turned out to be wrong. Another member of the team, Richard White, ran computer simulations which showed that the shock wave from the core bounce is not strong enough to accomplish all of this. Instead of the star imploding in a supernova explosion, then shrivelling to nothing, there is just a fizzle.

White studied physics at Pomona College in California, where he also took an evening course in the fledgling subject of computer programming. On graduating in 1956, he joined Livermore National Laboratory. At the age of twenty-two, he was the first person there with any formal training in computers. Computers were still in their infancy, and programmers and theoretical physicists at Livermore simply learned about them on the job. Working with Colgate at Livermore, he later remembered, was 'akin to having a close personal relationship with a whirlwind'.[6]

Colgate and White sat down to take a long hard look at supernovae. They set out to create models of stars on the verge of collapse. It was a very ambitious programme. It entailed constructing a mathematical description of the gas in the star, an equation of state that built in nuclear forces adjusted to prevent the star collapsing completely, as well as the mixtures of chemical elements that made up the star. The equation of state they were looking for was far more complicated than the perfect gas law or Chandra's equation for an 'ideal' white dwarf, both of which ignored interactions between particles of gas, and was also far more sophisticated than the one used by Wheeler's team. The whole equation of state had to be encapsulated in computer software that could model real stars. White began with a computer program that combined software used to design bombs with the most up-to-date equation of state for a star from astrophysics. At that time, only Livermore and Los Alamos had computers fast enough to perform such complicated calculations.

Working at a weapons laboratory, Colgate and White knew nothing about the supernova research that Zel'dovich and his group were carrying out. This was an era when Soviet scientists attended international meetings accompanied by 'minders', apparatchiks or even KGB agents with dubious scientific credentials. This is not to say there was no scientific communication between East and West at these events, but it was restrained. Zel'dovich, of course, was not allowed out of the Soviet bloc. On one rare occasion, according to an oft-told story, the organiser of the astrophysics section of a meeting of the American Physical Society decided to make a goodwill gesture and invited a member of the Soviet delegation, chosen at random, to give a scientific lecture. Unwittingly he chose a KGB agent.

At first, says Colgate, he 'didn't think he was doing anything particularly original'. As far as he knew, he and White were just creating models of scenarios other people had already suggested, like those in the B^{2}FH paper, but which no one had yet run through a computer. But things turned out otherwise. After Colgate and Johnson hypothesised that a supernova explosion is sparked by the shock wave from core bounce smashing into infalling nuclear ash from the collapsing star and igniting it, Fowler and Hoyle began to fill in the details. But Colgate and White, working together, were able to show that Fowler and Hoyle had made a mistake in analysing the physics of the explosion.[7]

The challenge was to work out how core bounce could create an explosion. At this point, Colgate had a flash of inspiration: 'Why not look into something that everyone else had neglected?' At the moment when the core of the star is crushed from the size of the Earth to less than 12 miles across, the electrons in the core are smashing into the protons in hydrogen atoms and into heavier nuclei, such as iron, and turning themselves into not only neutrons but trillions upon trillions upon trillions of neutrinos. After the bounce, all these subatomic particles fly outwards like a sonic boom at more than 6000 miles per second. Perhaps neutrinos were the clue. Perhaps they might heat the nuclear ash plummeting inwards so much that they powered an explosion.

'That did it.' Now Colgate was eager to learn about neutrinos and tie them into his and White's computer simulations to see whether they could help in producing an explosion. He set off for Caltech again to meet up with Christy. He had been impressed with Christy's work on Cepheid variables, which made use of atomic bomb design and for which he had been awarded the Royal Astronomical Society's Eddington Medal. Christy, in turn, was intrigued by Colgate's proposals. Colgate felt comfortable enough with Christy to air what he thought was an outlandish suggestion: 'Could neutrinos be treated as if they were a gas of radiation?' Christy replied, 'Of course. Just look at what Chandra did for an electron gas.' He was referring to Chandra's discovery of electron degeneracy pressure – that at the density of a white dwarf, a gas of electrons develops an outward pressure that prevents white dwarfs smaller than 1.4 times the mass of the Sun from collapsing. Like electrons, protons and neutrons, neutrinos can also develop a degeneracy pressure. Aha, thought Colgate: 'A vital piece of the puzzle had been neglected, namely, the huge degeneracy pressure expected from the flood of neutrinos. Excitement within the astrophysics community ran high. Colleagues generously offered advice and time. Science was really working at every step, no one was jealous.' Incredible results were emerging.

'Everyone talked about certifying a nuke, but I kept talking about how could you possibly ever certify a supernova?' says Colgate, with a glint in his eye. By 'certifying' a hydrogen bomb, Colgate meant that he had worked out the theory of the bomb and understood how it exploded. But this was still far from the case for supernovae.

By inputting models of stars into their computer and just letting it run, Colgate and White discovered far more than they had set out to find. It turned out that they had proved numerically, beyond any shadow of a doubt, what most astrophysicists, even Wheeler himself, had refused to believe, despite the evidence of his and Chandra's own mathematics – that stars really could undergo an ongoing and endless collapse.

Up until the time they ran their first computer simulation, even the most cutting-edge astrophysical research had been based only on simplified models of stars. Those investigations were valuable, but everyone knew that the early models were far removed from real stars, where there are many different factors that could prevent stars from collapsing or imploding. 'Prior to the Colgate–White research, conventional wisdom held that singularities were untenable,' White recalled; astrophysicists refused even to consider that a real star might begin a process of infinite collapse.[8] They took for granted that stars would always shed enough mass to shrink below the Chandrasekhar limit, so that when their fuel burnt out they would quietly expire as white dwarfs. The question of whether they might collapse completely did not even need to be considered. It was a mere mathematical conceit.

At that time, Wheeler often travelled to Livermore and was in regular contact with Colgate and White, along with Chandra, just as their results were pouring from their computers. Colgate recalls:

> Wheeler, and Chandra as well, were incredibly satisfied to see that when suitable numbers and details referring to the constitution of matter were taken into account – that is, equations of state and nuclear reactions . . . it is clear that the core really can collapse. It made it very clear to Johnny [Wheeler] and to Chandra that collapse was inevitable and then the beauty is that we can't really put our finger on what happens next.

As Wheeler exclaimed excitedly to Colgate, 'The big deal, Stirling, is that it makes a reality of these things!'[9] But to Colgate it was not a big deal at all: he had acquired his experience doing diagnostics on some of the largest hydrogen bombs ever produced by the United States, working on both theoretical and experimental physics. His work had always involved first working out the theory and then putting it into practice. He for one was convinced that the concepts of complete collapse and relativistic degeneracy were practical realities. For him it had always been real, never mere theory:

> The sense of the reality of the nuclear physics, of the radiation, of the hydrodynamics seemed to me absolutely beyond question. I had understood all the phenomena first, then designed the experiments, then saw that the results of the experiments confirmed in detail the design physics and the calculations. It was not that I was nonchalant, it was instead the conviction that physics works at a very very deep level.

As Kip Thorne, an eminent relativity theorist at Caltech, recalls, Wheeler was so excited about Colgate and White's work that after one visit he shelved the topic for the day in his relativity class at Princeton, where Thorne was a student, and gave an impromptu lecture on what he had just heard.[10] For Wheeler, it was a turning point in his research on the fate of the stars. It had finally been proved, conclusively enough to overcome even his most deep-rooted objections, that if a neutron core could not eject enough mass for it to get below the maximum mass to make it stable, it would have to collapse indefinitely, to a point of infinite density.

Colgate and White wrote up their work, and in 1962 submitted it to *Reviews of Modern Physics*. The editors 'could not figure out what to do with it within the context of physics, because it was really astrophysics', recalls Colgate. 'After two years of muddling around' it was passed to Chandra, who was the editor of the *Astrophysical Journal*. The paper was a mess. Despite his sessions boning up on astrophysics at Caltech, Colgate was writing about astrophysics as if it were physics. 'We were rediscovering astrophysics,' remembers Colgate. 'Chandra took one look at it and said, "This has got to be published. But, look here, Stirling, there's some astrophysics you need to learn."' Chandra must have realised immediately how important this research was in verifying his lifetime's work. The paper finally appeared in 1966. It had taken four years to publish what was to be one of the most important papers in astrophysics. Colgate is careful to point out that 'Chandra carefully edited it and in places rewrote it'. Among his alterations was to add a reference to a paper of his own alongside one of Eddington's.[11]

The essence of Colgate's discovery was that neutrinos could provide the key to understanding how stars go supernova. They might just be the catalyst that would allow the shock wave from the bouncing core to continue its way through the star without being stalled by nuclear ash falling towards it. The core could produce huge amounts of neutrinos which would be carried along by the outgoing shock wave. They would heat up the matter plummeting in from the onion-like layers, which would then turn around and sweep through the star, blowing off the star's outer layers and exposing the neutron star at its core. But, fascinating as all this was, neutron stars remained purely hypothetical. Astrophysicists were still sceptical as to whether they really existed at all.

Then, in 1967, two astronomers at Cambridge, Jocelyn Bell and Anthony Hewish, were studying radio signals emitted by certain stars. They noticed that some were pulsing at regular intervals, with a range of 0.25 and 2.0 seconds, like the beam of a lighthouse sweeping over the sky. What could they be? To fertile imaginations it sounded very much like a message being semaphored from a civilisation on another planet. The pulses were too close together to come from a pulsating star like a Cepheid variable, which changes its size over periods of weeks. Neither could they be a signal from a particularly hot spot on a rotating white dwarf star: if it was spinning that fast it would be ripped apart.[12] Then, the following year, pulsing radio waves were detected coming from the Crab Nebula, a remnant of a supernova.

It was a remarkable moment for astronomers. Whatever was emitting these waves, it was not a white dwarf. Until then astrophysicists had assumed that white dwarfs were the end of the line. For all the talk about neutron stars, they were still in the realm of hypothesis, while the idea of stars collapsing into nothingness had been mathematically verified but still seemed like wild speculation. People doubted that either could really exist. It now seemed that they were wrong. Only one thing could be emitting the pulsing radio waves: a neutron star. It was the first time that anyone had actually found concrete observational evidence that there was any way for a star to die other than as a white dwarf.

Everything fell into place. Being the core of a huge collapsed star, the neutron star was bound to spin much faster than the original giant star, just as an ice-skater spins faster and faster as she draws in her arms. Similarly, the magnetic field of a neutron star can be some 10 billion times that of the original giant star before its collapse. These spinning neutron stars became known as 'pulsars', from the original notion that something was pulsing.[13] Since their discovery, over a thousand have been found, and there are estimated to be a million in our Galaxy alone. Bell and Hewish had discovered what Zwicky and Baade first predicted in 1933; what Oppenheimer and Volkoff had provided a theoretical basis for in 1938; and what had been almost ignored ever since. Neutron stars existed.[14]

Although the Colgate–White model would ultimately be revised and refined in the coming decades as computing power increased, the basic idea remained sound. There was, however, another possible and even more dramatic way in which a massive star might die. In the years after Chandra discovered his maximum mass, astrophysicists had circled around it. Colgate and White's computations, which so excited Wheeler, had revealed that it was, in mathematical terms at least, a certainty. But even when the mathematics seemed to point to the inevitable, astrophysicists refused to believe that a star could dwindle not just to a white dwarf or a neutron star but to a state of almost unimaginable nothingness – a never-ending collapse beyond smallness to a colossal density and a gravity so intense that it seemed even to violate the laws of space and time. It was not until the early 1970s that the impossible finally began to seem possible.†

---

† See Appendix B for current developments in supernova research.

PART III

12

The Jaws of Darkness

Stirling Colgate and Richard White's computer simulations were the first step in convincing scientists that stars really could collapse completely. Mathematicians, meanwhile, had begun to accumulate results that helped to clear up the seemingly discordant views of the two observers in the puzzle proposed by Oppenheimer and Snyder back in 1939. To recap, a collapsing star is about to reach the size of its Schwarzschild radius. A nearby observer sees it shrinking faster and faster, almost at the speed of light, as it falls through its event horizon, the surface defined by its Schwarzschild radius.[1] To a distant observer, however, the collapsing star appears to freeze. This turned out to be an illusion: the star's gravity becomes so strong as it disappears behind the event horizon that light can no longer escape.[2] Solving the paradox required finding mathematical equations that would enable the two observers to compare their views of what happens to matter trapped in a region of space and time hugely contorted by gravity. It turned out that Eddington had first worked out how to do this way back in 1924, while he was examining Karl Schwarzschild's solution to Einstein's theory of general relativity, without realising that it could be applied to collapsing stars.[3]

But the real stumbling block was the concept of a star many times bigger than the Sun shrinking to an infinitesimally tiny point. Physicists were prepared to accept any number of scenarios, however strange, illogical, contradictory or downright impossible they might at first appear, as long as the mathematics bore them out. Astrophysicists were a far more hidebound breed, but little by little they began to accept the inevitability that a star could disappear into nothingness.

In 1965, a year before Colgate and White completed their computer simulations, a thirty-four-year-old English mathematician, Roger Penrose, at Birkbeck College, London, had had the brilliant idea of applying topology to stars. Topology is a branch of mathematics that studies the properties of objects and their surfaces that remain unchanged when the objects are distorted. A standard example is a coffee cup and a doughnut. They have different shapes but the same topology, because they each have a hole – the coffee cup's is in the handle. So they can be transformed one into the other without tearing their surface. Penrose used topology to study the surface of the event horizon, to try to work out a connection between the shape of an imploding object and the singularity it can form. He came up with a watertight proof, framed in the most abstruse and rigorous mathematical detail, that whenever a star disappears through its event horizon, it is bound to carry on collapsing until it becomes a singularity. This was the argument that clinched the possibility of complete collapse, in purely theoretical terms.

The next question was whether it might be possible to see this infinitely small and infinitely dense star. When general relativity is applied to the question of the collapse of stars, a singularity is the logical conclusion. But at a singularity the laws of physics break down and the course of the universe becomes unpredictable – highly disturbing to scientists. In 1969 Penrose put forward the 'cosmic censorship conjecture' – that there can be no naked singularities.[4] By definition, singularities and the bizarre physics they produce are hidden from view behind the event horizon, from which nothing can escape. But general relativity takes no account of quantum effects. It could be that a new physics – a quantum theory of gravity – is required to describe what goes on inside a black hole. There may not be any singularity to worry about, after all.

As it happens, Penrose's brilliant topological idea was not pure mathematical speculation. He had been inspired by an extraordinary discovery by Maarten Schmidt, a thirty-four-year-old Dutch-born astronomer at the California Institute of Technology, two years before Penrose published his paper. Schmidt had been deciphering data on the composition of light as broken up by a spectrograph into its different frequencies, which appear as spectral lines on a photographic plate.

Scientists have learnt practically everything we know about the stars – from their chemical make-up to their movements – by analysing their spectra. A prime example is the Doppler effect, which reveals how fast a star is moving in relation to the Earth. This is the effect, explained by the Austrian scientist Christian Doppler, that causes the pitch of, for example, an ambulance siren to change. As the ambulance moves away from us, the distance between successive peaks in the sound wave from its siren increases by the distance the ambulance has travelled between emitting the two peaks, and our ears interpret the longer sound wave as a lower pitch.

In 1842 Doppler predicted, on the basis of the wave theory of light, that light would behave in the same way – that the colour of a luminous body such as a star will vary depending on its speed relative to the observer. Twenty years later, astronomers confirmed this prediction. To measure a Doppler shift in light, scientists compare the wavelength of a spectral line from a particular element in the laboratory with the wavelength of the same spectral line in a star. Thus they can calculate the speed of the star in relation to the Earth. Stars moving towards the Earth (a star in a binary system in the part of its orbit in which it moves towards us, for example) appear bluer than they would otherwise because their spectral lines are shifted towards shorter wavelengths, at the blue end of the spectrum. This is known as a blueshift. Stars moving away from the Earth appear redder because their spectral lines are shifted towards longer wavelengths, at the red end of the spectrum. This is an optical redshift. It is different from the gravitational redshift discussed earlier, which is a consequence of the effect of gravity on light.

In 1929, the forty-year-old American astronomer Edwin Powell Hubble used the optical redshift to make a momentous discovery that contributed greatly to our understanding of the universe. For almost a decade he had been involved in the massive project of classifying galaxies in terms of their shapes and distances from the Earth, using the 100-inch Hooker Telescope at Mount Wilson Observatory near Pasadena, California – at that time the biggest and best telescope in the world. From his painstaking measurements he discovered that galaxies are redshifted – in other words, moving away from us – at a speed that depends on their distance. The farther away the galaxy, the faster it is moving. This is now known as the Hubble law. According to the Hubble law, the speed at which a galaxy is receding from us can be calculated by multiplying its distance from the Earth by the Hubble constant. A wealth of observations has now set the Hubble constant at around 70 kilometres (42 miles) per second per 19 million trillion miles (a megaparsec). In other words, for each 19 million trillion miles a galaxy is away from us, the speed at which it is receding increases by 70 kilometres per second.

Hubble's discovery was the first piece of observational evidence that the universe is expanding and so also for the Big Bang theory of how the universe was created, which astrophysicists began to formulate in the late 1960s. According to this theory, some 13 billion years ago space exploded out of a tiny, superdense, superhot state – a singularity in space and time – and has been expanding ever since, like the surface of a balloon. The galaxies are like spots· on the balloon's surface; they move farther and farther apart as the universe expands. In other words, the redshift is due not to the relative motion of the galaxies and the Earth, but to the expansion of the universe itself. It is what is called the cosmological redshift.[5]

Shorter wavelength radiation – gamma rays, X-rays, ultraviolet radiation and some infrared radiation – cannot penetrate the atmosphere, so astronomers observe it by using telescopes carried on orbiting satellites. Radio waves, however, at the opposite end of the spectrum, have long wavelengths and low energy and frequency. They pass through the Earth's atmosphere, so astronomers can study them from the ground. Radio astronomy began almost by accident in 1932, when Karl Jansky, working at the Bell Telephone Laboratory, was trying to identify some hissing static that was interfering with transatlantic telephone calls. It turned

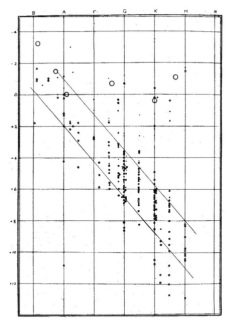

Figure 1(a). Henry Norris Russell's first HR diagram. Brightness is on the vertical axis, from dimmer (positive numbers) to brighter stars (negative numbers). The horizontal axis shows temperature, using the spectral classification OBAFGKM, with the highest temperature at the left. Most stars are on the diagonal strip – the main sequence. The lone star at the lower left is the white dwarf o² Eridani B, very hot yet very dim. (*Nature*, **252**, 1914)

Figure 1 (b). A modern HR diagram. Each dot represents a star whose spectral type and luminosity have been measured. Absolute magnitude (brightness) is on the vertical axis to the right. Luminosity, in units of the luminosity of the Sun (L$_\odot$), is on the vertical axis to the left. Surface temperature (degrees kelvin) is on the upper horizontal axis and spectral type on the lower one. The solid line defines the main sequence. White dwarfs (including Sirius B) are grouped on the lower left, giants (such as Arcturus) are on their own branch to the right of the main sequence, and supergiants (such as Betelgeuse) are above it. (Kaufmann and Freedman (1999), *Universe*, W. H. Freeman)

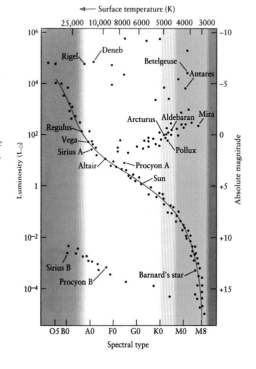

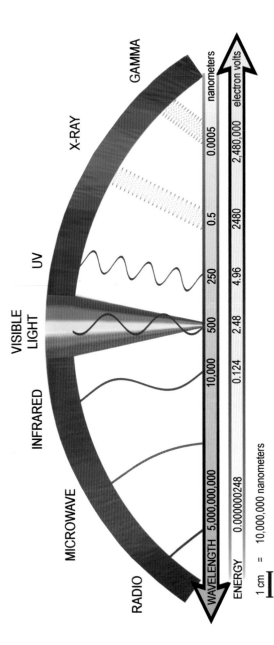

Figure 2. Electromagnetic spectrum showing the full spectrum of radiation. Radiation in the form of electric and magnetic waves travels through space at 186,000 miles per second. Short wavelength radiation, of high frequency and energy – gamma rays and X-rays – is on the right. Long wavelength radiation, of low frequency and energy – radio waves – is on the left. (NASA/Chandra X-ray Center/S. Lee)

RADIO   MICROWAVE   INFRARED   VISIBLE LIGHT   UV   X-RAY   GAMMA

| WAVELENGTH | 5,000,000,000 | | 10,000 | 500 | 250 | 0.5 | 0.0005 | nanometers |
| ENERGY | 0.000000248 | | 0.124 | 2.48 | 4.96 | 2480 | 2,480,000 | electron volts |

1 cm = 10,000,000 nanometers

Figure 3(a). Artist's rendition of the swirling gases that mark the event horizon of a black hole. The gas near the event horizon is ejecting a jet of high speed particles to alleviate the enormous build up of pressure. The event horizons of small black holes are several miles across; those of supermassive black holes are millions of miles across. (NASA/Chandra X-ray Center/A. Hobart)

Figure 3(b). Artist's rendition of a star (orange circle) that wandered too close to a supermassive black hole whose enormous gravity stretched it (arrow), tearing it apart. The white stream indicates the doomed star's mass swallowed by the black hole. Data from the Chandra X-ray Observatory, XMM-Newton and the X-ray satellite ROSAT, provided evidence of this occurring in galaxy RX J1242-11 in the constellation Virgo. (NASA/Chandra X-ray Center/M. Weiss)

Figure 4(a). Artist's version of the Chandra X-ray Observatory in orbit. (Marshall Space Flight Center)

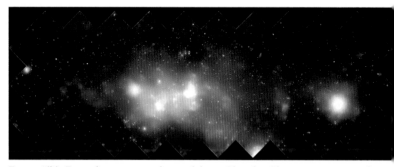

Figure 4(b). Deep sky survey, made up of thirty Chandra Observatory images taken in July 2001, showing an oblong section of the core of the Milky Way, 2400 trillion by 5400 trillion miles across, showing hundreds of white dwarf stars, neutron stars and black holes. Sagittarius A*, the supermassive black hole in the middle of the galaxy is inside the bright white patch in the centre of the picture. There are no true colours in the original. The colours are computer generated to differentiate the types of radiation emitted by the stars in a process called 'false colour' imagery. (NASA/Umass/D. Wang et al.)

Figure 5(a). Sirius A and Sirius B. Sirius A, the brightest star in the night sky, all but obscures its hot but dim companion, the white dwarf Sirius B, which can be seen faintly to the lower right. (R. B. Minton)

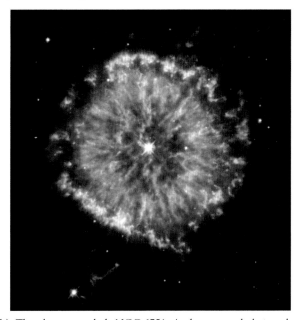

Figure 5(b). The planetary nebula NGC 6751. A planetary nebula is a glowing cloud of gas formed by a relatively low mass star like the Sun throwing off its outer layers towards the end of its life. NGC 6751 was formed several thousand years ago in the constellation Aquila, 40,000 trillion miles away, and is 5 trillion miles across, six hundred times larger than our solar system. The hot core will cool down and become a white dwarf, eventually becoming a diamond in the sky. (NASA/The Hubble Heritage Team (STScI/AURA))

Figure 6(a). The Crab nebula, created in a supernova explosion more than 6500 years ago. At the centre is a pulsar, or neutron star, rotating thirty times per second. This photograph was taken by the Very Large Telescope in Cerro Paranal, Chile. (FORS Team, 8.2-metre VLT, European Southern Observatory)

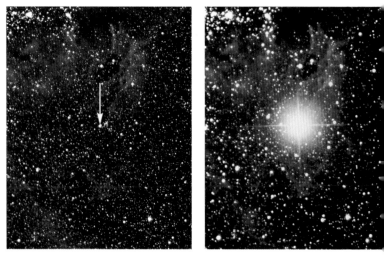

Figure 6(b). Supernova 1987A (SN 1987A). The picture on the left shows the blue supergiant B3 in the Large Magellanic Cloud just before it exploded and became Supernova 1987A. The picture on the right shows the same star after it went supernova. (Anglo-Australian Observatory)

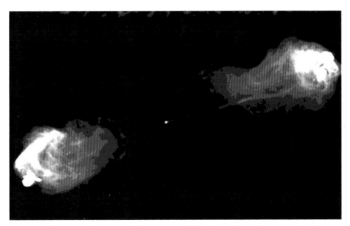

Figure 7(a). A radio map of the powerful radio galaxy Cygnus A taken by the Very Large Array, Soccoro, New Mexico. There is a compact radio source or quasar at the centre of the galaxy which ejects energetic jets of particles that feed two radio lobes three million trillion miles apart. The map shows a jet shooting towards the radio lobe on the right. (Perley, R. A., Dreher, J. W. and Cowan, J. J. (1984), *Astrophys. J.* **285**, L,35)

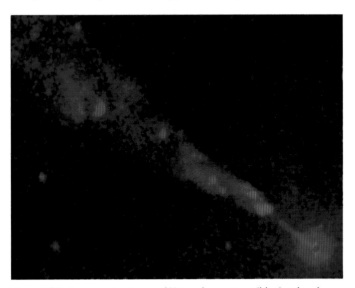

Figure 7(b). A composite image of X-ray observations (blue), taken by the Chandra X-ray Observatory and radio observations (red) taken by the Very Large Array, Soccoro, New Mexico, showing a 24,000 trillion mile long jet from the radio source Centaurus A, 70 million trillion miles away from us. The jet comes from the black hole at the lower right-hand corner of the picture and is moving at about half the speed of light. (NASA/Chandra X-ray Center/U. Bristol/M. Hardcastle)

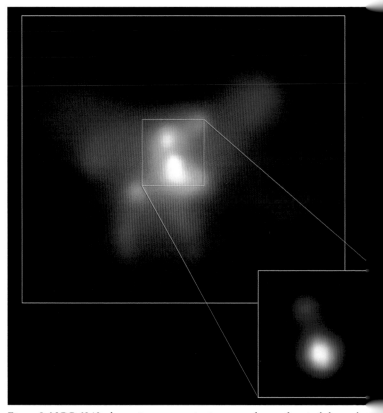

Figure 8. NGC 6240, shown in a composite image made up of optical data taken by the Hubble Space Telescope and X-ray data (blue) taken by the Chandra X-ray Observatory. The X-ray image shows powerful X-rays emitted by the gases around the two supermassive black holes at the centre of the galaxy. (NASA/Chandra X-ray Center/MPE/S. Komossa et al.)

out to be radio waves that came from the central portion of the Milky Way and were far stronger than any from the Sun. Initially radio astronomers used a concave dish that focused radio waves into its centre, where they were scooped up by an antenna. But after World War II, in which radar played a vital part, radio astronomy had become radically more sophisticated, with highly trained operators able to use much improved instruments.[6] Radio astronomy was to be responsible for one of the most ground-breaking discoveries of all.

When charged particles accelerate, they emit radiation. High-speed electrons spiralling through a galaxy's magnetic field emit long-wavelength radiation – radio waves. As a result, every galaxy emits radio waves. Some galaxies, however, are more powerful emitters than others and give out radio and light waves of equal intensity, which is highly unusual: most galaxies mainly give out light. The most intense radio waves from distant galaxies are a billion times stronger than the radio waves from the Milky Way. By the late 1940s, astronomers had detected three different sources of exceptionally strong radio wave emissions, far more powerful than any that had ever been detected from a galaxy. They pinned them down to sources in certain constellations. But they were a complete puzzle. No one had any idea what they could be.[7]

Walter Baade (Zwicky's one-time collaborator) and the fifty-six-year-old German-born astrophysicist Rudolf Minkowski decided to try to locate the optical counterparts of these mysterious sources.[8] In 1954, using what was now the largest optical telescope in the world, the 200-inch Hale Telescope at the Palomar Observatory on Palomar Mountain, they found an object which they realised could not be a star, whose location seemed to coincide with one of the sources of strong radio wave emissions, in the constellation Cygnus, outside our galaxy. They called it Cygnus A. Its redshift indicated that it was speeding away from us at 10,200 miles per second – 5 per cent of the speed of light. Applying the Hubble law showed that it was an astounding 4500 million trillion miles away.[9] Sirius, the brightest star that can be seen with the naked eye in the night sky, is 52 trillion miles away, while the Sun is a mere 93 million miles. If Cygnus A were closer and radio waves could be seen, it would have a radio 'luminosity'

10 million times that of the entire Milky Way. It is in fact the strongest source of radio waves in the northern sky.

Studying their photographic plates, Baade and Minkowski suggested that the source of this barrage of radio waves might be two galaxies colliding. But galaxies would need to be colliding again and again, with enormous violence, to maintain such a huge output of radio waves, which made the idea dubious.[10] Around that time, astronomers using radio telescopes at Jodrell Bank in England noticed that the radio emissions came from two widely separated lobes. It appeared that Cygnus A was made up of an extremely high-energy, central 'star-like' object with a radio lobe on either side.

By 1963 astronomers had made two other oddball sightings using radio telescopes. They were catalogued as 3C 48 and 3C 273.[11] As with Cygnus A, astronomers found optical counterparts by using optical telescopes. But these were far more mysterious even than Cygnus A. Certain of the lines in their spectra made absolutely no sense: they did not represent any known elements.

Perhaps this meant that there were entirely new elements in the universe. Maarten Schmidt and another astronomer at Caltech, Jesse L. Greenstein, a contemporary and former colleague of Chandra's at Yerkes, often discussed this possibility with Willy Fowler, Caltech's expert on the chemistry of stars. Usually scientists discuss over coffee. With Fowler it was martinis. He kept saying that this 'new element business' made no sense.[12] It was Schmidt who realised that the mysterious spectral lines in 3C 273 could be lines from the light emitted by hydrogen atoms but with their wavelengths redshifted by an incredible 16 per cent. If this were true, it would mean that 3C 273 was 12 billion trillion miles away from us, even more distant than Cygnus A, and receding at 25,000 miles per second, 15 per cent of the speed of light.[13] He put forward his theory rather tentatively, saying only that explaining this phenomenon in extragalactic terms seemed the most 'direct and least objectionable'.[14]

Within days, Greenstein and Thomas A. Matthews at Caltech had worked out that the apparently indecipherable spectral lines of 3C 48 were redshifted by 37 per cent. This would mean that

3C 48 was receding from us at 46,500 miles per second, 25 per cent of the velocity of light, and, from the Hubble law, was 24 billion trillion miles away.[15] In other words, it would take light from these stars over a billion years to reach us. Looking at these stars was like glimpsing the universe over a billion years ago.

Given that they were so far away and yet could still be photographed, 3C 48 and 3C 273 also had to be enormously bright. Indeed they were. 3C 273 turned out to be a thousand times brighter than its entire galaxy, and completely obscured it. It was as bright as 25 trillion Suns. This star-like object turned out to be right in the nucleus of a fast-moving galaxy, and was a hundred times brighter than the luminous galaxies surrounding radio sources such as Cygnus A. Schmidt also noticed a wisp of a jet, as he put it, extending from the star-like object towards one of the lobes. A close connection between the jet and the star-like object, he mused, would be 'intriguing'.[16]

These extraordinarily bright star-like objects have as much energy as would be produced by applying Einstein's equation, $E = mc^2$, to the mass of a billion Suns, a galaxy's worth of stars. Astronomers measured variations in their output, and by 1965 had discovered that 3C 273 generated its energy – produced its 'light', as it were – within a region a trillion miles across. The radiation from star-like objects such as 3C 273 differed dramatically from that of ordinary stars. It consisted almost entirely of very high-energy gamma radiation and X-rays, emitted from a material hotter than 100,000 degrees kelvin.

One possible explanation was gravitational redshift, in which intense gravity can increase the wavelength of light, making it look redder. Greenstein and Schmidt examined this hypothesis and concluded that it was very unlikely that the incredible redshifts for 3C 48 and 3C 273 could be produced by gravitational means. The stars would have to be so hyperdense to produce such extremely high gravity that it would strain the laws of physics, at least as they were known at the time.[17] They also came up with a name for these star-like objects. They dubbed them 'quasi-stellar objects', which they shortened to 'quasars'. But they had not completely discounted the gravity hypothesis. Perhaps there was some other condition under which a gravitational redshift might take place. And there was no agreement at all on the source of the quasars' enormous power.

Perhaps, exaggerating Baade and Minkowski's original suggestion, the energy was generated by several galaxies colliding. But this would require an impossibly high efficiency rate for the conversion of mass into energy. Or could it be that 100 million closely packed stars on the verge of going supernova, each thirty to fifty times the mass of the Sun, suddenly exploded in a chain reaction? In 1967, Stirling Colgate formulated a model along these lines and demonstrated that, incredible though such a celestial fireworks display might sound, it agreed fairly well with the data. It was an ingenious idea, but it was fiendishly difficult to work out the details.[18]

Hoyle and Fowler, meanwhile, had been working on a simpler but very dramatic model.[19] Since the radio waves seemed to emanate from the centre of the galaxy, perhaps a quasar was some sort of supermassive star. This star, formed from the dense cloud of gas in the centre of a galaxy in the usual way, might have become so massive that the pressure from its radiation exceeded the pressure of gravity pushing inwards. At this point it would have to stop taking in gas or it would explode. Fowler and Hoyle suggested that these supermassive stars must be 100 million times the mass of the Sun. After each period of burning, the star would collapse under the pressure of its own gravity. Perhaps, then, supermassive stars undergoing gravitational collapse were the source of the quasars' prodigious energy. It was these two ideas – Schmidt's theory of quasars and Hoyle and Fowler's suggestion that the source of their energy might be a massive star which had collapsed to close to the size of its Schwarzschild radius – that inspired Penrose to look more deeply into the mathematics of a star's collapse.

One day, walking across the Caltech campus, Fowler bumped into his colleague Richard Feynman. Feynman was a brilliant forty-five-year-old physicist who, two years later, in 1965, was to share a Nobel prize for figuring out a cure for the ills of quantum electrodynamics, which had eluded Dirac, Heisenberg and Pauli (the theory had predicted, among other things, that the electron has an infinitely large mass). Feynman cut straight to the chase.

'Willy, you know those supermassive objects that you and Fred have been working on are unstable,' he said. 'They'll collapse due to general relativity.' Hoyle and Fowler had been using Newton's theory of gravity in their model. But the general theory of relativity gives gas molecules a far stronger gravitational attraction than Newton's theory does. This meant that as a supermassive star contracted while letting off energy in a cosmic belch, it would not be able to stop contracting but would carry on collapsing indefinitely. Fowler ran back to his office. He 'immediately put general relativity in [and] found out rightaway that Dick was right, and the damn things would collapse'.[20] He tried to save the theory by varying the amount of nuclear fuel burned and making the star more realistic by making it rotate, 'but it really wasn't enough'.[21]

At this point, Chandra had turned his attention to relativity theory, a field he had long avoided. His decision came in 1960, at the end of a long period studying hydrodynamic and hydromagnetic stability, which he had closed with another of his famous tomes, published in 1961, and titled, not surprisingly, *Hydrodynamic and Hydromagnetic Stability*. Chandra was not happy with the way things had turned out. The calculations had been long and hard, and he felt that he had wasted ten years working on something that was ultimately inconsequential. He decided to turn to 'more contemplative matters'. But it was more than that – it was 'something to do with vanity'. At this point he wanted to try something that for years had been on his mind: working on pure theoretical physics, like Dirac. 'There's not a single table in any of his papers,' Chandra remembered thinking.[22] But he was hesitant. Could he do it? Could he come up with significant ideas? He confided in his friend and colleague in the physics department, Gregor Wentzel. Wentzel urged him to take the leap. After all, he told him, 'they can't fire you'.[23]

In 1962, to bring himself up to date on current developments in the field, Chandra attended a meeting in Warsaw on the general theory of relativity. There he discussed astrophysics with Zel'dovich, whom he had not seen since his trip to Russia twenty-eight years earlier. He came away feeling that there was a place for someone with extensive experience in astrophysics who could formulate problems 'whose solution might be useful'.[24]

It was at that moment that Willy Fowler's predicament about the stability of supermassive stars came up. When a physicist of Feynman's stature comments on a topic on the edge of science, news spreads immediately. Chandra was familiar with Fowler's computer calculations in which stars 100 million times the mass of the Sun burn up their nuclear fuel, then collapse. He had also heard about Feynman's intuitive response and Fowler's desperate attempts to rescue his theory. Chandra set to work. The topic struck a chord in him, taking him back to his days in Cambridge thirty years earlier. It was like the task that Eddington had set himself in 1916: to explore the stability of stars going through periods of bulging followed by compression. Chandra simply applied the general theory of relativity, and thus, he claimed, obtained results Eddington could have found decades earlier.

In 1964, a year after Feynman's flash of intuition, Chandra published his first papers on the general theory of relativity.[25] He proved with mathematical certainty that if a star 100 million times the mass of the Sun began to pulsate, it would become unstable. Once it had pulsed down to the estimated size of a quasar, 100 billion miles across, it would have to collapse completely. Thus Hoyle and Fowler's postulated supermassive star could not be the source of the quasars' energy. Chandra was delighted with this result. He was, he declared, 'pretty convinced' that singularities – stars of infinite smallness and infinite density – really could exist, and that enormous stars, above the upper limit he had established for the mass of a white dwarf, could actually shrivel away to nothingness and disappear into a niche in space and time.[26]

That December, all the major players in quasar research gathered for the Second Texas Symposium for Relativistic Astrophysics at the University of Texas in Austin. Information and ideas were flooding in. Session followed session where Fowler and his coworkers 'presented another talk with yet another theory to fit the rapidly expanding data base. Some of their theories put the quasars at the edge of the solar system, others at the edge of the universe.'[27] One day redshifts were gravitational; the next, cosmological. There were different theories for different data at different times of day.

Fast jets served Dallas, but delegates had to travel from Dallas on to Austin by prop-driven plane. They called it the 'milk run'. Heading back to Dallas on one of the planes after the end of the meeting, Richard White found himself among such luminaries of astrophysics as Chandra, Fowler, Wheeler and Schmidt. Soon after take-off, when the seat-belt sign went off, Fowler stood up, cupped his hand over his brow in a dramatic gesture, scanned the passengers in their cramped seats and announced in his booming voice, 'If this plane were to crash, we could get a new start on this quasar problem.'[28]

The mystery of quasars and what made them so enormously bright had still not been solved. Then, in 1964, Edwin Salpeter at Cornell University and Zel'dovich and his co-workers in Russia proposed a scenario in which an object greater than a million times the mass of the Sun might move through an interstellar gas cloud, scooping up gas particles. Eventually this supermassive star would undergo gravitational collapse through its event horizon, collapsing until it was an unimaginably tiny dense point. Then it would drag space and time in around it, continuing to move along, scooping up gas particles like a cosmic vacuum cleaner. Accelerated to speeds close to that of light as they were sucked over the event horizon, particles near the edge would crash into each other, heat up and emit enormous amounts of radiation in the form of X-rays. Salpeter and Zel'dovich had discovered a way in which huge amounts of radiation could be generated, far more than in a nuclear reaction, as much as the energy radiated from an entire galaxy.

All the different lines of research in different fields were converging. A year later, in 1965, Penrose produced his elegant topological proof that if a star disappears through its event horizon, it is bound to keep disappearing for ever. Then, in 1966, Colgate and White used computer simulations to investigate the actual mechanisms by which stars could collapse. It was John Wheeler who, in 1967, found the perfect and deeply memorable term to describe the region of space into which collapsed stars fall: a 'black hole'.[29] In other words, quasars were powered by black holes. At the centre of a quasar was a black hole, with particles on the verge of tumbling over its event horizon emitting enormous amounts of radiation. No one could doubt any longer that black holes existed. The tempo of research quickened.

In 1969, Donald Lynden-Bell, an English astrophysicist at the Royal Greenwich Observatory in Sussex, took Salpeter and Zel'dovich's model and applied it to the centre of a galaxy. He suggested that a massive star that completely collapsed through its event horizon would become such a strong source of gravity that it would drag everything in its vicinity into a disk that would spin around it. It suddenly became obvious that black holes too must be spinning. As massive stars blow off mass and grow smaller, they spin faster and faster. If this were the case, then black holes were not merely receptacles for dead stars – they had structure. As it happened, the mathematics of a spinning black hole had already been proposed, in 1963, by a twenty-nine-year-old mathematician from New Zealand, Roy Kerr. He had done a Ph.D. at Cambridge and was on the staff of the University of Texas in Austin.

Kerr applied Einstein's general theory of relativity to space and time around a spinning spherical object such as a star.[30] His solution provides a way of measuring distances and times in the space distorted by the object and is referred to as the Kerr metric, a metric being a measure. The metric we use in everyday life is based on Euclidean geometry, the geometry we learn in school, which assumes that space is flat and three-dimensional.[31] The Kerr metric tells us the structure of the curved, four-dimensional geometry described by the general theory of relativity, of the three dimensions of space plus time. Schwarzschild's solution of Einstein's general theory of relativity – the Schwarzschild metric – described space and time around a non-spinning spherical object and, as Oppenheimer and Snyder realised in 1939, is actually the metric for the region around a non-spinning black hole. Thus spinning black holes are referred to as Kerr black holes, and non-spinning ones as Schwarzschild black holes. Kerr's metric becomes Schwarzschild's when the black hole's rotation is set to zero.

The power of Kerr's solution was that it applied to every possible black hole. Its beauty lay in its simplicity, in that its description of every black hole requires only two parameters, the black hole's mass and its spin. The mass of a black hole is the mass of the star and of the material that has fallen into it, straightforward to measure if the black hole is in a binary system; its spin is the amount of times it rotates per second, thousands of times if it results from the collapse of a huge star.

But how was such a simple classification possible? After all, each spinning, collapsing star had different surface characteristics. Yet after the star collapsed through its event horizon, the event horizon always turned out to be perfectly smooth, like water after a pebble has fallen into it. Every physical characteristic of the star that fell through, momentarily deforming space and time, vanished except for its mass and spin. These were left behind, a bit like the Cheshire Cat's smile. This astounding discovery was made in the 1960s, primarily by Zel'dovich and his group. Once again, Wheeler found memorable words for it: 'black holes have no hair'.[32] No matter what sort of star fell through its event horizon, it left no trace of its previous idiosyncrasies.

Penrose's cosmic censorship hypothesis went on to place a limit on a black hole's mass and spin. If the black hole was spinning too fast, it would make the rim 'roll down' to reveal the singularity within, which was unacceptable according to the laws of physics. Thus black holes, among the most massive and powerful objects in the universe, turn out to be elegantly simple.

Chandra was eager to join the party. In autumn 1971 he took research leave to spend three months with one of the most active groups in the field. This was headed by Kip Thorne, by then thirty-one and Professor of Theoretical Physics at Caltech. Thorne had been involved in black hole physics right from the start, and was now at the leading edge of research. He had studied under Wheeler, and was one of the authors of *Gravitational Theory and Gravitational Collapse*. Six feet tall, rail-thin and wiry, with long hair and a beard, Thorne exudes the assurance of an expert, moving with ease from the complex mathematical theory of black holes to highly technical aspects of the equipment used to detect them. He imparts to his students a 'we're all in it together' sense of teamwork.

Thorne remembers Chandra affectionately. Always immaculately dressed in his conservative 'Chandra grey' suit, he would eagerly join Thorne and his scruffy graduate students for lunch in the student cafeteria (the 'Greasy'), rather than dine at the much more formal and elegant Athenaeum, Caltech's famous faculty club where eminent guests such as Einstein stayed. He was always in his office bright and early with his door open, ready to talk to any astronomer who passed by.

Two of Thorne's graduate students, William H. Press and Saul Teukolsky, drew Chandra's attention to Kerr's result. He was stunned by the power of its mathematics:

> In my entire scientific life, extending over forty-five years, the most shattering experience has been the realisation that [Kerr's] exact solution of Einstein's equations of general relativity provides the *absolutely exact representation* of untold numbers of massive black holes that populate the universe. This 'shuddering before the beautiful', this incredible fact that a discovery motivated by a search after the beautiful in mathematics should find its exact replica in Nature, persuades me to say that beauty is that to which the human mind responds at its deepest and most profound.[33]

Teukolsky went on to look for a way to calculate how spinning black holes interact with their surroundings when, for example, they are hit by electromagnetic waves (i.e. light waves) or by gravitational waves.[34] The latter are the ripples in space and time caused by the movement of objects in it; they are predicted by general relativity, but have yet to be observed directly.[35] Cutting through the complications of Teukolsky's work, Chandra sensed the glimmer of a connection between the structure of black holes, electromagnetism and gravitational waves. Teukolsky recalled Chandra's intense determination to understand this connection in as complete a way as possible. It set Chandra off on a long journey which involved him in some of his most complex calculations ever.

By then, the perception of black holes had been transformed. Far from being rejected as ugly, ungainly objects that spoiled a harmonious universe, they had come to be seen as the 'most perfect macroscopic objects there are'.[36] It was an utter reversal of the way things had been on 11 January 1935. As Chandra put it when he began his research, quoting Eddington's famous dismissal with mingled bitterness and irony, 'we are today considering seriously situations that were brushed aside as *reductio ad absurdum* not so long ago'.[37]

On 5 August 1971, the spaceship Apollo 15 was about to start its long journey home from the Moon. This was the first manned scientific expedition into outer space. Its crew had already walked on the Moon and circled the Earth outside their spaceship, carrying out repairs. But as well as being pioneers of exploration, their task was to carry out a variety of scientific experiments. That day, Colonel David R. Scott and his crew were tracing the sources of X-rays passing through space. They had identified rapid irregular bursts of X-rays coming from Cygnus X-1, an X-ray star 20,000 trillion miles away that was the companion star of a blue supergiant catalogued as HDE 226868, in the constellation Cygnus.[38] Their results were to be incorporated into observations made since the previous year in a scientific spin-off of the cold war.[39]

After the Soviets tested the massive Czar bomb in 1961, the United States launched surveillance satellites that would detect X-rays from Soviet hydrogen bomb tests. Circling the Earth, these satellites began to pick up all sorts of surprising signals and information, such as intense X-rays which seemed to be coming from far outside our galaxy. On 12 December 1970, NASA launched SAS-1 (the first of three Small Astronomical Satellites) from the San Marco Platform in Kenya to continue steady surveillance of these X-rays. It was renamed Uhuru, the Swahili word for 'freedom'. It homed in on the source of the X-rays, which turned out to be Cygnus X-1. But whereas Uhuru could study this source for only two minutes at a time, the Apollo 15 crew could monitor Cygnus X-1 for an hour.[40]

What first drew scientists' attention to Cygnus X-1 was how extremely variable its X-ray bursts were. Some continued for months, others were as short as a thousandth of a second. Since an object flickers for exactly the length of time it takes light to travel across it, and the speed of light is always the same, scientists can work out the size of an object from the longest duration of light it emits. They calculated that the X-rays were coming from a compact object about 1800 miles across at most – less than a quarter the size of the Earth. What tiny object could be generating so much power? From comparison with other blue supergiants, astrophysicists had estimated that HDE 226868 was about thirty times the mass of the Sun. As Cygnus X-1 was its binary companion, taking account of the shape of their orbit and the mass of HDE 226868, they deduced that it was at least seven times the mass of the Sun. That made it too big to be either a white dwarf (maximum mass, 1.4 times the mass of the Sun) or a neutron star (maximum mass, two to three times the mass of the Sun). With growing excitement they realised that Cygnus X-1 had virtually all the necessary characteristics of a black hole.

Today, most agree that Cygnus X-1 is indeed a black hole, relentlessly devouring its companion star. The 'mouth' of the black hole itself, its event horizon, is just 26 miles across. Deep inside is the collapsing star, continuing its eternal demise. The matter that the black hole sucks from its blue supergiant companion forms an 'accretion disk', a broad flat disk of matter revolving around it like a vast compact disk in a CD player. The X-rays come not from the black hole – no radiation can escape from there – but from the gas in the accretion disk. The inner parts of the disk rotate faster than the outer ones, and as the inner particles rub against their slower neighbours, they heat up from the friction and give off X-rays. Matter from the accretion disk is continually being dragged into the black hole like water down a drain, from the inside outwards. By observing the spectrum of the X-rays, astrophysicists calculate that it comes from a gaseous accretion disk some 4 million miles across. The hottest part – the portion emitting the X-rays – is some 300 miles across. Today we take this scenario for granted. But for decades, research was impeded because astrophysicists simply refused to believe that stars could disappear.

Chandra was aware that black hole physics would not be an easy field to enter. It was already full of brilliant and experienced young men – added to which, by the time he started writing papers on the subject in 1975, it seemed that the most exciting discoveries had been made and the golden years of black hole physics were over. All that remained was to clarify and systematise the details. This was, however, exactly what Chandra liked best. Instead of

juggling numbers, he would be working in pure mathematics, dealing with symbols without involving 'a single table' of figures, as in
Dirac's work.

Chandra began by streamlining the subject and creating a
direct and simple derivation of Kerr's metric. It was quite an
accomplishment. In the opinion of experts, including Kerr himself, it could not have been done without horrendously complex
calculations.[41] The next step was to establish a method of
analysing what might happen when a rotating black hole is hit by
electromagnetic or gravitational waves, and how that black hole
might affect a particle moving near it. Scientists usually begin by
taking only small changes into account and initially discarding
larger ones. Chandra took the reverse path and 'just bull-dozed in',
as he had described Bethe's approach to physics problems.[42]

In the 1980s, further mathematical developments of Kerr's solution and better data from observations of quasars finally enabled
astrophysicists to come up with a unified description valid for all
sources of radio waves, from weaker radio galaxies to quasars. It
was a stunning marriage of general relativity and theoretical and
observational astrophysics. Quasars, it seems, are powered by a
supermassive spinning black hole, accompanied by its accretion
disk, which is also rotating. (The black hole powering the quasar
3C 273 is a colossal billion times the mass of the Sun.) As the disk
spins, it compresses the gases that make it up. This, together with
the gas particles rubbing against their slower neighbours, generates
unimaginably high temperatures. The highest-energy radiation
comes from the innermost part of the accretion disk, which spins
fastest and is thus hottest. It is about a trillion miles across. As its
particles are sucked into the black hole in 3C 273, which is about
a billion miles across, it gives off gamma rays and X-rays.

In recent years, astronomers have been able to make many
observations of quasars using radio, X-ray and infrared devices.
There are, it appears, two lobes of gas on either side of an active
galaxy, or quasar, at whose centre is a supermassive black hole
surrounded by an accretion disk. Shooting out of the inner part of
the disk are two back-to-back streams, narrow jets of electrons
travelling at speeds close to that of light and emitting radio waves.
These jets can extend over vast reaches of space – a hundred

thousand trillion miles, five to ten times bigger than the entire galaxy in which the quasar is located. As the jets ram into the lobes and excite the atoms in the gas, the lobes in turn emit radio waves. The biggest quasars are so intensely luminous that they obscure the stars in the host galaxy so that there appears to be just a single star, one quasi-stellar object.[43]

Quasars are found only in older galaxies, far, far away in both space and time. Over time, the accretion disk around a spinning black hole uses up all its gas particles, and the quasar ceases to emit light – it shuts down. We can only conclude that there must be millions of black holes out there. Thanks to these developments, beginning in the 1960s, Chandra's discovery of 1930 was finally vindicated, and Eddington proved wrong.

13

'Shuddering Before the Beautiful'

On 6 February 1960, Babuji died of a sudden heart attack. He was seventy-five. Chandra had not seen his father since 1951, and that had been on his first visit home in fifteen years. Babuji had always regretted that his brilliant son never returned to live in India, which led to periods of great strain in their relationship. It had all come to a head in 1953, when Lalitha and Chandra were pondering whether or not to become naturalised American citizens. It was a huge decision for them because it would mean giving up their Indian citizenship. There were practical reasons for it, such as Chandra's difficulties in leaving and returning to the United States on an Indian passport. On a deeper level, having noticed the poor research conditions in India during his visit and been reminded of the savage academic rivalries there, he had become convinced that it would be very hard to return to the country of his birth.

America was becoming a more liberal place. In 1952, Adlai Stevenson, a brilliant and enlightened Democrat, campaigned for the presidency on a platform that included ridding the country of McCarthyism. The Chandrasekhars were caught up in the liberal fervour. They threw themselves into Democratic party activities in Williams Bay. Then, in December, the Immigration and Naturalisation Act was passed, allowing a small quota of people of Asian origin to become US citizens. After much deliberation the Chandrasekhars decided to go through the arduous procedure, and in October 1953 they became Americans.

For Babuji the news was a slap in the face, both for himself and for India. He accused Chandra of turning his back on his homeland and his Hindu heritage. Deeply hurt, the Chandrasekhars pointed out that they had lived in the United States for seventeen years and intended to remain there. They had discussed the matter with the Fermis, the Kuipers and the Schwarzschilds, all of whom had chosen to become naturalised. Chandra and Lalitha intended to keep close ties with India, just as their friends did with their native countries. Nothing helped – Babuji was deeply offended. There was no further communication with him for several years.

And now he was dead. For Chandra, the news must have come as a shock. No doubt he thought back to his lonely days at Cambridge, when Babuji had lent a sympathetic ear. Perhaps he remembered too the emotional evening in London when Babuji was about to return to India, when Chandra prayed to his God to make him a worthy son. As soon as he could, a year after Babuji's death, he made the long journey to India. Back at Chandra Vilas he sought out all the letters he had sent to his father, right back from when he was at Presidency College.

Besides the pain of Babuji's death, Chandra also anguished about changing subjects. He was now concentrating his efforts on general relativity. But the field was darkened for him because one of the dominant figures in relativity studies had been his demon, Eddington. Then, one Sunday morning in September 1974, when he was sixty-three, he felt an enormous tightening in his chest. Lalitha rushed him to the hospital, where he had an angina attack. Chandra recalled the doctor's assessment of his condition: 'On a scale of 5, 5 being fatal, mine was $3\frac{1}{2}$.'[1] Had he not been in the intensive care unit at that moment, he would have died. Forcefully reminded of his own mortality, Babuji's death came back to him. Three years later, Chandra had another heart attack and underwent an extremely risky triple bypass operation. The medical staff suspected there would be post-operative problems, and indeed they came to a head early the next morning. Excessive bleeding had caused a dangerous build-up of blood in his lungs, and an emergency procedure was necessary. Assuming that everything was safely in hand, Lalitha had already gone home, and Chandra had to be roused to sign the permission slip himself. A lengthy convalescence followed. 'It was extremely painful,' he remembered with anguish.[2] He did not fully recover his health until 1980. His work on black holes kept him going, but the years of gloom and disappointment were taking their toll. Chandra had always kept an informal diary of the problems he was working on and his scientific progress. Now his notes took on a more painfully introspective tone.

On 20 January 1976, he copied out some lines from Gustav Mahler's *Das Lied von der Erde* ('The Song of the Earth'), which must have seemed to resonate with his own deepest feelings:

> *My fortune was not kind to me in this world!*
> *Whither I go? I go, I wander in the mountains,*
> *I seek rest for my lonely heart!*[3]

That November, he was agonising over his research into black holes. He was trying to disentangle some lengthy equations, but they did not yield.[4] Dispirited, he wrote:

> I feel utterly discouraged, despondent, and lonely. Life has lost its savour. And I am desperately tired. I wish that I could overcome the present impasse in my work: I have been up against a stone wall for nearly three months . . . How I long for the time when the book will have been written; and I with a sigh of relief leave this university!; and go into exile like the saints of old. I wish, I wish but so many vain things continue to haunt me.[5]

Chandra was beginning to assemble his black hole results into a book which he would entitle *The Mathematical Theory of Black Holes*. But he still tormented himself over whether it was worth it. Could he have lost his passion for research? By then he had received every possible important scientific award except the highest of all, a Nobel prize. His work on white dwarfs had been recognised in 1974, when the American Physical Society presented him with the Dannie Heinemann Award for excellence in mathematical physics, always his 'first love'. He had also received the prestigious Gold Medal of the Royal Astronomical Society, and the Astronomical Society of the Pacific's Bruce Medal. But he still complained that these were neither fulfilling nor encouraging 'in the kind of work that I considered most relevant' – his discovery of the upper limit to the mass of a white dwarf, which was given only brief mention in the award citations.[6]

Four years on, Chandra's mood continued to be dark. Just before his seventieth birthday, on 18 October 1980, he was writing, 'Why has the optimism of the forties, and even the remnants of the sixties, dissipated? How is it that my outlook has become dark and pessimistic: with no hope, no joy . . .'.[7] The next day he meticulously wrote a couple of lines in his neat handwriting on the same page:

> Oh!
> *Let me not imagine vain things*
> *Let my consolation be my book.*

Chandra found it impossible to hide his deep sadness. Once, in the 1980s, a physics department meeting finished earlier than expected. The professors relaxed, and one by one began to reminisce. When it came to Chandra's turn, his eyes filled with tears. He felt, he confessed, only regrets about his life. Perhaps it had all been in vain. 'I am afraid that by giving to the pursuit of science the highest priority, one necessarily distorts one's personal life,' he said. 'That includes my marriage, in the sense that life has been very hard for my wife . . . I had, in fact, no time for anything else [besides science].'[8] He often referred to Lalitha as his 'noble wife', as indeed she was. She had consciously made the decision to give up her own career in physics to make life as comfortable as possible for her husband, so that he could concentrate entirely on his research. She took care of all the household tasks except shopping, which he enjoyed. As she says, rather sadly, gazing wistfully into the distance, 'so many things could have been different'.[9] She had dreamt of working with her husband and keeping up to date with developments in science. But that proved impossible.

'What is the point of it all? I feel so cheated,' Chandra scribbled in his diary.[10] He seemed convinced of the futility of existence. The only thing was to carry on working obsessively, even though he might question whether ultimately what he was doing had any meaning. Work became his salvation, helping him through the bouts of severe depression that followed his heart attack. He seemed to believe, like Picasso, that as long as he kept busy he would not die. Then, late in 1982, he finished his book on black holes, and the futility of existence overwhelmed him once again. He recorded the emotional ups and downs of the last half century in a long diary entry:

> During the 40s, 50s, and 60s, I did have occasional spells of depression and gloom. But the overriding anxiety to be scientifically productive subdued those periods; besides, certain hopes I had were realised. And on the whole, I maintained a feeling of wholesome well being. But since 1969, I have worked and lived with essentially no encouragement; and whatever hopes I may have had, were shattered . . . And nothing assuaged the enveloping gloom. Still the persistence to be productive kept me going. The only bright interludes were brief, few, and far between – as during May 1976 when I separated Dirac's equation and February 1978 when I decoupled the Newman–Penrose equations governing the perturbations of the Reissner–Nördstrom black hole. I escaped the gloom of the 70s during 1980–82 by my preoccupation with the book. I cannot imagine what those years might have been had I not had the book constantly weighing on me. Now that the anodyne that kept me going during the past three years has been exhausted, I am at a complete loss; and I do not have even the desire to be productive. The book seems to have taken revenge and destroyed my motivations. The future is bleak and there is no prospect even for a moment of joy. I am against a dark impenetrable wall; and I foresee no release.[11]

The weight of memory began to take its toll. It was not just the usual malaise that every writer feels on completing a book. Something else was preying on his mind. Perhaps in his study of black holes he had hoped to fulfil his dream of becoming a theoretical physicist, like Dirac. But his work had been inconclusive and not well received, and recognition had come late in life.

In *The Mathematical Theory of Black Holes*, Chandra often writes that he is seeking the 'simplest solution'. But in the end both style and simplicity eluded him. This is especially evident in the monumentally complex calculations by which he solved the problem of what happens to a black hole hit by gravitational waves. It tormented him that he was unable to arrive at the solution by simpler means. Perhaps in the future, he wrote, the 'complexity will be unravelled by deeper insights'. In the meantime, 'the analysis has led us into a realm of the rococo: splendours, joyful, and immensely ornate'.[12] For the detail of the calculations he refers the reader to the 600 folio-sized pages and six notebooks which he put on deposit in the Joseph Regenstein Library at the University of Chicago. They are awe-inspiring to behold.

His contemporaries complained that Chandra's excessive use of mathematics tended to obscure the underlying physics. But that was his way of solving the sort of problems he liked to work on. No doubt Eddington would have gleefully ripped apart *The Mathematical Theory of Black Holes*, as he had Chandra's second book, on stellar dynamics. Colleagues joked that Chandra would rise at the crack of dawn and would have written two hundred equations by 7.00 a.m. At a colloquium, he displayed equations so long that they did not fit on a single transparency. Nevertheless, as he reminded everyone, 'You may think I have used a hammer to crack eggs; but I have cracked eggs!'[13] In describing *The Mathematical Theory of Black Holes*, the eminent British astrophysicist and Astronomer Royal Sir Martin Rees quoted the nineteenth-century polymath William Whewell's description of Newton's presentation of his theory of motion: 'We feel as when we are in an ancient armoury where the weapons are of gigantic size; . . . we marvel what manner of men they were who could use as weapons that we can scarcely lift as a burden.'[14] This was Chandra's style, a classical style very different from the elegant, modern yet powerful topological approach introduced into black hole literature by Roger Penrose, whom Chandra regarded as 'of course, at the pinnacle'.[15]

On 19 October 1983, Chandra celebrated his seventy-third birthday. That same day it was announced that he had been awarded a Nobel prize. Finally, he had received the greatest honour science has to offer. Congratulations poured in. But Chandra's own response was ambivalent. At the ceremony in Stockholm, the citation specified that the award was in recognition of 'one of Chandrasekhar's most well known contributions, his study of the structure of white dwarfs'.[16] Chandra was aggrieved. As far as he was concerned, the body of work he had produced over his whole life had been disregarded. He published an autobiographical statement in which he listed the seven periods of his career and the epic publications that marked each: stellar structure (1929–39), stellar dynamics (1938–43), radiative transfer (1943–50), hydrodynamic and hydromagnetic stability (1952–61), equilibrium of ellipsoidal figures (1961–68), general relativity and relativistic astrophysics (1962–71) and the mathematical theory of black holes (1974–83). 'However, the work which appears to be singled out in the citation for the award of the Nobel prize is included in the following papers,' he continued.[17] Almost all were from the 1930s, when he was in his youth. The award statement, he argued, distorted his life's work because all it acknowledged was that early insight of his into white dwarfs. His friends tried to convince him that in reality the award was honouring a lifetime of research.[18] At last his discovery of the maximum mass for white dwarfs had been fully recognised. But he was still dissatisfied.

Chandra was awarded his Nobel prize jointly with Willy Fowler. Fowler was immensely popular throughout the astrophysics community, while Chandra, though highly respected, was a more retiring figure. On the day the prize was announced, a symposium on nuclear astrophysics was in progress at Yerkes. Fowler gave a stirring yet self-effacing speech in response to the momentous news. David Schramm, a prominent astrophysicist at the university, phoned Chandra to congratulate him. Lalitha answered. The Nobel was long overdue, she said. Her exact words were, 'It's about time.' As Chandra had launched his career at Yerkes, suggested Schramm, perhaps he would like to come out and join the celebration. After all, Willy was there. Chandra politely declined.[19]

Peter Vandervoort, a former student of Chandra's, was a colleague at the University of Chicago. To mark the occasion he gave a lecture to his graduate class in astrophysics on Chandra's maximum mass. To his surprise, when he slid up the front layer of blackboards, there was a derivation of it already written there. Vandervoort told Chandra of this coincidence, to which Chandra replied ungraciously, 'I bet five years ago no one in the department could have done that derivation.'[20]

Soon after the award was announced, Chandra was interviewed. It would have been different, he said, if his work had been recognised in the 1930s and the damaging controversy with Eddington had never happened. Instead, he had received no encouragement to go any further. He recalled attending several of Russell's lectures on stellar structure. There was never any reference to his work. Then there was the shabby treatment he had received at Yerkes when he arrived there.

The burning desire to have one's work recognised with a Nobel prize can have a damaging effect on a scientist's life. Some who have done important work may wait for the phone to ring in October, only to wait and wait. Only three scientists can win it in any one field.[21] Some become bitter because they feel they have been unjustly overlooked. By the time Max Born received a Nobel prize in 1954, he had become paranoid about it. Chandra recalled being in the audience with Born in 1933 at a ceremony at the Cavendish Laboratory in Cambridge, to celebrate the award of Nobels to Dirac and Heisenberg. When the guests of honour filed in and everyone stood up to applaud, Born's eyes filled with tears. 'I should be there, I should be there,' he said.[22]

Some Nobel prizewinners are so overwhelmed with lucrative speaking engagements and invitations to move up into higher administrative positions that they have to give up their research careers. For Chandra it was just the opposite: it was an incentive for him to step into new fields of research. He decided to look into what happens when gravitational waves collide with each other and with stars, making the stars they hit change shape and 'breathe' ('non-radial oscillation'). This then generates new outgoing ('scattered') gravitational waves, in a process very much like waves in the ocean hitting a rock and rippling back to intersect with the incoming waves. His *Selected Publications* lists over twenty-seven articles on these subjects. His productivity never flagged.

Valeria Ferrari, now a professor of physics at the University of Rome, worked alongside Chandra in his research on colliding gravitational waves. They met in Rome in the summer of 1983. She impressed him sufficiently for him to invite her to spend some time in Chicago, perhaps later that year. Excited about the invitation, she carefully prepared for this new project. Three days after the Nobel prize had been announced she was at his office, ready to start work. Chandra was still reeling from the publicity and was stunned to see her. Ferrari's telexes had not arrived, a situation reminiscent of Chandra's first unheralded appearance in Cambridge as a student. She still recalls his graciousness amid all that was going on in his life and, most of all, his 'humility of starting learning something new, digging into some difficult problem and trying to unravel its structure and order. It was certainly a quest for harmony and symmetry that led Chandra to think that the theory of colliding waves could be constructed along the same lines as the theory of black holes.'[23] It was an enormously difficult task. Yet a year later their first paper appeared. Chandra always referred to it as 'the Valeria paper'.[24]

In the mid-1980s, now well into his seventies, Chandra set to work on yet another project. He had been invited to speak at a conference to celebrate the 300th anniversary of the publication of Isaac Newton's *Philosophiae naturalis principia mathematica* (usually known as the *Principia*) in June 1987. As an expert on the general theory of relativity, he agreed to speak on Newton and on gravitational theory after Newton. With his usual seriousness, he began to delve into Newton's magisterial work. He was profoundly moved by the elegance of Newton's presentation and of his results, which had laid the foundations of modern science and brought about the Age of Reason. Chandra entitled his lecture 'The aesthetic base of the general theory of relativity'. But he was not satisfied with it and did not submit a text for the conference proceedings.[25] He described what he tried to do as follows: 'I decided to try to read [the *Principia*], but that I estimated would take a year, so I took 15 propositions, tried to prove them on my own, and then compare them with Newton's proofs. I was astounded at his originality.'[26] Even on a historical subject, he still took a scientific approach.[27]

He was hooked. Newton became the love of this last phase of his life. He was now an old man, but he planned his work for the future:

> There are two things that I should like to do before I die:
> 1. To write to my satisfaction and send to Press my *Lobegesang* [song of praise] on Newton.
> 2. To be able to give a carefully prepared Lecture on a suitable occasion on 'The serial paintings of Claude Monet and the landscape of general relativity'.*
> [*But I shall write the essay regardless – even if the occasion does not arise.][28]

For the next two years he devoted himself to these two projects. Eight years after the Cambridge conference, *Newton's Principia for the Common Reader* was finally published. Uncharacteristically, Chandra had written it hastily, as if racing against time, sending each chapter off to the publisher as it was completed. He also wrote jottings to himself, entitled 'How I came to writing my book on the *Principia*'. He noted that he intended his method to be very different from that of the historians of science who dominated Newton studies. While they, as he put it, took 'the easy way of talking in large, soft and vague terms', he chose to analyse Newton's work using the heavy machinery of twentieth-century mathematics.[29] If anyone could do this, it was he. At this point he was probably the world's leading expert on Newtonian mechanics.

Chandra stated his approach in straightforward terms. He intended that this book should be assessed 'as an undertaking by a practising scientist to read and comprehend the intellectual achievement that the *Principia* is'.[30] Historians of science might argue that this approach would obscure Newton's thinking. He was, after all, a man of the seventeenth century, whose principal interests were biblical chronology and alchemy. When asked why he had chosen to ignore practically all the historical work on Newton, Chandra replied, 'Why shouldn't I? Was Newton wrong?'[31] The reviews of *Newton's Principia for the Common Reader* spoke highly of its technical merits. Robert Westfall, one of the high priests of Newtonian studies, described it as 'required reading', although he added that 'historians of science will find it a disturb-ing book'.[32] Chandra never saw this review, which was published in 1996.

He also completed his essay on Monet and the landscape of general relativity, which he had promised himself he would write regardless of whether the occasion should ever arise to deliver it as a lecture. Perhaps he felt that his life was drawing to a close. There were things he wanted to say – even though they were in areas he had never touched on before, such as aesthetics.

In fact he did have the opportunity to present his thoughts on Monet as a lecture, several times: first at the Inter-University Centre for Astronomy and Astrophysics in Pune, India, in 1992, then at Presidency College the same year, and in his last collo-quium at Chicago's department of physics, on 6 October 1994. His colleagues there were taken completely by surprise. What on earth was Chandra going to come up with next? How could abstruse equations of general relativity, occupying an entire page, be beau-tiful? Chandra had his own ideas.

Chandra had already written several articles on the relationship between art and science, collected into a book with the elegant title *Truth and Beauty: Aesthetics and Motivations in Science*, which had been published in 1987. In it he explored grand themes against a background of personal reflections. Chandra loved read-ing biographies of famous scientists. Seeking out the differences and similarities in the creativity of artists, writers and scientists, he found that artists and writers may continue to develop over many years, during which time they create ever better works, even blos-soming in old age. But highly creative scientists are different. At forty-two, for example, Newton gave up serious scientific research; after the age of thirty-five, James Clerk Maxwell, unifier of elec-tricity, magnetism and light, never did anything to match his earlier achievements; and Einstein spun no more great theories after the age of thirty-six. Chandra compared them with Lord Rayleigh, who, throughout his long career, sought to bring logical consistency, clarity and mathematical rigour to subjects ranging from electricity and magnetism to light, heat and sound. But from this he drew what must have been a rather bitter conclusion – that he himself was one whose most important discovery had been his

first.[33] This, indeed, had been the opinion of the Nobel committee. What about all his other contributions, which were as solid as
those of Lord Rayleigh? But they had emerged from a lifetime of
hard work, while his theory of white dwarfs had come to him in a
flash of youthful brilliance.

When all was said and done, in Chandra's view, what was important was to 'have added something to knowledge'.[34] He always bore
in mind what G. H. Hardy had told him in Cambridge:

> If anybody asks you why you are an astronomer and what use
> is astronomy, then to such a philistine the correct answer is,
> 'Astronomy may not be useful, indeed my work may not be
> of "importance" in any sense, but I am convinced that my
> work in astronomy is at least the most important part of my
> personality. That I am married, have a salary, am brown in
> colour, loved by others are of no significance except to
> myself. But my creative work has a value to others (however
> few these others may be) which has a significance apart from
> and therefore of the greatest value in myself.'[35]

For Chandra, the quintessential loner who never considered himself
'part of the astronomical establishment', it was an appropriate
expression of purpose.[36]

Chandra's Monet essay is deep and original. He had shaped it so
as to convince colleagues of the importance of his discoveries in
black hole physics via another avenue, the interplay of art and science. Since Kerr had proved conclusively that every black hole
can be characterised by just its mass and spin, many things had
fallen into place for Chandra. Perhaps part of the cause of his
soul-searching was that he had taken the leap into theoretical
physics so late in his life. Why had he not done so earlier?

From his earliest letters home, it is clear that Chandra loved art
and literature. He believed that an appreciation of the arts 'might
help one to do science better'.[37] In his essay on Monet, Chandra
takes scientists to task for writing about the beauty of Einstein's
general theory of relativity, 'though they are mostly silent or vague
on the elements that constitute its beauty'.[38] He demanded that
they define their terms. He asked himself 'how one may evaluate

scientific theories as works of art in the manner of literary or art criticism',[39] and offered two criteria. The first was from the seventeenth-century English philosopher Francis Bacon: 'There is no excellent beauty that hath no strangeness in the proportion!' The second was from Heisenberg, his boyhood hero: 'Beauty is the proper conformity of the parts to one another and to the whole.'[40] In Chandra's opinion, Heisenberg's criterion was applicable to 'both the arts and the sciences'.[41]

The general theory of relativity demonstrates 'strangeness' in that it relates things that were previously assumed to be completely unrelated, namely the structure of space and the matter in it. Large masses and very dense objects, such as white dwarfs and neutron stars, distort the space around them. Stephen Hawking brought more surprises and more strangeness when he introduced quantum concepts into black hole physics and revealed that black holes will eventually evaporate, though over trillions and trillions of years, longer than the expected lifetime of our universe.[42]

For Chandra, Kerr's solution of Einstein's general theory of relativity was another example of the beauty to be found in scientific forms, and he analysed it in his essay on Monet and general relativity. Kerr's solution began as a mathematical exercise in analysing space and time around a rotating sphere. But further investigations revealed the astounding consequences of assuming the sphere to be a spinning star inside a spinning black hole. That black holes could be identified by just their mass and spin stunned everyone. Chandra's response was, 'It seems to me that there are a number of instances in which what the human mind perceives as beautiful has counterparts in nature. [This] is to me in many ways a very sobering thought.'[43]

Monet was Chandra's favourite artist. Looking at his paintings of haystacks and Rouen Cathedral with his mathematician's eye, Chandra saw the immutable objects behind the surface glitter of light and colour. In Monet's paintings of Rouen Cathedral, the immutable object is the cathedral, which exists as a 'thing in itself', regardless of the light illuminating it. 'For me a landscape does not exist in its own right: it is the surroundings that bring it to life,' he wrote, quoting Monet.[44] What brings the landscape of general relativity to life? According to general relativity, the landscape is spacetime, determined by its geometry and described by the complicated equations that shape it. Just as painters have in their mind's eye colours, contours and symmetries, the scientist uses geometry and equations as a way to describe the reality behind appearances.

Already deep into his black hole research, Chandra had been inspired by a paper by Penrose in which he explored the problem of colliding gravitational waves.[45] Surprisingly, the mathematics indicated that after colliding, the waves bounce apart, then collapse and form a singularity which is mathematically the same as a black hole. Like that day on the Arabian Sea over fifty years before, Chandra noticed something everyone else had missed. His deep knowledge of the mathematical complexity of general relativity 'led him to think that the theory of colliding waves could be constructed on the same lines as the theory of black holes' – particularly Kerr's.[46] He could bypass the difficult mathematical equations that deal with gravity waves because the structure of spacetime after gravity waves collide is the same as around a rotating black hole.

Monet's series of paintings struck Chandra as an excellent way of describing the unexpected symmetry between black holes and colliding gravitational waves. He presented the fearfully complex equations describing these two phenomena as two 'tableaux', showing how the Kerr metric describes a black hole and how gravitational waves collide. Superficially they are quite different, like two of Monet's paintings of Rouen cathedral. From there he went on to tease out the immutable object, but with none of the complex mathematical details of his scientific papers.

With a dazzling mathematical sleight of hand, he showed that both sets of equations derive from the same equation, $\varepsilon = p\eta + iq\mu$, representing the distortions of spacetime caused by colliding gravitational waves as well as around a rotating black hole. 'It is a remarkable fact,' he wrote, 'that the entire set of equations of both tableaux can be reduced to the same equation.'[47] The meaning of the symbols aside, suffice it to say that ε was Chandra's 'immutable object', the reality behind the glittering surface.[48] It ties two apparently distinct parts of astrophysics together: the properties of black holes and the scattering of gravitational waves become different aspects of the same general theory, encapsulated in the mathematical quantity ε. Here was a precise way of talking about the beauty of general relativity. Chandra was extremely excited about this unexpected simplification. In his Monet essay he quoted a wider description by Heisenberg of what constitutes beauty in science: 'If nature leads to mathematical forms of great simplicity and beauty – to forms that no one had previously encountered – we cannot help thinking that they are true and that they reveal genuine features of Nature'.[49]

Some colleagues did not share Chandra's enthusiasm. Valeria Ferrari recalled, 'I think that he was mainly disappointed because people did not fully appreciate and understand the deep physical insights that had been achieved in unifying in a coherent view these theories [black holes and colliding gravitational waves], a result that he considered a major achievement in his work.'[50]

In his earlier book, *The Mathematical Theory of Black Holes*, Chandra reported a discussion he had with the sculptor Henry Moore about whether sculptures should be viewed from close up or far away. Moore replied that they should be seen from every distance because their beauty is revealed on every possible scale. Like a Michelangelo statue, to Chandra 'the mathematical perfection of black holes of Nature is, similarly, revealed at every level by some strangeness in the proportion in conformity of the parts to one another and to the whole'.[51] Theirs is a frightening yet beautiful strangeness.

Eddington wrote of falling into a black hole as falling into 'nowhere'. But is it really 'nowhere', at least as we understand that term? At the edge of the singularity – the star that has shrunk to an infinitesimally small, infinitely dense point – the laws of classical physics fail, as does the general theory of relativity: they yield singularities – infinities – that create untenable contradictions. But quantum physics, a more realistic description of nature, can handle the infinities at which classical physics balks. Deep inside the black hole, the laws of quantum gravity take over. Physicists conjecture that in such an extreme region, space and time are ripped apart and there is no causal order of events, no before or after. The laws of quantum physics signal the onset of strangeness and ambiguity. Space and time are unglued from each other. Space loses definite form, leaving a fluctuating 'quantum foam', like an amorphous mass of soap suds. It is a world where there is no certainty and probability reigns.

It is thrilling to try to imagine what further surprises may be waiting in the wings. In mathematical terms the singularity deep inside this bottomless gravitational pit is uncannily similar to the singularity that generated the Big Bang. Is it possible that it might spawn 'baby universes' that will evolve into universes capable of supporting life? Again and again, mathematics has turned out to be a key to unlocking nature's deepest secrets, as Chandra perceived – and marvelled at – towards the end of his life.

Chandra formally retired from the University of Chicago in 1980. Until 1985 he continued to hold the title of Morton D. Hull Distinguished Service Professor as a full-time, paid, post-retirement appointment. He then took up a non-salaried position, funded by a research grant, so that he could invite his favourite young collaborators, such as Ferrari, to work with him.

On 21 August 1995 he was awakened by severe chest pains. He did not want to disturb Lalitha, so he quietly dressed and slipped out of their bedroom. Perhaps he took a quick glance at the mirror on their dresser, where there was a postcard-sized reproduction of Monet's painting of a little girl. 'That is you', he often told Lalitha. He took the lift downstairs, got into his car and drove to the university clinic. As he was entering, he collapsed. He had had a massive heart attack. He died later that day, with Lalitha at his side.

Later she sprinkled some of his ashes in the university grounds. Some she threw into the air, some she sprinkled into Lake Michigan and some she took back to India, 'so he is here with us since he is everywhere'.[52]

14

Into a Black Hole

Four years after Chandra's death, a little after midnight on 23 July 1999, the space shuttle *Columbia* lifted off from Cape Canaveral. It was carrying a state-of-the-art X-ray telescope – the Chandra X-ray Observatory.[1] Lalitha was guest of honour at the launch. Would Chandra have been pleased? Lalitha thinks not: 'I can imagine him coming home and telling me, if he should be alive today, "They built an X-ray satellite and named it after me. So what?"' Perhaps it would have violated his intense sense of privacy, as he felt his Nobel prize had. As Lalitha puts it, 'all he cared for was his work'.[2] He would have been more interested in what the satellite revealed about the universe than in its name.

Appropriately, the mission of the Chandra Observatory has been to seek out black holes. In this it has been enormously successful, succeeding in clearing up mysteries which have troubled astronomers and astrophysicists for years. One of the more intriguing of these mysteries concerned the galaxy NGC 6240.

NGC 6240 lies in the constellation Ophiuchus, an almost unimaginable 2400 million trillion miles, or 400 million light years, away (25 trillion times the distance from the Earth to the Sun; we see NGC 6240 as it was 400 million years ago). Previous X-ray observatories had identified a mysterious source of X-rays right in the centre of the galaxy. Then a combination of radio, optical and infrared observations picked up a second source. The Chandra Observatory, with its super-sharp resolution, was able to identify that the X-rays were being emitted by two active galactic nuclei powered by supermassive black holes.[3] Astrophysicists deduced that NGC 6240 is the product of two galaxies colliding, each with its own active black hole. It is a so-called starburst galaxy – a star nursery in which stars are forming, evolving and sometimes exploding in the aftermath of a galaxy merger estimated to have happened some 30 million years ago. The two supermassive black holes are 18,000 trillion miles apart, and each has an enormous gravitational field. They will gradually drift closer and closer, form a binary system, spiral around each other and finally merge to form an even larger black hole. This lends support to the idea that supermassive black holes at galactic centres are formed by smaller black holes merging.

The 'merge' will take about 400 million years. It is a three-part process. First the two black holes spiral around each other, then they merge, clicking together in the final note in this cosmic tango, a phase known as ringdown. This final 'ping' comes when the opening in spacetime snaps out of its deformed shape, caused by the two black holes combining. Astrophysicists predict that the result will be a single rotating black hole with a smooth opening – since black holes have no hair – leaving no evidence that there were originally two. The fingerprint of the spiralling and the merger will be imprinted in the gravitational waves emitted, and future astrophysicists will be able to extract it using the mathematics of the general theory of relativity. The strongest part of the signal will probably come from the merge and the ringdown. Depending on the character of the gravitational waves, it should be possible to test whether the final black hole, the product of the merge, is actually as predicted by Kerr's theory of rotating black holes. Astrophysicists estimate that there are several cataclysmic merges of this sort every year in the observable universe – which is just as well, as it will be 400 million years before we can detect the NGC 6240 merge from Earth.

In the last few years, scientists have designed ground-level instruments to detect the gravitational waves rippling out from ringdown and other violent events in the cosmos. One is the Laser Interferometer Gravitational Wave Observatory (LIGO), in which a laser beam is split into two by a beam-splitter. Each beam races back and forth along a perpendicular path, two miles long; the two paths form an 'L'. Where the beams recombine they form a characteristic interference pattern that can be accurately monitored. Any change in the distance that the laser light travels will disturb this pattern. LIGO is so sensitive that it can detect a change of one part in a trillion billion in one of the laser beams. Ripples in the fabric of spacetime caused by cosmic events such as the spiralling of two black holes or two neutron stars towards each other will hopefully alter the length of at least one of these paths, disturbing the interference pattern and enabling astrophysicists to detect gravitational waves.

There is also a very exciting project being developed jointly by NASA and the European Space Agency to launch an orbiting detector, the Laser Interferometer Space Antenna (LISA), sometime around 2011. It will consist of three spacecraft equipped with laser devices, located at the three points of an equilateral triangle, 3 million miles apart. The craft will direct laser beams at one another. Out in space, LISA will be far more sensitive to the long wavelengths of low-frequency gravitational waves. There is too much background noise from events such as earth tremors to make such measurements from the ground, and the LIGO arms are too short to detect long wavelengths. LISA will be able to detect binary stars which orbit each other in hours and even minutes, whereas LIGO will be sensitive to much more rapid phenomena such as the final spiralling-in of binary black holes.[4]

Other projects are under way. One is a programme that will be run on a supercomputer at the Lawrence Berkeley National Laboratory. The plan is to calculate the sort of gravitational waves that two rotating Kerr black holes with no particular alignment of their axes (i.e. no 'north–south' alignment) are likely to generate at the moment when they merge. The gravitational waves created at ringdown have momentum, like breaking waves at the water's edge which can knock you down. As the two black holes collide and merge, the gravitational waves emitted by the newly formed black hole as it snaps into its smooth shape should cause it to recoil, perhaps even knocking it out of the centre of its host galaxy. The core of the galaxy where this takes place should be scarred by this cosmic cataclysm and have an unusual radiation profile. Martin Rees has suggested something even more spectacular – a recoil so violent that the black hole breaks free from its galaxy and hurtles through space, as if from a catapult.[5]

Amazingly, one of Chandra's companion observatories, the Hubble Space Telescope, has observed a black hole streaking across our galaxy at 70 miles per second. Catalogued as GRO J1655–40, it is the first known example of a black hole exhibiting evidence of runaway motion imparted by a kick from the supernova explosion of a massive star. It can be tracked because it is travelling with a companion star which it is gradually gobbling up. Astronomers believe that the binary companion is the survivor of the same supernova explosion that gave birth to the black hole. The companion star orbits GRO J1655–40 every 2.6 days. Fortunately, GRO J1655–40 is a safe distance from us, around 36,000 to 54,000 trillion miles.

The Chandra Observatory has also homed in on jets emanating from quasars. Observations of radio waves show that the inner portion of the jet shoots out from around the black hole at the centre of the quasar at half the speed of light. Chandra has picked up X-rays generated farther from the centre, at the point where the jet ploughs through gas in the galaxy and comes to a halt. Astronomers hope that images taken by Chandra will help them to understand how these jets affect the regions around them.

The Chandra Observatory has even 'heard' the sound of a black hole. This cosmic performance took place in the supermassive black hole in the Perseus galaxy cluster, some 1500 trillion trillion miles away. More than 53 hours of observation using special image processing techniques revealed wave-like features around the black hole that appear to be sound waves. Scientists think that they result from explosive activity near the black hole caused by large amounts of gas falling into it from smaller galaxies that it is cannibalising. The sound is pitched at a steady B flat, but we cannot hear it because it is a low grumble, fifty-seven octaves below middle C. To give some idea of just how low this is, a piano keyboard spans a mere seven octaves. The hum of a black hole is over a million billion times deeper than can be heard by the human ear, the deepest note ever heard in the universe.

One of the most exciting discoveries of all is right in the middle of our own galaxy, the Milky Way, a mere 168,000 trillion miles away. By observing at radio, infrared and X-ray wavelengths, astronomers have penetrated the gloom of the thick interstellar dust that surrounds the Galaxy's core. We have known for some time that, right at the centre, there is an incredibly strong source of radio waves, far too strong to come from a star, a pulsar or an exploded fragment of a supernova. Astrophysicists hypothesised that every galaxy has a black hole at its centre and suggested that this was a massive black hole feeding off its accretion disk. They named it Sagittarius A* ('Sagittarius A star', or Sgr A* for short).

In 1996, astronomers started to construct a precise three-dimensional map of the orbits of star clusters rotating around the centre of the Galaxy. In October 2002, observations began to come in. From data collected over ten years using the highest-resolution equipment available, including the European Southern Observatory's Very Large Telescope at Cerro Paranal, Chile, they were able to confirm that the star clusters were moving around a non-stellar source of radiation, and recorded a dark object at the exact location of the compact X-ray source Sgr A*.[6] The star closest to Sgr A*, known as S2, orbits it once every 15.2 years at a distance of 10 billion miles, from which astronomers deduced that Sgr A* is 3 to 4 million times the mass of the Sun and has an event horizon 14 million miles across – smaller than the Earth's orbit around the Sun. It spins on its axis once every 11 minutes.

Something that massive yet that compact could only be a supermassive black hole – right in the middle of our own Galaxy. It is the only hypothesis that fits the facts, and the most precise and clear-cut evidence yet for a black hole. Simplicity and beauty reign, as Chandra believed.

Astronomers are continuing their study of Sgr A*, collecting data they hope will reveal how a supermassive black hole in the centre of a galaxy is born and evolves. Even when it flares up, Sgr A* emits X-rays of rather low intensity, which indicates that it is a starved black hole, perhaps created by a recent supernova explosion which cleared away most of the interstellar gas around it. Thus black holes are not, it transpires, cosmic vacuum cleaners, voraciously and incessantly gobbling up galaxies.

Chandra would undoubtedly have been fascinated by the deep

sky survey shown in the astronomical picture section of this book (Figure 4(b)). This is a mosaic of thirty separate Chandra Observatory images taken during July 2001 and shows an oblong section of the core of the Milky Way, some 2400 by 5400 trillion miles across. If you look hard you can see hundreds of white dwarfs, neutron stars and the flaring X-rays marking the location of black holes, all bathed in an incandescent fog of gas, many millions of degrees kelvin in temperature. Sgr A* is to the right of the bright patch at the centre. The escaping gas is rich in elements forged in the furnaces of supernovae, which will eventually be deposited in the Galactic suburbs and perhaps elsewhere in our universe.

All black holes must have spun at one time. As the star inside collapses it spins faster and faster, and the whirling accretion disk adds even more spin. But this theory had never been verified in the case of an actual black hole. Then, in September 2003, the Chandra Observatory detected strong evidence. As a black hole hurtles through space, it sucks in gas from its companion star, building up its accretion disk. Holes that lack a companion draw in surrounding interstellar gas to form the disk. As the gas is dragged towards the black hole, it heats up to tens of millions of degrees kelvin. Iron atoms in this gas produce distinctive X-ray signals that reveal the orbits of particles around the black hole. How close a particle orbits a black hole depends on the curvature of spacetime near the hole. If the black hole is rotating it drags spacetime along with it, twisting it up into a funnel whose event horizon is smaller than if it were not rotating. Thus a particle orbits closer to a rotating black hole than to one which is not.

The Chandra Observatory can detect this smaller orbit because the X-rays emitted by the particle orbiting the rotating black hole have a longer wavelength and different characteristics from those orbiting a non-rotating black hole. There is also a stronger gravitational redshift because the particle is closer to the event horizon. Astrophysicists studying X-rays from iron atoms orbiting black holes have found that some iron atoms orbit deep within the gravitational well, as close as 20 miles from the event horizon. Particles orbiting Cygnus X-1, however, get no closer than 100 miles from its event horizon, from which astrophysicists infer that Cygnus X-1 is a non-rotating black hole. These conclusions depend on complicated mathematical models of accretion disks. Tempting as such black holes are by their beauty, there is at present no litmus test that can tell us whether a black hole is rotating, although evidence strongly suggests that some are, and that their properties are exactly as proposed by Kerr's theory.

Some black holes see their companion star expire before they have had time to acquire much spin. These holes will eventually run down. Cygnus X-1 may have been one of them, although astrophysicists have yet to agree that it is definitely not rotating. And why do some black holes spin faster than others? To make it more complicated still, a black hole's rotation can result from the spinning collapsing star inside, the spin of its companion, or the chaotic flow of hot gas from the accretion disk, and the spins may not always have been in the same direction.

The Chandra Observatory has witnessed all sorts of extraordinary cosmic events. In one of the most lopsided gravitational battles ever observed, a star with about the mass of the Sun was accidentally knocked off course by another star and ended up in the neighbourhood of the supermassive black hole at the centre of a galaxy catalogued as RX J1242–11, in the constellation Virgo. With a mass 100 million times greater than the Sun's, this black hole ripped the intruder to shreds, creating one of the most powerful X-ray bursts ever detected in that galaxy. Chandra picked up this catastrophic destruction on 9 March 2001. Its data were combined with those from the European Space Agency's XMM-Newton and earlier observations by Rosat, the German–UK–US X-ray observatory. This phenomenon had been predicted from the earliest days of black hole theory, but it had never before been confirmed.[7]

Black holes have opened the minds of scientists to incredible speculations and provided all of us with a breathtaking perspective on the universe of which we are a minuscule part. Will we one day be able to produce black holes in the laboratory? As it happens, physicists are already making concrete plans for just such an attempt. It will be done not by crushing matter to its Schwarzschild radius – the technology does not yet exist to do that – but by using machines that accelerate elementary particles to near the speed of light, then smash them into one another.

The collision will produce an enormous concentration of energy, hugely hot yet minuscule in volume, far smaller than a nucleus, thereby mimicking conditions a fraction of a second after the Big Bang, when the temperature was about ten thousand trillion degrees kelvin and black holes were being created. The experiment is scheduled to be carried out in 2007, smashing together oppositely directed beams of protons, at the Large Hadron Collider at the European Centre for Nuclear Research, near Geneva. Physicists are already talking about 'black hole factories' in which they hope to create a black hole every second.[8] The black holes will be a million times smaller than a nucleus (10^{-13} centimetres across), with a mass around that of a proton (10^{-24} grams). Hawking predicts that they will evaporate in a fraction of a second, leaving behind a telltale flash of Hawking radiation produced by each black hole evaporating. From this, physicists hope to pry information about how black holes are formed as well as the structure of space at its tiniest level.[9]

According to the mathematics of Hawking's theory, the shape of the radiation spectrum of the Hawking radiation depends on the number of dimensions of the space through which it undulates. Thus Hawking radiation may also reveal information on the character of space around a black hole. Theories of the early universe suggest that at that time space had more than the four dimensions we now experience. The other dimensions did not expand into the world as we perceive it, but remain 'rolled up' and make their presence felt only on a scale so small that it is way beyond microscopic, a tenth of a trillionth of a trillionth the size of an atom (a length of 10^{-33} centimetres), where scientists expect quantum gravity to take over from general relativity.[10] This is the so-called Planck length, the smallest length conceivable in physics. It is constructed from three fundamental constants of physics: the gravitational constant G, representing the large scale of nature; the Planck constant, representing the atomic scale; and the velocity of light. To get some idea of its smallness, if an atom could be expanded to the size of the entire universe, the Planck length would be around four feet long.

The Planck length may be unimaginably small, but it is not zero – it defines a region of space with a volume of 10^{-99} cubic centimetres. Incredibly, scientists can explore nature at this minuscule scale in the laboratory. When they use radiation to investigate the structure of an object, they choose a wavelength about the same size as the structure. Radar uses wavelengths at least as large as an aeroplane. To investigate atomic properties they use very small wavelengths, and trillionths smaller for looking into nature on the scale of the Planck length. Physicists can produce radiation of such a minuscule wavelength – and therefore enormously high frequency and energy – in the Large Hadron Collider. Hopefully they will succeed in reproducing conditions in the early universe, and the Hawking radiation given off by the black holes created in the Collider will tell us something about the additional dimensions that are believed to be there. Perhaps deep inside a black hole – at the level where quantum gravity takes over – there is no endlessly shrinking star. Instead, extra dimensions come into play. Surprises are in store.

Scientists are also daring to explore fantastic scenarios which until recently had been dismissed as science fiction – such as whether black holes might be used as portals for travel to distant galaxies. The most daunting obstacle is the incredibly strong gravity of the eternally collapsing star, guaranteed to reduce any rash space travellers and their craft to atoms and beyond. To overcome this, scientists have come up with the cunning idea of finding a black hole where there is a second, 'benign' singularity not too deep inside and are currently working on mathematical models of this.[11] They also hypothesise that some supermassive black holes may have a hybrid structure, with parts where the gravity is less intense than at others. It might be possible to navigate a spacecraft towards these sections, thus giving the intrepid astronauts a greater chance of survival.

As a spacecraft plunges into one of these black holes, distortions in the fabric of space and time around the benign singularity or the low-gravity area jostle it violently, like hitting a series of speed bumps. It accelerates with jarring speed, then bumps to a halt. When the astronauts look out of their porthole, they see a glaringly strange, unearthly yet tranquil terrain. Checking their instruments, they discover that they have landed on a planet in a galaxy a million trillion miles away. They have travelled this vast distance in just a few minutes by their watches. While light coursing along the surface of spacetime would take a million years to make the journey, the folds in spacetime near the singularity bumped them into a shortcut, like travelling from the North Pole to the South by going straight through the Earth instead of around it.

The shortcut is actually a mathematical solution to Einstein's general theory of relativity, and is called a wormhole. It is a rip in spacetime that permits the astronauts to tunnel through instead of travelling across the surface. A wormhole can open up anywhere, and close just as suddenly, because of the huge force of gravity in the 'throat' linking the two distant regions, which is so powerful that it prevents even light from passing through. One way to avoid this would be to find or develop an exotic material capable of pushing against the crush of gravity – a material with negative gravity, a phenomenon predicted by quantum physics – which would hold open the throat of the wormhole long enough for a traveller to come and go. Perhaps in the distant future a super-advanced civilisation will work out how to do this.

It is virtually impossible to find a wormhole; but, as we have seen, we do know the locations of black holes. Sagittarius A*, the nearest supermassive black hole to Earth, is a good candidate for space travel, though unfortunately it is 150,000 trillion miles away. No doubt scientists will eventually crack the problem of placing humans in a state of suspended animation in order to survive interstellar travel, so that perhaps one day we really will be able to send astronauts into a black hole and from there bounce them into a wormhole, launching them on a journey through time and space.[12]

Black hole research has involved taking the mathematics of general relativity to its logical conclusion, thereby producing results which for decades seemed totally absurd and divorced from reality. There are yet more deductions to be made from special relativity, the implications of which have yet to be taken seriously. It is now known for certain – as predicted by special relativity – that a clock in motion marks time more slowly than a clock at rest.[13] In the future, astronauts might be able to make a journey to a distant star travelling at a speed close to that of light. According to their calendars, they would have travelled – and aged – only a few years. But on Earth a huge amount of time would have passed. Their family would have long since died and generations would have come and gone. When they returned to Earth, these astronauts – whether they liked it or not – would find themselves in the future. Wormholes are another mathematical manifestation of relativity. Scientists have only just begun to look into the many bizarre implications they may offer, such as reversing time. If our astronauts could make use of antigravity material to hold wormholes open, they might be able to travel to a star many trillions of miles away or even arrive before they had left.[14]

All this may seem mere mathematical conceit – but so did Chandra's flash of inspiration on the Arabian Sea, one sunny August afternoon in 1930. No one believed him either, and for many years. Yet it was Chandra's proof that white dwarf stars can shrink to an infinitely tiny point of infinite density that paved the way for the discovery of black holes – mysterious chasms in the fabric of space and time, millions of miles across, vast enough to gobble up a star. It was a concept that forever changed the course of astrophysics and profoundly transformed our view of the universe.

Chandra was a man who dealt in complexities, in a world of intricate calculations, both numeric and symbolic. He may have helped to reveal some of the secrets of the universe, but there were mysteries in his own life that he never really resolved. He was confident of his own brilliance, yet permanently bitter at never having received the recognition he thought was his due. After making his great discovery, he became more conservative in his approach to his work, focusing on making fields such as hydrodynamics and radiative transfer more rigorous and consistent, rather than innovating. He was always afraid to reach too high, until right at the end of his career. Then he finally saw that he could achieve what he had always wanted – to be a theoretical physicist, albeit a very mathematically inclined one. He knew deep down that he could achieve his life's goal only by giving up astrophysics. Chandra often liked to refer to Eddington's image of Daedalus and Icarus. Scientists in the Daedalus mould play safe with their theories, applying them only where they are confident of success. Icarus types are more adventurous, stretching their theories to the breaking point. Chandra yearned to be an Icarus – he preferred to take chances. But by the time he did, it was too late: the wonder years of black hole research were over. Maybe he was thinking of this when, many years later, he transcribed some lines from one of Rabindranath Tagore's Cantos: 'I forget, I ever forget, that I have no wings to fly, that I am bound on this spot evermore.'[15] Yet he managed to achieve something exciting, interesting and deep, incorporating notions from art and science to communicate his views, and using the fluid prose and wide expanse of knowledge he had reserved till then for his personal correspondence.

In India, Chandra had been a golden youth, publishing complex papers from the age of seventeen, achieving the highest grades at university and being very popular as well. He went from success to success, meeting and befriending world-class scientists such as Werner Heisenberg and Arnold Sommerfeld. India's greatest scientists, C. V. Raman and Meghnad Saha, foresaw a brilliant future for him. Then he arrived in England, bursting with high hopes, expecting to be welcomed as the new Ramanujan. He brought with him a discovery about nothing less than the fate of the stars. Instead, he found himself not only a stranger in a strange land, but pitted against the most powerful astrophysicist in the world, one of the most important scientists of his day. Rather than comparing himself with other students, he set himself up against the likes of Dirac, Fowler and Rutherford; although he was always in awe of Eddington. Eddington was his intellectual hero, from whose books he had learned his trade.

Whatever the complexity of factors driving Eddington's urge to undermine the young Indian scientist, what is certain is that for Chandra it became an obsessive quest to prove that he was as good an astrophysicist as Eddington. By the late 1930s his international reputation was soaring. He could easily have moved on and put the humiliation of 11 January 1935 behind him. But, like the Ancient Mariner, he never ceased to recall the events of that day. Eddington's image was forever with him.

In the 1980s, Chandra had begun to make frequent visits home, laying the groundwork for Lalitha to return permanently after his death. She thought otherwise. She decided to stay on in the flat they had lived in for so many years, surrounded by memories. Chandra's study remains exactly as it was on the day he died. It is dominated by a massive wooden desk. Opposite is a large window; he could look up and see the campus. In the corner, to his right, is the chair where Lalitha would sit and watch him work. Sometimes he would glance at her, gazing adoringly at him, and ask her if everything was all right, whether she wanted anything. 'Just love me,' she would say, and they would embrace. For fifty-nine years they were soulmates. Usually Indians are bad listeners, says Lalitha, but she was a good one; she opened Chandra's eyes and his mind. She knew him through and through. He would be pleased and surprised at her insights about him. 'You have made a discovery about me,' he often said.[16]

Chandra loved to drive. Once, on a long journey along the Ohio Turnpike, Lalitha suggested turning on the radio to listen to some music. The drive the day before had been boring and tiring. But the radio was not working. So she suggested going through the alphabet, choosing a subject to talk about beginning with each letter. They never got past 'A' for astronomy. When she asked Chandra to tell her some stories about astronomers, he said immediately, 'Why don't I dictate?'

'That will be a fine idea,' Lalitha replied. 'About whom?'

Eddington's name came up instantly. 'The words came steadily while I wrote them down,' remembers Lalitha. 'Time passed quickly.'[17]

Lalitha has her own take on Chandra's relationship with Eddington. As she sees it, while Jeans and Milne allowed their lives to be ruined by Eddington, Chandra's was a life of steel, forged in an Eddingtonian furnace. 'Chandra had the character to overcome the shock of his experience with Eddington, and turn to other fields of investigation,' she wrote:

What Milne and Jeans did was the easy way: they kept their hurt alive and fretted over it. What Chandra did was the harder way: He set the whole matter aside, and told himself, 'No, it will not help to brood over this. I must push it aside and proceed to do other things if I am to survive.' And not only did he analyse the situation and come to the above decision: he was also able to act on it and was able to discover a whole new range of subjects he could explore, and in true gratitude be able to say, 'I owe all this to Eddington!'[18]

Chandra seldom discussed his mental turmoil with friends and, surprisingly, not even with his soulmate, Lalitha. One person he did try to confide in was his close friend of those days, William McCrea. But while McCrea listened patiently to Chandra, he refused to stand up in public and defend him. In an obituary of his old friend written for *The Observatory*, McCrea reviewed Chandra's achievements but dismissed the suggestion that the events of 11 January 1935 had had any significance for him. McCrea based his remarks on his memory of events sixty years before; the two had lost touch after Chandra emigrated to America. He wrote, 'These days in the literature it is sometimes alleged that some astrophysicists at the time concealed their agreement with Chandra lest they should offend Eddington; this is nonsense historically since Eddington was even then regarding himself as the odd man out.'[19] This is most unlikely, of course. In 1936, when Shapley asked astronomers to list their peers by rank, Eddington was number one on every list. He would certainly not have considered himself the odd man out. McCrea went on, 'Personally I cannot recollect details of contacts I had with Chandra over the rest of 1935, but had he ever betrayed any stress resulting from the meeting I am fairly certain I should have been aware of it – I never was.'[20] Having had no further contact with him over the rest of 1935, McCrea was not the bosom pal he portrays himself to be. Although he offered to write to Eddington in 1935, he never did. Instead he had chosen to sit on the fence so as not to offend Eddington.

Chandra's close friend and colleague at the University of Chicago, James W. Cronin, who received a Nobel prize in 1980, responded to McCrea's obituary. Cronin was one person Chandra had confided in, during the many relaxed walks they took around the picturesque woodlands of Williams Bay. 'It was absolutely clear that this confrontation had a lifelong effect,' he wrote in a letter to *The Observatory*. 'It is true that he never expressed a personal bitterness towards Eddington, but to imply that the incident had little effect completely contradicts the many conversations we had on this subject.'[21] McCrea replied that Cronin had been duped by the 'mythology about Eddington which has come into being over more than half a century since the RAS meeting'. No evidence had ever been unearthed that 'Eddington did anything whatever that was in any way detrimental to Chandrasekhar's advancement'. Chandra, of course, had never made any such claim.[22]

Cronin and others at Chicago felt that Chandra was never happy or content with his life.[23] The melancholy he suffered as a boy in India was exacerbated at Cambridge when he tried to measure up to established scientists instead of comparing himself with his fellow students. Obsessed with work, never congratulating himself on doing a good job, predisposed towards gloom and fits of depression, things only worsened for him after 11 January 1935. That day led to an explosion of self-doubt. Chandra spent the rest of his life attempting to prove himself, most of all to Eddington's ghost, which haunted him like the ghost of Hamlet's father. Sometimes he could not control his distress. It surfaced in outbursts of emotion in front of colleagues and when he spoke tearfully to Lalitha, regretting the toll his work had taken on his personal life. By the 1980s he had received every prize he could possibly have won, far more recognition than Eddington received. Yet Eddington's shadow continued to hang over him.

As we have seen, Chandra loved reading the biographies of great physicists, though he was sometimes critical of the lifestyles they revealed. Reading Heisenberg's reminiscences of his golden years, Chandra commented that he doubted if he himself had had any at all. He was disturbed to read about the many women in Erwin Schrödinger's life. As far as he was concerned, clean living was an essential component of scientific achievement.[24]

Lalitha was an accomplished singer in the Indian tradition and often sang for her husband. Right after their marriage, when she entered Chandra Vilas for the first time as Chandra's wife, she was invited to sing. She sang a song about the inevitability of dying and of being thrown into the ditch of death. Everyone was taken aback. Babuji muttered how inappropriate the song was and abruptly left the room. Years later, she realised that as a newlywed she should not have sung at all. But on that day she had felt perfectly happy, and Chandra had expressed delight in her singing. He saw a parallel between the song and being thrown into a ditch by Eddington. He needed to pull himself out and be reborn. That song came to have a special relevance for them; they called it 'Our Song'.[25]

Lalitha speaks about Chandra's 'generous nature to forgive, to "overlook", and continue his friendship with Eddington'. Chandra himself said that 11 January 1935 'did not affect our personal relations'.[26] Eddington continued to take Chandra to cricket matches and to the tennis at Wimbledon. They sometimes went cycling together. Eddington confided to Chandra his private code which he used to drive himself to cycle further and further. The letter n referred to the number of miles cycled that day. The idea was to cycle n miles n times and to up the figure regularly. He assiduously maintained it. In 1938 he wrote to Chandra, 'My cycling is still 75. I was rather unlucky this Easter as I did two rides of 74¾ miles which do not count.'[27] In 1943 he noted that his 'n is now 77. I think it was 75 when you were here.'[28]

Chandra often told Lalitha that 'it is because of Eddington that I became the sort of scientist changing my field periodically from one to another. I had to change my field after the controversy.'[29] It was not until the 1960s that he returned to the fate of the stars. He eased himself back in by attacking the same problem that Eddington had chosen when he began work on the subject, the stability of Cepheid variables.[30] He changed his area of study every five years or so. Whenever he did this, he always asked Lalitha to sing 'Our Song'.[31] Those periods he considered to be fallow times. To guard against depression he would begin work in the new area of study while he was still completing the previous one.

Recognition was slow in coming. To his dismay, Chandra discovered that physicists of some repute referred to the upper limit for the mass of white dwarfs as the Landau limit, not the Chandrasekhar limit. He recalled that one colleague who gave lectures on astronomy to undergraduates in the mid-1960s was unaware of Chandra's work on the upper limit. Even Struve used to confuse Chandra's limiting mass with the Chandrasekhar–Schönberg limit, the maximum amount of helium which a star can produce by burning hydrogen before it begins to bloat and become a red giant.

Eddington was famous for being brutally abrasive when he spoke at the Royal Astronomical Society. But Chandra was the same. His chair in the department colloquia was his sacred territory. He would stand over anyone that had the temerity accidentally to sit in it. He chimed in whenever he felt like it. People had to be aware beforehand that they might be in for rough treatment from Chandra. One speaker remembers giving a lecture. At the beginning of the discussion session, Chandra made a single brusque comment and then strode out, straight across the middle of the room. His colleagues reassured the speaker that this was actually a compliment – if Chandra did not like a lecture, he would walk out well before the end.[32]

Intimidating though he was in public, Chandra was, like Eddington, attentive and supportive in one-to-one sessions with his Ph.D. students. While Eddington had few research students, postdoctoral students or collaborators, Chandra had many, around forty-five Ph.D. students alone. Among his postdoctoral students were Mario Schönberg, and his collaborators included von Neumann and Fermi. After 1954, Chandra became even more deeply committed to the physics department, and all his students came from that department. In 1964, when he moved to the University of Chicago campus, so did his students. Previously they had commuted with him back and forth from Yerkes.[33] For twenty years, directors at Yerkes kept Chandra's office there vacant, hoping he would return. He never did.

Chandra chose his students carefully. A definite prerequisite was mathematical proficiency. The students' research always reflected whatever areas Chandra was currently interested in, and no one ever failed to complete a thesis. Chandra never gave a student a problem that he could not conceive of solving himself. Former students remember him as approachable, attentive and concerned about their future. He arranged positions for them and followed their progress. He could be a harsh taskmaster if they did not measure up to his standards of research. But no one complained in later years; all agreed that his toughness had been for their own good. Chandra always went out of his way to be cordial to the families of colleagues and students and was especially kind to their children.

Those who worked with him recall Chandra's sense of humour as rich, complex and sometimes bordering on bitter. He could take a joke, though, and his private persona often differed from the one he presented in public. On one occasion, late in the 1950s, John Sykes, a mathematician, cooked up a parody of Chandra's work with the help of some of Chandra's co-workers at Yerkes. Modelled closely on one of Chandra's papers, 'The instability of a layer of fluid heated below and subject to the simultaneous action of a magnetic field rotation. II', it was entitled, 'On the imperturbability of elevator operators: LVII', by S. Candlestickmaker of the Institute for Studied Advances, Old Cardigan, Whales. It is a hilarious take-off of Chandra's mode of writing, with hugely complex detail backed up by nested footnotes to many of his previous publications on the topic. Sykes actually sent it to the *Astrophysical Journal*, supposedly for the issue of 19 October 1910, the date of Chandra's birth. The secretary, seeing it was a joke, showed it to Chandra, assuming he would not be amused. In fact he was delighted, and took enormous pleasure in showing it to others. He also agreed to have it printed in the style of the *Astrophysical Journal* in a reprint format. The 'authors' took up a collection to pay for the printing.[34] Sykes had caught Chandra's style so well that Chandra seriously recommended it as an example of how to write a scientific paper.

Jesse Greenstein, the astronomer at the California Institute of Technology who did important work on quasars, left Yerkes for the West Coast in 1948. He sent Chandra a photograph of himself and his family enjoying a beautiful sunny southern California day. In return Chandra sent him a photograph of a typically grey day at Yerkes. He added a note: 'Dear Jesse, This shows how Williams Bay looked on that day. I worked the entire day, and I am now one day ahead of you.'[35]

Eddington made it all look easy, moving effortlessly between astrophysics, astronomy, the general theory of relativity, cosmology, quantum mechanics, abstract mathematics, philosophy and religion, which he combined with trekking, cycling and voracious reading. Chandra, on the other hand, struggled with a small subset of these endeavours. Yet what really separated him from Eddington was the way he was able to pour out his heart, to Babuji, Balakrishnan and friends such as Rosenfeld and Cronin. With Lalitha he went to great lengths to spare her the torment he felt, his private suffering, his yearning to be a theoretical physicist instead of the reluctant astrophysicist he was for so much of his life.

Always looking inwards, Chandra examined every moment of his life and his own feelings, providing us with glimpses of the kind of complex human stories that lie behind the greatest scientific discoveries. For even though science deals with abstractions and grand cosmic issues that dwarf our small human lives, those who carry out scientific research and spend their lives absorbed in calculations and theories are human beings. They may be driven by irrational impulses, as was Eddington when he refused to believe a result which did not square with his view of the universe; and, as they go about their intensely solitary work, they may be haunted by passions, jealousies, fears, ambitions and disappointments. Chandra's eternal quest for personal peace could never be fulfilled.

Appendix A

The ongoing tale of Sirius B

To astrophysicists in 1939, Sirius B, the white dwarf companion of Sirius A, presented some particularly knotty problems. By the time of the Paris conference, new measurements had shown that it was actually brighter than previously thought. Astrophysicists now believed that its temperature was one to two thousand degrees kelvin higher than the figure Eddington had used in 1926, and had reclassified it as an A-type star, in the light of which they recalculated its radius. The new figure, combined with the star's mass, which they still assumed to be correct, gave a gravitational redshift of about 20 miles per second, almost double as much as had been measured by Adams in 1925. Gerard Kuiper described all this at the 1939 Paris conference, adding that it was somewhat irksome but could probably be resolved once it became possible to make more precise measurements.[1]

There was a bigger problem. Although Chandra's theory was supported by astronomical observations of Sirius B, it clashed with some exciting new developments on the constitution of stars that had begun to emerge from nuclear physics. Chandra's theory predicted a relationship between the mass and the radius of a white dwarf from which astrophysicists determined its mean molecular weight. On the basis of the most recent assessment of the mass, radius and surface temperature of Sirius B, the theory produced the result that Sirius B was 40 per cent hydrogen.[2] But only the previous year, Hans Bethe had developed a theory supporting

Eddington's original suggestion that the basic source of energy in stars was the burning of hydrogen by fusion. If the temperature inside white dwarfs was high enough to spark the fusion of hydrogen, why was Sirius B not enormously much brighter? Perhaps the mass was actually less than was currently assumed, in which case Chandra's theory would give the result of a smaller amount of hydrogen inside Sirius B. This might be the case if there were a third star in the orbit of Sirius A and B. But that was unlikely.

Kuiper proposed another way out. What if there were no hydrogen at all in the interior of white dwarfs? They were, after all, believed to be at the very end of their lives. In that case, according to Chandra's theory, Sirius B would have a smaller radius, a surface temperature of 25,000 degrees kelvin and a gravitational redshift of 50 miles per second, thus moving the star into spectral class B. But in the end Kuiper rejected all such suggestions as 'equally impossible', because the radius of Sirius B as measured by astronomers, as well as its gravitational redshift, was accepted by everybody in the field.[3]

It turned out that this extraordinary scenario was indeed close to the truth. Improved methods for measuring the radius and brightness of Sirius B made it possible to eliminate the overwhelming brightness of Sirius A. This revealed that the spectrum of Sirius B that Adams and others had found was hugely contaminated with lines from Sirius A showing iron, chromium and magnesium, which are not actually present in Sirius B at all. F-type stars are made up of large amounts of these elements, which had led Eddington mistakenly to classify Sirius B as an F-type star, and thus to deduce the very low temperature of 8000 degrees kelvin. From this he worked out the gravitational redshift of Sirius B, having calculated its mass as 0.85 times the mass of the Sun and its radius as 11,280 miles, about three times the radius of the Earth.

In 1971, however, three astronomers – Jesse Greenstein, John Beverley Oke and Harry Shipman – took precise measurements of Sirius B using modern instrumentation and methods. They showed that it actually has a surface temperature of 32,000 degrees kelvin, placing it in spectral class O, a radius 0.85 times that of the Earth (far smaller than Eddington had assumed) and a mass about the same as the Sun's, giving a gravitational redshift of some 55 miles per second. Their results have since been verified and refined, and remain largely unchallenged.[4] Current measurements suggest that Sirius B's surface temperature is 27,000 degrees kelvin (making it a B-type star), its radius about 3000 miles, a little smaller than the Earth, and its average density a million grams per cubic centimetre, a hundred times greater than the figure Eddington used.

George Gamow took Bethe's theory – that stars shine as a result of the fusion of hydrogen – more seriously than the astrophysicists did. He analysed it thus: since a white dwarf is a burnt-out star, it has no hydrogen left. Its average molecular weight is therefore 2, which makes the Chandrasekhar limit 1.4 times the mass of the Sun,[5] the figure usually quoted in texts. In 1940 Robert Marshak, a Ph.D. student working under Bethe at Cornell University, took the first step towards resolving in detail the problem of hydrogen in Sirius B. Using the most up-to-date results from nuclear physics, he calculated that the temperature at the core of Sirius B was at least 10 million degrees kelvin, and the density a million grams per cubic centimetre. Under such conditions hydrogen would fuse very quickly, making Sirius B much brighter than astronomers had observed. To explain the discrepancy he suggested that hydrogen showed up so strongly in the spectral lines because there was a thin layer of hydrogen floating on the surface of the star, at most 1 per cent of its total constitution, held in place by gravity.

If we assume that there is no hydrogen inside Sirius B, Chandra's theory results in a much smaller radius than the figure obtained by adopting Kuiper's assumption that the star was 40 per cent hydrogen. This leads to a sharply increased gravitational redshift, four times that measured by Adams. Ten years later, T. D. Lee substantiated and elaborated on Marshak's results in conversations with Chandra at Yerkes. Lee proved that white dwarfs which contained any more than the minuscule figure of one tenth of one per cent of hydrogen in their interior would be bound to explode, and that energy was carried through the degenerate core of the star primarily by the movement of the electrons, as nuclear reactions were no longer possible there. This was all consistent with what was expected to happen at the end of a star's lifetime.[6]

Greenstein recalled the intensity of the debates that raged from the 1930s on 'the "mysterious case of the gravitational redshift of Sirius B" . . . Few participants were right, except for Chandrasekhar, who remained aloof.' Kuiper withdrew early on. Frustrated by a lack of proper instrumentation, he abandoned white dwarfs and went into planetary astronomy: he 'dropped out in favour of planets'.[7]

Appendix B

Bringing supernovae up to date

In 1966, Stirling Colgate and Richard White formulated the theory that a star goes supernova when it is blown up by the sudden release of an enormous number of neutrinos from its collapsing core. As White points out, this was 'the earliest application of digital computers to the study of stellar hydrodynamics. No one previously had treated the dramatic gravitational collapse that leads to a supernova.'[1] At the time, for all its complexities, it was very convincing; but eventually it turned out to be a glorious failure. Many astrophysicists challenged its assertions.

All the computing power at Colgate and White's disposal was still not enough. They were limited to modelling a star as if it were a circle, instead of a sphere, and studying its properties only along its radii – in one dimension. They also had to ignore the way in which elements mix as a result of currents caused by temperature differences in a star, which need to be studied in at least two dimensions. The stellar models they were creating were unrealistic.

'Colgate and White started an industry,' recalls Stan Woosley, a leading astrophysicist at the University of California, Santa Cruz, who specialises in models of supernovae. As Woosley explains, Colgate and White's deep insight was to take the 'gravitational energy of the neutron star, convert it into neutrino energy, then take part of that neutrino energy to blow up the star in a neutrino-powered explosion'.[2] As to why the cataclysmic core collapse takes place, Woosley points out that 'massive stars are gravitationally confined thermonuclear reactors – they are like gravity bombs'.[3]

One of the problems with Colgate and White's model was their suggestion that a supernova explosion could be caused by neutrinos heating up the mass of nuclear ash falling into the star's centre. Fowler and Hoyle, among others, discussed other possibilities. One was that the star's rotation and its magnetic field might provide the push needed to set into motion the shock wave that had been stalled by the falling ash. However, computers in the 1960s were simply not fast enough to perform the calculations necessary to test modern theories of supernova explosions.

But there was something even more fundamental to Colgate and White's failure. This did not emerge until the electroweak theory was formulated in the late 1960s by Sheldon Glashow at Harvard, Steven Weinberg at MIT and Abdus Salam, a Pakistani physicist at the International Centre for Theoretical Physics in Trieste, Italy, who all shared the 1979 Nobel Prize for Physics for discovering it. The electroweak force is a union of the electromagnetic force with the weak nuclear force that controls beta decay, the process of radioactive decay in which a neutron turns into a proton, an electron and a neutrino. Among other things, it predicts ways that neutrinos can interact.

Neutrinos are loners. They whiz around through space, through the Earth and through us, each on its lonely course, as if we were not there. They interact so weakly that they can fly through space for 3 trillion miles unhampered. According to the electroweak theory, when a neutrino and an electron collide, it is like two billiard balls rebounding off each other. Neutrinos collide in a similar way with neutrons and protons. These reactions are relatively rare under most conditions, but in the cramped, hot innards of stars they are common. Even though neutrinos interact weakly, when they become trapped inside the incredibly dense collapsing core of a star they have to fight their way out, banging against electrons, nuclei, protons and neutrons. Just as light struggles to emerge from a star, so do neutrinos. The crowd of barging neutrinos creates what is called neutrino opacity. Eventually the star explodes and the neutrinos burst out of their prison, generating neutrino luminosity or 'shine'.[4]

There was still no observational evidence for this neutrino scenario. Then, on 24 February 1987, Ian Shelton, a Canadian astronomer working at Las Campanas Observatory in Chile, noticed a spot on his photograph of the Large Magellanic Cloud, a satellite galaxy of the Milky Way. It was startlingly bright, so much so that it had to be a supernova. As it happened, just the previous day he had photographed a star catalogued as Sanduleak –69° 202, twenty times the mass of the Sun and some hundred thousand trillion miles away. Shelton went outside and looked up. It was now a thousand times brighter: Supernova 1987A (SN 1987A), the death throes of Sanduleak –69° 202, had just exploded. It was the first supernova to be seen with the naked eye since 1604.

A few years earlier, two neutrino detectors – Kamiokande II, near the city of Gifu, right in the middle of Japan, and IMB, near Fairport, Ohio – had been built deep underground and had gone into operation with the aim of testing predictions of certain unified theories of elementary particles. As they were on twenty-four hours a day, scientists were able to go back over their records. What they found was that three hours before SN 1987A lit up the sky, the Japanese detector had spotted and recorded twelve neutrinos, while the American one had recorded eight, practically simultaneously. These could only have come from the supernova.[5] It was even possible to trace some of the Kamiokande neutrinos to the region where SN 1987A had erupted. Finally, here was conclusive proof that supernovae emitted neutrinos. And knowing that the neutrinos had arrived three hours before the supernova lit up gave vital clues about how the explosion worked.

Three years after SN 1987A exploded, NASA sent up the orbiting Hubble Space Telescope. One of its tasks was to collect more precise data on the supernova. Studying the characteristics of the light it emitted, astrophysicists discovered that there had been a great stirring-up of elements, like alphabets in a soup. They had moved around in all dimensions, up, down and sideways; in order to study them, scientists would need to make computer models of stars in at least two dimensions. By now there were powerful desktop computers that could handle this sort of analysis. Everyone got to work. One much debated scenario is as follows.

The shock front hurtling outwards at 6000 miles per second smashes into burnt-out nuclear ash collapsing inwards at 36,000 miles per second, and stalls. But instead of being beaten back, the shock front is held in suspension by the combined degeneracy pressures of electrons (which still exist in some abundance in the outer regions of the star, where temperatures are not high enough for them to be badly depleted through combination with protons) and neutrinos (from reactions in which nuclei in the core are smashed by high-energy radiation). Interacting with other particles in the core, the neutrinos are trapped and compressed so much that they develop an outward degeneracy pressure, in much the same way as electrons. This is called the delayed shock process, to distinguish it from the 'prompt shock' in earlier theories to do with core bounce. As the falling nuclear ash passes through it, the suspended shock front, 360 miles in diameter, accumulates matter inside and outside.

At this point the star's iron core is a mere 50 miles across, and is made up of electrons, trapped neutrinos trying desperately to escape, neutrons and heavier nuclear matter. Between the spherical shock front and the highly dense core at its centre is a shell of matter waiting to blast its way out. It is blocked by the stalled shock front. As more and more neutrinos ooze out, the core loses its rock-hardness and in a fraction of a second shrinks to a dense neutron star just 18 miles across with a mass 1.4 times that of the Sun. The matter heated by the neutrinos energises the shock front so much that it starts to move again, pushing against the falling nuclear ash like a giant plunger. The remaining neutrinos burst out, driving the shock front and making it move faster.

As the shock front reaches more diffuse regions of the star it becomes supersonic, plowing into the star's outer layers. Speeding through the burned out nuclear ash, the shock front compresses the ash so much that it sets off a new round of thermonuclear reactions in a last gigantic ignition, creating all the elements heavier than iron, among them zinc, silver, tin, gold, lead, mercury and uranium. These reactions require so much energy that they cannot take place before a star goes supernova. The fact that these elements are all found on Earth shows that everything that makes up the world we live in, including ourselves, is a product of massive

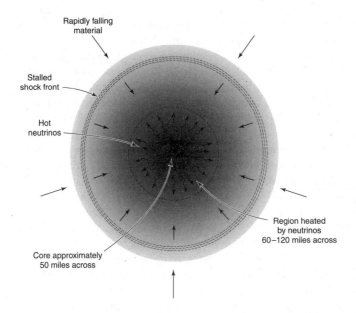

Rapidly falling
material

Stalled
shock front

Hot
neutrinos

Region heated
by neutrinos
60–120 miles across

Core approximately
50 miles across

The iron core, initially 1000 miles across, is squeezed by the nuclear ash falling
in on it until it rebounds – 'core bounce'. This generates a shock front which
sweeps out through the infalling matter, leaving behind a very hot core, about
50 miles across. The front is then stalled by the infalling matter. Neutrinos begin
to ooze out of the very hot core, heating up the layer of matter between the core
and the shock front. Eventually there is so much pressure against the stalled
shock front that it bursts into motion again.

supernova explosions. We are indeed children of the stars.

As the ash blasts into space there is a dazzling flash of light.
The energy released is huge – hundreds of times greater than all
the energy the Sun has produced the whole lifetime of Earth, 4.6
billion years. The star has become a supernova.

This theory of a delayed shock finally seemed to be a viable scenario. It created a model of a supernova that fitted with what
astronomers had observed, suggesting that the material around

the stalled shock front was suspended for long enough to be 'cooked' by neutrinos. Astrophysicists created two-dimensional simulations to analyse the light emitted by SN 1987A. These showed that when Sanduleak −69° 202 exploded, its onion-like structure was not intact. There was a great deal of swirling and turbulence inside the star's interior, like when smoothly flowing water breaks into eddies or smoke rings become fuzzy. This was what produced the percolation of heat that reinvigorated the shock wave so that it could blast through. The resistance to neutrinos passing through a star (neutrino opacity) is less than the resistance to light and decreases as the neutrinos arrive near the star's surface, which was why neutrinos from SN 1987A arrived three hours earlier than its light. SN 1987A actually exploded 160,000 years ago, because that is how long it takes light to reach us from there.

Neutrinos reveal a great deal about the way stars die. A straightforward calculation shows that in just ten seconds SN 1987A emitted trillions upon trillions upon trillions of neutrinos, a hundred times more energy than the Sun will emit in its entire lifetime.[6] Unlike sunshine, however, we cannot see this 'neutrino shine'. But for a few seconds the luminosity of the neutrinos from SN 1987A was greater than the light from every star in the observable universe.

Every researcher into supernovae emphasises that much of what we think we know about these exploding stars is still controversial, and much remains to be discovered. One unsolved problem is that the mass of a neutron star as predicted by current models is lower than the masses measured in binary systems in which one of the two stars is a neutron star. The figure depends critically on the equation of state describing the neutron star's interior. No one is sure what constitutes the core of a neutron star, where densities are more than ten times that of nuclear matter. Perhaps neutrons are stripped down to their constituent quarks, or even beyond, to unknown forms of matter. The core is so dense that a piece the size of a basketball would weigh a trillion tons on Earth. A neutron star's gravitational pull is 190 billion times that of the Earth. An object dropped from three feet above the surface of one would smash into the ground at 70,000 miles per second; on Earth it lands at a gentle two-thousandths of a mile per second.

As Colgate put it, in words that could have been taken from the pages of *The Observatory* during the debates between Eddington, Jeans and Milne, 'Intellectual honesty demands that we remove the approximations in our models, so that we become confident that our simulations duplicate reality and are not merely exercises in differential equations'.[7] Astrophysicists will need to make more precise simulations to model something as complex as the convection processes that trigger a supernova explosion. At the moment the theory of convection in supernovae is rudimentary and does not take magnetic fields and rotation sufficiently into account. It is also possible that some supernovae may be powered not by neutrinos, but totally or partially by bursts of gamma rays.

Stars that go supernova as a result of gravitational collapse in the way we have been considering are what astronomers call Type II supernovae, or 'gravity bombs' to use Woosley's term. White dwarfs in binary systems, such as Sirius B, meet their end in a different way. In these Type I supernovae, the white dwarf accretes matter from its companion, growing until it exceeds the Chandrasekhar limit, then blowing itself to pieces.[8] Under certain conditions it can eject the excess mass in a mild supernova explosion and become a white dwarf again. Astrophysicists are finally groping towards an answer to the question that Milne often put to Chandra: what would happen if a speck of dust were added to a white dwarf that had reached the Chandrasekhar limit? The answer seems to be that it will go supernova.[9]

Astrophysicists believe they have most of the physics right about white dwarfs, but details are still missing. It is impossible to simulate the conditions in white dwarfs in terrestrial laboratories. Instead scientists have to make theoretical calculations of the properties of the various elements that make them up, predominantly carbon and oxygen. Just as neutron stars have magnetic fields, so do most white dwarfs. Both are the result of a much larger star being hugely compressed. White dwarfs can have a magnetic field over ten thousand times greater than normal stars. While neutron stars rotate very fast, white dwarfs rotate much more slowly, taking hours or even days. White dwarfs can also pulsate. All these properties have to be taken into account to understand these stars properly. They also affect Chandrasekhar's maximum mass. Chandra calculated his upper limit for an idealised white dwarf, a sphere of highly compressed electrons that do not interact with one another. In his day, it was a stunning leap of the imagination.

BIOGRAPHICAL SKETCHES

Adams, Walter Sydney (1876–1956) One of the top American observational astronomers of his day, who, among much else, made ground-breaking measurements of the white dwarf star Sirius B. At the time these seemed so outrageous that Eddington dismissed them as 'absurd'.

Ambartsumian, Viktor Amazaspovich (1908–96) Armenian astrophysicist whom Chandra met during a visit to Leningrad in 1934.

Anderson, Carl David (1905–91) American physicist who worked under Robert Millikan at Caltech studying cosmic rays. In 1932 Anderson discovered the positron, the anti-particle of the electron, which Dirac had originally predicted in his theory.

Anderson, Wilhelm Robert Karl (1880–1940) Russian-born physicist of German extraction whose claim to fame was his realisation that a full understanding of white dwarf stars required the theory of relativity.

Artin, Emil (1898–1962) Viennese-born mathematician whose speciality was the theory of numbers.

Aston, Francis William (1877–1945) English physicist who was awarded the 1922 Nobel Prize for Chemistry for his precision measurements of the masses of isotopes.

Atkinson, Robert d'Escourt (1893–1981) Welsh-born astrophysicist who, with Fritz Houtermans, worked on early theories of thermonuclear reactions in stars, following a suggestion of George Gamow.

Ayyar, Chandrasekhara Subramanya (1885–1960) Chandra's father, whom Chandra often referred to affectionately as 'Babuji'.

Baade, Walter (1893–1960) German-born astronomer who came up with the concept of the neutron star in 1933, with Fritz Zwicky.

Balakrishnan (1914–98) One of Chandra's younger brothers, to whom Chandra was particularly close. He later became an outstanding paediatrician and author, whose writings in both Tamil and English were read throughout southern India.

Balakrishna, Sitalakshni (1891–1931) Chandra's mother.

Bethe, Hans Albrecht (1906–2005) German-born American physicist who discovered the thermonuclear reactions that produce energy in stars. He played an important role in the development of the atomic and hydrogen bombs, and eventually returned to astrophysics.

Bohr, Niels Henrik David (1885–1962) Danish physicist who formulated the first quantum theory of the atom.

Born, Max (1882–1970) German physicist who did important work on quantum mechanics, and coined the term.

Chandrasekhar, Lalitha (b. 1910) Chandra's wife and soulmate for fifty-nine years.

Chandrasekhar, Ramanathan (1837–1906) Chandra's grandfather, a professor of mathematics and author of several mathematics books.

Chandrasekhar, Subrahmanyan ('Chandra') (1910–95) Indian-born American astrophysicist who discovered the maximum mass for a stable white dwarf, the Chandrasekhar limit.

Chowla, Sarvadaman (1907–95) Indian mathematician who was Chandra's close friend at Trinity College, Cambridge, during 1930 and 1931.

Colgate, Stirling (b. 1925) American physicist who applied his expertise in developing hydrogen bombs to formulate the first realistic models of supernovae. He is still working at Los Alamos.

Compton, Arthur Holly (1892–1962) American physicist who discovered the Compton effect, the way in which X-rays change their wavelength when they are deflected by electrons, which provided proof of the particle nature of radiation.

Cowling, Thomas George (1906–90) English astrophysicist who was Milne's student and a friend of Chandra's.

Critchfield, Charles Louis (1910–94) American nuclear physicist who was a graduate student of George Gamow. His early results on how stars like the Sun shine influenced Hans Bethe. After working on the Manhattan Project, Critchfield taught intermittently in universities but could not resist the lure of Los Alamos, where he spent the rest of his career.

Cronin, James (b. 1931) Nobel prizewinning American physicist who was a close friend of Chandra's at the University of Chicago.

Dingle, Herbert (1890–1978) English physicist, editor of the *Monthly Notices of the Royal Astronomical Society*, with whom Milne corresponded at the height of his dispute with Eddington. Dingle is best remembered for his controversial criticism of Einstein's special theory of relativity.

Dirac, Paul Adrien Maurice (1902–84) English physicist celebrated for his eccentricities, and best known for the Dirac equation, which gives an accurate description of electrons. At Trinity College, Cambridge, Chandra idolised him and often sought his advice.

Douglas, Alice ('Allie') Vibert (1894–1988) Canadian astronomer who studied with Eddington in 1921. Vibert completed her Ph.D. in astrophysics at McGill University, making her one of the first women to win a Ph.D. in North America. She was professor of astronomy at Queens University from 1946 to 1964, and was awarded the MBE. She was Eddington's biographer.

Eddington, Arthur Stanley (1882–1944) The world's greatest astrophysicist in his day, and Chandra's arch-rival and nemesis.

Eddington, Winifred (1878–1954). Eddington's sister.

Einstein, Albert (1879–1955) Indisputably the greatest physicist of the twentieth century. Einstein discovered, among much else, the special and general theories of relativity and the iconic equation $E = mc^2$. Yet he refused to accept the most spectacular consequence of his life's work – black holes.

Fermi, Enrico (1901–54) Italian-born American physicist who discovered, among much else, Fermi–Dirac statistics. He was a close colleague and friend of Chandra's at the University of Chicago.

Feynman, Richard Phillips (1918–88) American physicist legendary for his antics and brilliance dating back to the Manhattan Project.

Finlay-Freundlich, Erwin (1885–1964) German astrophysicist and a colleague of Einstein's at the University of Berlin. He was Director of the Astrophysical Observatory in Potsdam and editor of the journal *Zeitschrift für Astrophysik* when Chandra visited in 1932.

Fowler, Ralph Howard (1889–1944) English physicist who was the first to apply quantum mechanics to astrophysics. Dirac and Chandra were among his most distinguished Ph.D. students.

Fowler, William Alfred (1911–95) American astrophysicist who studied supernovae and how these superbrilliant stars produced the chemical elements found in the universe.

Fuchs, Klaus (1911–88) German-born nuclear physicist. He worked on the British atomic bomb project and then the Manhattan Project at Los Alamos, while passing key information on atomic and hydrogen bomb development to the Soviets.

Gamow, George (1904–68) Flamboyant Russian-born American nuclear physicist who initiated modern research on the thermonuclear processes that power stars.

Ganesan, Angarai Seshia (1900–86) The husband of Chandra's sister Rajam, Ganesan completed his Ph.D. in physics in 1926 at Imperial College London. He went on to do research with C. V. Raman, leading to the discovery of the Raman effect.

Greenstein, Jesse Leonard (1909–2002) American astrophysicist who was a colleague and friend of Chandra's at the University of Chicago before moving to Caltech, where he did important work on quasars.

Hardy, Godfrey Harold (1877–1947) English mathematician and colleague of Chandra's at Trinity College, Cambridge. He was responsible for discovering the enigmatic Indian genius Srinivasa Ramanujan.

Hawking, Stephen William (b. 1942) The world's best-known astrophysicist today, Hawking has played a key role in exploring the properties of black holes. Among his many major advances is his discovery that black holes can evaporate by emitting radiation – Hawking radiation.

Heisenberg, Werner Karl (1901–76) Major German physicist whose many discoveries included quantum mechanics and the uncertainty principle. He visited Madras when Chandra was a student there, and greatly encouraged him.

Hill, Archibald Vivian (1886–1977) English biologist who won a Nobel prize in 1922 for his adroit use of physics to discover how muscles function.

He played an important role in defence research during World War I, and influenced Fowler and Milne.

Houtermans, Friedrich ('Fritz') Georg (1903–66) German physicist who did pioneering work with Gamow and Atkinson in the late 1920s on the thermonuclear reactions that power stars.

Hoyle, Fred (1915–2001) English astrophysicist who advanced our knowledge of how heavier elements are generated from hydrogen and helium.

Hubble, Edwin Powell (1889–1953) American astronomer who discovered that galaxies are receding at speeds related to their distances from the Earth.

Hutchins, Robert Maynard (1899–1977) As president of the University of Chicago he played a key role in hiring Chandra and keeping him there.

Jeans, James Hopwood (1877–1946) English astrophysicist, also famous as a mathematician, theoretical physicist and populariser of science, and Eddington's arch-rival.

Kerr, Roy (b. 1934) New Zealand physicist who in 1963 discovered a solution to Einstein's general theory of relativity that turned out to describe the properties of rotating black holes – although, as he later recalled, he did not realise the significance of his result at the time.

Kuiper, Gerard Peter (1905–73) Dutch-born American astronomer whose early observations of white dwarfs supported Chandra's theory. The two were colleagues at Yerkes Observatory.

Landau, Lev Davidovich (1908–68) Russian physicist who discovered the Chandrasekhar limit in 1932, independently of Chandra. The two met in the Soviet Union in 1934.

Lee, Tsung-Dao (b. 1926) Chinese-born American physicist whose research on the constitution of white dwarfs was carried out partly at Yerkes Observatory, where he discussed the subject with Chandra.

Lewis, Gilbert Newton (1875–1946) American physical chemist who coined the term 'photon' for the smallest particle of light.

Littlewood, John Edensor (1885–1977) English mathematician who, along with G. H. Hardy, made important advances in the theory of numbers. He spent his entire academic career at Cambridge and lived in Neville's Court, Trinity College, for fifty-one years.

McCrea, William Hunter (1904–99) English astrophysicist and Chandra's friend in the 1930s.

Marshak, Robert Eugene (1916–92) American physicist who included in his Ph.D. thesis Bethe's new theory of the thermonuclear processes that make stars shine.

Millikan, Robert Andrews (1868–1953) American physicist known for his talent as a researcher and creator of physics departments, at both the University of Chicago and Caltech.

Milne, Edward Arthur (1896–1950) English astrophysicist whose theory of the constitution of stars clashed with Eddington's and then with Chandra's.

Møller, Christian (1904–80) Danish physicist who wrote a paper with Chandra in support of his theory of white dwarfs and against Eddington's.

Oppenheimer, Julius Robert (1904–67) American physicist who co-authored critical papers on neutron star formation and collapsing stars. He went on to direct the Manhattan Project.

Osterbrock, Donald E. (b. 1924) American astrophysicist and a student of Chandra's. He has made important contributions to our understanding of nebulae and quasars, and is an eminent historian of astrophysics.

Pauli, Wolfgang (1900–58) Famously ironic Viennese physicist who discovered the Pauli exclusion principle and postulated the existence of the neutrino.

Payne-Gaposchkin, Cecilia Helena (1900–79) A student of Eddington's at Trinity College, Cambridge, she obtained a Ph.D. in astrophysics at Radcliffe College, and went on to become Harvard's first female professor.

Peierls, Rudolf Ernest (1907–95) German-born English physicist who wrote papers in support of Chandra's theory of white dwarfs.

Penrose, Roger (b. 1931) English mathematician who applied topology to black holes.

Plaskett, Harry Hemley (1893–1980) Canadian-born pioneer in observational astrophysics and a close friend of Milne.

Pryce, Maurice Henry Lecorney (1913–2003) English physicist who, along with Dirac and Peierls, wrote a paper in support of Chandra's theory of white dwarfs.

Raman, Chandrasekhara Venkata (1888–1970) India's most illustrious physicist, who had a rocky relationship with his famous nephew, Chandra.

Ramanujan, Srinivasa (1887–1920) Indian mathematician whose brilliance in number theory and Cinderella-like career were an inspiration to Chandra's generation.

Rees, Martin John (b. 1942) English astrophysicist and Astronomer Royal who made essential contributions to the study of the formation of black holes.

Robertson, Howard Percy (1903–61) American researcher on the general theory of relativity who frequently discussed his work with his colleague at Princeton, Einstein. At Caltech he gave valuable advice to Oppenheimer and Snyder in their research on collapsing stars.

Rosenfeld, Léon (1904–74) Belgian physicist who was a confidant of Chandra's during the 1930s.

Russell, Henry Norris (1877–1957) Dean of American Astronomers and a powerhouse in the American scientific scene during the first half of the twentieth century.

Rutherford, Ernest (1871–1937) New Zealand-born English physicist who made many important discoveries relating to radioactivity and in atomic and nuclear physics. He was a major force at the Cavendish Laboratory in Cambridge.

Saha, Meghnad (1894–1956) Indian astrophysicist who used Bohr's atomic theory to relate a star's spectral lines to the chemical constitution of its outer layers.

Sakharov, Andrei Dmitrievich (1921–89) Russian physicist who did significant work on the Soviet hydrogen bomb, along with Yakov Boris Zel'dovich.

Schmidt, Maarten (b. 1929) Dutch-born American astronomer who deciphered the puzzle of quasars.

Schrödinger, Erwin (1887–1961) Viennese-born physicist whose wide and diligently pursued interests included biology, literature and women. He is best known for the equation and paradox named after him.

Schwarzschild, Karl (1873–1916) German astronomer and physicist who provided the first exact solution to the equations of Einstein's theory of general relativity in 1916. What the scientific community long considered to be a flaw in his solution turned out to be a black hole.

Scott, David Randolph (b. 1932) American astronaut who flew three space missions, the last one as Commander of Apollo 15, which landed on the Moon, and was the first expedition into outer space with astronauts specifically trained for scientific exploration. He writes and broadcasts on space travel, and advised on the film *Apollo 13*.

Serber, Robert (1909–97) American nuclear physicist who co-authored with Oppenheimer an early paper on neutron cores inside stars. He went on to play an important role in the Manhattan Project.

Shapley, Harlow (1885–1972) Director of the Harvard Observatory who offered Chandra a research position there.

Snyder, Hartland (1913–62) American physicist who, with Oppenheimer, wrote the first paper to explore what happens when a collapsing star becomes smaller than its Schwarzschild radius. During World War II he worked on the development of radar and in the 1950s made valuable contributions to the design of the most powerful particle accelerator of its day, the Cosmotron at the Brookhaven National Laboratory, Long Island, NY.

Sommerfeld, Arnold Johannes Wilhelm (1868–1951) German physicist renowned for his research and teaching.

Steigman, Gary (b. 1941) American astrophysicist at Ohio State University who has made important contributions to our understanding of the early universe.

Stoner, Edmund Clifton (1899–1968) English physicist who, in 1924, came close to discovering the Pauli exclusion principle. He then took a break from atomic physics and looked into the nature of white dwarf stars. He and Chandra arrived at similar results simultaneously and independently.

Stratton, Frederick John Marrian ('Chubby') (1881–1960) English astronomer known for his work in spectroscopy. He spent his professional career at Cambridge University and played an active role in the Royal Astronomical Society, where he was president from 1933 to 1935.

Strömgren, Bengt Georg Daniel (1908–87) Danish astrophysicist who was a friend of Chandra's in the 1930s and his colleague at the University of Chicago. In 1957 he become the first professor of astrophysics at the Institute for Advanced Study in Princeton. He returned to Copenhagen in 1967 to assume the prestigious honorary position first occupied by Niels Bohr.

Strutt, John William (Lord Rayleigh) (1842–1919) English scientist who researched in both experimental and theoretical physics. Chandra admired his mathematical rigour.

Struve, Otto Ludwig (1897–1963) Russian-born American astronomer who put Yerkes on the map by hiring dynamic young researchers such as Chandra.

Teller, Edward (1908–2003) Hungarian-born American physicist whom Chandra met in the 1930s. They became colleagues at the University of Chicago, but fell out over politics.

Thomson, William (Lord Kelvin) (1824–1907) Irish-born Victorian-era scientist whose research spanned electricity, magnetism, hydrodynamics, mathematical physics, geology and the theory of heat (thermodynamics). Many of his discoveries were highly practical as well as deeply theoretical. He played a key role in establishing the first successful transatlantic telegraph cable. Thomson formulated a temperature scale which relates to the laws of thermodynamics, known as the kelvin scale.

Thorne, Kip Stephen (b. 1940) American physicist known for his work on black holes, gravitational wave detection and time machines. Chandra spent part of 1971 with his group at Caltech.

Trimble, Charles John Agnew (1883–1958) Classmate of Eddington's at Trinity College, Cambridge, and his lifelong friend and companion. Trimble taught mathematics, and his students remembered him as 'a great trainer of minds'.

Ulam, Stanislaw (1909–84) Polish-born American mathematician who contributed important ideas to the development of the hydrogen bomb.

Volkoff, George Michael (1914–2000) Canadian physicist remembered for the famous paper he wrote with Oppenheimer on neutron stars.

Von Neumann, John (1903–57) Hungarian-born American polymath who discussed astrophysics with Chandra at Trinity College, Cambridge, during 1934 and 1935.

Waerden, Bartel Leendert van der (1903–96) Dutch scientist who did important work in mathematics, physics and the history of mathematics. In 1931 he became a colleague of Heisenberg's at the University of Leipzig.

Wali, Kameshwar C. (b. 1935) Indian-born American theoretical physicist who wrote the first biography of Chandra.

Wentzel, Gregor (1898–1978) German-born physicist who was a close friend of Chandra's at the University of Chicago.

Wheeler, John Archibald (b. 1911) American physicist who made important contributions to nuclear physics, the development of the atomic and hydrogen bombs, and general relativity, particularly black hole physics.

White, Richard Harold (b. 1934) American physicist at Livermore National Laboratory whose expertise with computers was essential to early research into supernovae with Stirling Colgate.

Whittaker, Edmund Taylor (1873–1956) English physicist who arranged for the posthumous publication of Eddington's book on his fundamental theory.

Woosley, Stanford ('Stan') Earl (b. 1945) American astrophysicist who studies how massive stars evolve from placid beginnings to violent deaths as supernovae. He carries out his extensive research at the University of California at Santa Cruz.

Yang, Chen Ning (b. 1922) Chinese-born American physicist who befriended Chandra as a young instructor in the physics department at the University of Chicago.

Zel'dovich, Yakov Borisovich (1914–87) Russian scientist who did critical work in the Soviet hydrogen bomb programme and made important contributions to astrophysics.

Zwicky, Fritz (1888–1974) Quirky Swiss-born American astronomer. One of his flashes of brilliance was to link supernovae with neutron star formation. He published this work with Walter Baade.

GLOSSARY

accretion disk A broad, flat disk of gas rotating around a black hole. The inner parts of the disk rotate faster than the outer ones, and as the inner particles rub against their slower neighbours they heat up from the friction and give off X-rays. The disk is continually being dragged into the black hole like water down a drain, from the inside outwards.

alpha particle The nucleus of a helium atom, made up of two neutrons and two protons.

annihilation The process by which a particle and an antiparticle come together and disappear in a blaze of light.

annihilation hypothesis The idea that the energy that powers stars is generated by the mutual annihilation of protons and electrons. This was proposed before the discovery of antimatter.

antimatter Matter made up of antiparticles.

antiparticle Particles of the same mass as ordinary particles except for quantities such as charge. Thus, for example, the *positron* is the antiparticle of the electron.

astronomer A scientist who uses telescopes to study cosmic objects.

astrophysicist A scientist who uses mathematical equations representing laws of nature to analyse and understand measurements made by astronomers.

atom The smallest part of an element that can exist and still characterise that element. An atom is made up of a nucleus with a positive charge and enough electrons outside the nucleus to give the whole a net electrical charge of zero.

atomic bomb (fission bomb) A bomb whose explosive energy comes from a chain reaction of nuclear fissions involving uranium or plutonium nuclei.

beta decay The transformation of a neutron into a proton, electron and neutrino.

binary system Two objects that rotate around a common centre, such as two stars, a star and a black hole, a star and a neutron star, a star and a white dwarf, two neutron stars or two black holes.

black hole The well in spacetime into which a collapsing star falls when it becomes extremely dense and its gravity is so enormous that it pulls space and time in around itself. The gravity inside a black hole is so strong that not even light can escape from it.

Bohr's theory of the atom The earliest successful theory of the atom based on quantum theory as it was known before the discovery of quantum mechanics. It is based on the visualisation of the atom as a minuscule solar system, made up of a central nucleus surrounded by electrons in specified orbits.

Bose–Einstein statistics A theory for calculating the probabilities for how values of position, velocity and spin are distributed among elementary particles whose spin values can be only whole numbers, including zero; photons are an example of particles which obey Bose–Einstein statistics.

centrally condensed configurations Milne's term for all stars except white dwarfs.

chain reaction A sequence of nuclear fissions in which neutrons are produced and go on to trigger future fissions, releasing further neutrons, and so on.

Chandrasekhar limit The maximum mass of a stable white dwarf, about 1.4 times the mass of the Sun.

Chandrasekhar–Schönberg limit The maximum mass that a star's helium core can attain without collapsing during the process of hydrogen burning.

charge of the electron The electrical charge carried by a single electron. Until the postulation of quarks in the 1960s, this was considered to be the basic unit of electrical charge. But quarks can have $\frac{1}{3}$ or $\frac{2}{3}$ of the charge of the electron.

classical physics (Newtonian physics) The physics of matter moving at speeds much less than the speed of light, and to which the special theory of relativity therefore need not be applied.

classical statistics The manner in which values of position and velocity are distributed among particles in a perfect gas.

cold degenerate matter Highly compressed matter no longer radiating light or heat.

collapsed configurations Milne's term for white dwarfs. In order to avoid the conclusion that white dwarfs collapse completely, Milne made the ad hoc hypothesis that, since all gas laws may be invalid deep in the interior of white dwarfs, there must be a core made up of some material that cannot be compressed any further under any circumstances.

core bounce A recoil that occurs when the core of a massive collapsing star reaches its limit of compression. At this point the core stiffens and recoils like a spring.

cosmic rays High-energy particles raining down on the Earth from outer space. They were discovered in 1911 and named by Robert Millikan in 1925.

cosmological redshift The lengthening of wavelengths of light from stars, caused by the expansion of the universe.

critical mass The minimum amount of uranium or plutonium needed to make an atomic bomb explode.

degeneracy The limit at which densely packed electrons cannot be compressed any further. They retaliate by generating an outward degeneracy pressure and becoming rock-hard.

degeneracy pressure The pressure exerted by the electrons in a gas in a condition of degeneracy. Degeneracy pressure is what makes metals hard and maintains the stability of white dwarf stars whose mass is below the Chandrasekhar limit.

degenerate core The rock-hard core that results from the degeneracy pressure of densely packed electrons.

density, average The mass of an object divided by its volume.

density of particles The number of particles divided by the volume of the enclosure that contains them.

deuterium An isotope of hydrogen whose atoms each consist of a single electron and a nucleus with a proton and a neutron. Deuterium, also known as 'heavy hydrogen', has identical chemical properties to ordinary hydrogen.

deuteron A nucleus of deuterium.

dimensionless number A quantity that describes a physical system or process and is a pure number, with no associated unit. An example is the fine-structure constant.

Dirac equation The mathematical equation that describes the properties of an electron.

Doppler shift The optical Doppler shift is the change in the wavelength of radiation due to the relative motion of the source and an observer. When a star moves away from Earth, its light appears of longer wavelengths, that is, it is redshifted; when it moves towards Earth, its light is blueshifted. This is important for measuring the speed of a star relative to Earth.

dwarf star A star that, like the Sun, lies on the main sequence of the Hertzsprung–Russell diagram.

Eddington's paradox Although white dwarf stars continually lose heat and contract, they lack enough energy to cool down and expand to the density of terrestrial matter.

electromagnetic force The force that governs electricity and magnetism.

electromagnetic spectrum The gamut of wavelengths and frequencies of light, or radiation, from the most energetic radiation (short-wavelength gamma rays), to the weakest (long-wavelength radio waves).

electron An elementary particle with one unit of negative electric charge and a spin of ½.

electroweak theory A unified theory of the electromagnetic force and the weak nuclear force.

element A substance that cannot be made into simpler substances by chemical means. Atoms with the same number of protons or electrons are atoms of the same element. Each element has a place in the periodic table. An atom's electrons determine its chemical properties, i.e. how it combines with other atoms.

elementary particle A subatomic particle, such as an electron, a proton, a neutron, a neutrino or a quark.

E-numbers Eddington's version of Dirac's spinors.

equation of state An equation that relates pressure to density.

event horizon The rim of a black hole. Anything falling past the event horizon cannot escape.

Fermi–Dirac statistics A theory for calculating the probabilities for how values of position, velocity and *spin* are distributed among electrons in a quantum gas or an atom.

fine-structure constant A pure number of great importance in atomic physics, related to the distance between closely spaced spectral lines, that is a measure of the strength with which electrons interact with one another. It incorporates the Planck constant, the speed of light and the charge of the electron.

fission bomb See atomic bomb.

fundamental theory Eddington's name for the theory he was working on that would unify the general and special theories of relativity with quantum theory.

fusion bomb See hydrogen bomb.

gas laws Mathematical descriptions of the properties of collections of particles. The particular gas law used depends on the density of particles in the gas. For very high densities, quantum mechanics must be taken into account.

gas pressure The net outward pressure exerted by the gas particles within a star as calculated from the gas law pertaining to that star, which may behave as a perfect gas or as a quantum gas.

general theory of relativity A widening of Einstein's special theory of relativity to include gravitational phenomena. It is essential to the understanding of black holes and neutron stars.

giant–dwarf theory of stars Russell's idea that dwarf stars do not obey the perfect gas law because of their high density, while low-density giants do. Eddington disproved this for stars on the main sequence of the Hertzsprung–Russell diagram.

giant star A very large, bright star lying above the main sequence of the Hertzsprung–Russell diagram.

gravitational pressure The net inward pressure caused by the gravitational attraction of the molecules of gas that make up a star.

gravitational redshift The shift of light towards longer wavelengths caused by increasingly strong gravitational fields.

gravitational waves Ripples in spacetime caused most dramatically by movements of objects with extremely intense gravitational fields such as black holes and neutron stars.

group theory The branch of mathematics that studies the symmetry properties of equations.

half life The time in which half a sample of radioactive nuclei decays. The half life of a nucleus is a property of the nucleus itself and does not depend on how many nuclei are packed together or on the temperature of the sample.

Hawking radiation Radiation emitted by a black hole through processes related to quantum physics.

helium The second most abundant element in the universe, with a nucleus containing two protons and two neutrons, surrounded by two electrons.

Hertzsprung–Russell (HR) diagram A graph that plots the brightness of stars against their temperature.

Hubble constant The rate at which the speed of recession of the galaxies increases with their distance from the Earth, following Hubble's law.

Hubble's law Galaxies recede from us at a speed proportional to their distances from the Earth.

hydrogen The simplest chemical element in the universe, with a single electron and a nucleus that is a single proton.

hydrogen bomb (fusion bomb, nuclear bomb) A bomb whose explosive energy comes from nuclear fusion, i.e. by combining light nuclei such as hydrogen and deuterium. An atomic bomb supplies the heat necessary to trigger these thermonuclear reactions.

isotopes Atoms of the same chemical element which differ in the number of neutrons in their nuclei. Examples are deuterium and tritium, which are isotopes of hydrogen, and ^{238}U and ^{235}U, both isotopes of uranium. Since the nuclei of different isotopes have the same number of protons, they are virtually indistinguishable in chemical reactions.

jet A beam of fast-moving gas emitted from the vicinity of a black hole.

Kerr black hole A spinning black hole defined only by its spin and mass.

law of conservation of energy The energy before and after a process occurs must be the same.

LIGO Laser Interferometer Gravitational Wave Observatory, a device for detecting gravitational waves.

LISA Laser Interferometer Space Antenna, an orbiting detector for gravitational waves planned for launch in 2011.

main sequence The diagonal line on the Hertzsprung–Russell diagram on which most stars appear. These stars are at the most robust stage of their life.

Manhattan Project The massive US effort during World War II to build an atomic bomb.

mass A measure of the amount of matter a body contains. A body's mass is the same wherever it is in the universe, whereas its weight depends on its gravitational environment. Since the masses of stars and other celestial bodies are so huge, they are often expressed as multiples of the mass of Sun.

mass–energy equivalence Mass is equivalent to energy and vice versa, as expressed in the iconic equation $E = mc^2$.

mass–luminosity relation The relationship between the mass and luminosity of stars on the main sequence of the Hertzsprung–Russell diagram.

mean molecular weight The average number of free electrons per nucleus after atoms are stripped of their electrons by the high-energy radiation inside stars. The exact figure depends on the mix of chemical elements in the star.

metric A measure of distance in spacetime.

neutrino An elementary particle with zero electrical charge. It was originally hypothesised to have no mass; today we know that it has a minute amount.

neutrino opacity The resistance of a star to the escape of neutrinos from its interior.

neutron An elementary particle with no electric charge and a mass about that of the proton.

neutron star A highly compact, dense star made up almost entirely of neutrons, with a mass about that of the Sun but only about 12 miles in diameter.

Newtonian physics See classical physics.

non-relativistic Describing events in which the speeds of objects are well below that of light, so that the special theory of relativity does not need to be applied.

non-relativistic degeneracy A degenerate gas of electrons in which the pressure is too low for electrons to move at speeds close to that of light.

non-relativistic degeneracy pressure The pressure in a degenerate gas of electrons in which the electrons are moving at speeds much less than that of light.

nova A star that undergoes a sudden increase in brightness to about a thousand times that of the Sun.

nuclear bomb See hydrogen bomb.

nuclear fission Splitting a nucleus. When a neutron hits a uranium nucleus, it can split it, releasing an enormous amount of energy as predicted by the mass–energy equivalence. Atomic bombs and nuclear reactors are powered in this way.

nuclear force The attractive force between the particles that make up the nucleus, such as neutrons and protons. Unlike an electric force, in which oppositely charged bodies attract, in the nuclear force there is attraction between a neutral neutron and a positively charged proton.

nuclear fusion The binding of light nuclei such as protons, deuterons and alpha particles under the high temperatures and densities found in the interior of stars. These thermonuclear reactions generate the energy that makes stars shine and hydrogen bombs explode.

nucleus The massive central part of the atom, with a positive charge. The nucleus contains protons and (except in the case of hydrogen) neutrons.

opacity A measure of the resistance that light encounters as it tries to escape from matter.

Pauli exclusion principle No two electrons in an atom or a dense gas can simultaneously have the same position, velocity and spin.

perfect gas A hypothetical collection of particles whose density is low enough for its pressure, volume and temperature to be predicted by the perfect gas law.

perfect gas law The equation of state for how the particles in a perfect gas behave. If the temperature of a perfect gas is held constant, an increase in pressure results in a decrease in volume, and vice versa; if pressure is held constant, an increase in volume results in an increase in temperature, and vice versa.

periodic table A table of the elements ordered by increasing number of electrons, which allows comparison of their chemical properties.

photon (light quantum) The particle of light, corresponding to an electromagnetic wave according to wave–particle duality.

Planck constant A universal constant of nature at the basis of quantum theory.

planetary nebula A luminous shell of gas ejected from a dying red giant star. What remains at the centre is a white dwarf.

positron The antiparticle of the electron. Electrons and positrons differ only in that the positron has a positive charge.

proton An elementary particle with a positive electric charge. It is the nucleus of the hydrogen atom.

pulsar A rotating neutron star that emits radio signals at precise intervals.

quantum electrodynamics A theory that combines quantum mechanics and special relativity to study the interaction of light with electrons.

quantum gas A gas that has to be studied using quantum physics because the density of its electrons is high enough that their wave–particle duality has to be taken into account.

quantum mechanics The theory that describes how elementary particles such as electrons and protons interact.

quantum physics The study of light and matter under conditions in which phenomena such as wave–particle duality become important.

quantum statistics The manner in which values of position and velocity are distributed among particles in a quantum gas.

quark Elementary particle currently considered to be the fundamental building block of matter; neutrons and protons are made up of quarks.

quasar The bright centre of a young, distant galaxy, powered by a super-massive black hole.

radiation Energy travelling in the form of photons or as a stream of particles.

radiation pressure The outward pressure exerted by the light produced within a star.

radiative transfer The study of how radiation moves through matter.

radioactive dating Since the half life of a radioactive substance is not affected by temperature or pressure, radioactive samples decay at a predictable rate. Thus reasonable estimates of the original amount of uranium in a rock provide a way to measure the time that has elapsed since the rock was last molten and so to ascertain the age of the Earth.

radio galaxy A galaxy that emits strongly at radio wavelengths.

radio lobes Huge, radio-wave-emitting clouds of gas stirred up by jets of high-velocity electrons emitted by a quasar.

Raman effect A means to explore the structure of molecules by investigating the frequency of the light they scatter.

red giant A bloated star ten to a hundred times the diameter of the Sun and in the twilight of its life.

redshift The shift of spectral lines to redder (longer wavelengths) in the spectra of astronomical objects apparently moving away from the Earth, such as quasars or galaxies.

relativistic Relating to objects moving at speeds close to the speed of light, as described in the special theory of relativity.

relativistic degeneracy The limit of compression for a gas of electrons of sufficiently high density that they are moving at speeds close to that of light.

334 EMPIRE OF THE STARS

relativistic degeneracy pressure The way in which a densely packed gas of electrons travelling at speeds close to that of light retaliates to being squeezed. The electrons generate an outward pressure that makes them rock-hard, which explains the stability of white dwarf stars whose mass is below the Chandrasekhar limit. (See relativistic degeneracy.)

relativity See general theory of relativity, special theory of relativity.

Saha's equation An equation, based on the quantum theory of the atom, which relates the spectral lines from the light of a star to the elements that make up the star's surface layers.

Schwarzschild black hole A black hole that is not spinning.

Schwarzschild radius The size of a black hole, akin to its event horizon. (The Schwarzschild radius is a distance, the event horizon is a surface.) Every object has a Schwarzschild radius.

Schwarzschild singularity At the Schwarzschild radius, Schwarzschild's solution to Einstein's general theory of relativity yields a singularity. The Schwarzschild radius is the size of the opening in spacetime (event horizon) into which a collapsing star falls on the way to becoming a singularity. In other words, the Schwarzschild radius (event horizon) is not the measure of a singularity, but of the opening of a black hole. The star within the black hole becomes a singularity when it collapses to a point of infinite density and zero volume.

shock front The front of a shock wave; a narrow band of matter denser than the surrounding medium. A shock front can result from a violent pressure change, such as that caused by an explosion.

shock wave A wave made up of particles travelling at a speed that exceeds the speed of sound in the gas through which the wave is travelling. The sonic boom which a supersonic aircraft makes is produced by a shock wave.

singularity The point in a physical theory at which a quantity becomes infinite.

spacetime The three dimensions of space and one of time considered as a single four-dimensional entity.

special theory of relativity The theory used to study matter moving at speeds comparable to that of light.

spectral lines When light of different frequencies is passed through a spectrograph, a spectrum emerges made up of lines each corresponding to a certain frequency.

spectrograph A device used to break up light into its component frequencies.

speed of light Einstein's special theory of relativity asserts that the speed of light is constant no matter what the motion of the source of the light. It is usually taken to be 186,000 miles per second. Thus the speed of light is the cosmic speed limit, and time is a relative quantity. Since light takes time to travel across space, we observe the universe as it was in the past – the more so the farther away we look.

spin A property of elementary particles that emerged from the analysis of spectral lines. Its mathematical representation has a form analogous to what one would expect for a spinning top – hence the term spin.

GLOSSARY 335

spinors Mathematical quantities, intermediate between vectors and tensors, used in the Dirac equation.

standard model Eddington's theory of the constitution of the stars.

Stoner–Anderson equation Equation of state for a relativistically degenerate gas very similar to that deduced by Chandra. Chandra's later theory of white dwarfs was more detailed and complete than Stoner's.

supergiant A rare type of star, typically a thousand times the diameter of the Sun and more than ten times the diameter of a red giant.

supermassive black hole A black hole millions of times the mass of the Sun.

supernova The explosion of a massive star powered by the gravitational energy released by the collapse of the star's core, a billion times as luminous as the Sun.

supernova relic The neutron star left behind after a supernova explosion.

supernova remnant The expelled remains of a supernova, visible as a small nebula.

thermonuclear reactions Nuclear fusion reactions that occur in the dense interior of stars and also provide the power for hydrogen bombs.

topology The branch of mathematics that studies the properties of surfaces when they are distorted.

tritium An isotope of hydrogen, each of whose atoms consists of a single electron and a nucleus with a proton and two neutrons. This is an even heavier version of hydrogen than deuterium.

Type Ia supernova The explosion of a white dwarf in a binary system which has accreted enough mass from its companion to exceed the Chandrasekhar limit.

Type II supernova The death throes of an old, massive star undergoing gravitational collapse.

uncertainty principle In quantum theory, position and velocity cannot be measured simultaneously with the same accuracy: the more accurately the position of an elementary particle is measured, the less accurately its velocity is known.

universal constant of nature Numbers at the foundation of scientific theories, including the charge of the electron, the Planck constant and the speed of light.

wave–particle duality Elementary particles can behave as either a wave or a particle.

weak nuclear force The force that governs radioactive processes such as beta decay.

white dwarf An old star that has burnt up all of its nuclear fuel and contracted to about the size of the Earth, with a mass less than the Chandrasekhar limit.

NOTES

PROLOGUE

1. Eddington (1936b), p. 391.

CHAPTER 1

1 Details of the 11 January 1935 meeting are in *The Observatory*, **58**, 37–9. Chandra's lecture was based on his (1934b), and a summary is in Chandrasekhar (1935a).
2. Chandrasekhar (1977), p. 33.
3. Quoted in Wali (1991), p. 124.
4. Chandrasekhar (1980a), p. 3.
5. The lecture room dates from 1874, when the RAS moved to Burlington House. In 1969 it was partitioned and reconfigured to make two offices.
6. Quoted in Kanigel (1991), p. 277.
7. Chadwick (1961).
8. Letter from Chandra to his father, 8 June 1933.
9. Chandrasekhar (1979b).
10. Chandrasekhar (1979a), p. 5.
11. Chandrasekhar (1977), p. 14.
12. Einstein (1923), p. 482.
13. Chandrasekhar (n.d.2), p. 5. This passage is based on Chandrasekhar (1934b), p. 377.
14. Eddington (1935a), p. 38.
15. Chandrasekhar (1977), p. 33.
16. Eddington (1935a), p. 38.
17. Chandrasekhar (1977), p. 31.
18. Letter from McCrea to Chandra, 16 January 1935.
19. Chandrasekhar (1980c), p. 35.
20. Chandrasekhar (1977), p. 34.
21. Chandrasekhar (1977), p. 29.
22. Chandrasekhar (1977), pp. 27–8.
23. Chandrasekhar (1977), p. 34.
24. Chandrasekhar (1977), p. 34.

25. Milne (1935a).
26. Chandrasekhar (n.d.4).
27. Chandrasekhar (1980a), p. 2.
28. McVittie (1978), p. 18.
29. Quoted in Wali (1991), p. 128.
30. Chandrasekhar (1979b), p. 8.

CHAPTER 2

1. Chandrasekhar (1977), p. 4.
2. Chandrasekhar (n.d.1), p. 16.
3. Chandrasekhar (n.d.1), p. 17.
4. Chandra's six sisters, in order of age, were Rajalakshmi (Rajam) (the oldest), Balaparvathi (Bala), Sarada, Vidya, Savitri and Sundari. His three brothers were Visvanathan, Balakrishnan and Ramanathan (Ramanath).
5. Quoted in Venkataraman et al. (1983), p. 13.
6. Unless indicated otherwise, quotations from Lalitha Chandrasekhar in this chapter are from interviews by the author in Chicago, 1 November 2002.
7. Gandhi (1983), p. 40.
8. Quoted from Wolpert (1997), p. 295.
9. Balakrishnan (1983a), p. 18.
10. I am indebted to Professor Takeshi Oka, of the University of Chicago, for allowing me to quote from a fax message sent to him by Ramaseshan on 11 June 1998.
11. Balakrishnan (1983b), p. 2.
12. Ayyar (n.d.), p. 7.
13. Balakrishnan (1983b), p. 3.
14. Quoted in Venkataraman et al. (1983), p. 16.
15. Chandrasekhar (1977), p. 3.
16. Chandra's recollection quoted in Wali (1991), pp. 56–7.
17. Letter from Chandra to Balakrishnan, 5 June 1931.
18. Letter from Chandra to Balakrishnan, 12 May 1931.
19. Chandrasekhar (1977), p. 7.
20. Wolpert (1997), p. 313.
21. Eddington (1926), p. 393.
22. Eddington (1926), p. 1.
23. Eddington (1936b), p. 388. Professor Harlow Shapley, who was in the audience when Eddington delivered these vivid remarks, was a famous astronomer of the time and director of the Harvard College Observatory, which operated the Oak Ridge reflector telescope located in Harvard, Massachusetts. He will put in an appearance in later chapters.
24. Russell (1945), p. 134.
25. Einstein suggested that light could behave as particles in 1905. In the most astounding father-and-son act in the history of physics, J. J. Thomson discovered the electron as a particle, for which he received

the Nobel Prize for Physics in 1907; George, J. J.'s son, discovered the wave nature of the electron and won a Nobel in 1937.
26. Sommerfeld (1928a,b).
27. Chandrasekhar (1928).
28. Quoted in Wali (1991), p. 63.
29. Chandrasekhar (1977), p. 10.
30. Letter from Chandra to his father, 4 June 1929. Chandra refers to his (1929a); to 'The equipartition principle and the new statistics', which never appeared; and to a note to *Nature*, 'The Einstein method of deriving Planck's formula and the new statistics', which also was never published.
31. Later published as Chandrasekhar (1929b).
32. Chandrasekhar (1929b).
33. Chandrasekhar (1930). Despite its title, the 'Phil. Mag.' is actually a physics journal.
34. Letter from Chandra to his father, 4 June 1929.
35. Rosenfeld (1967), pp. 118–19.
36. Heisenberg (1962), p. 6.
37. Balakrishnan (1983b), p. 6.
38. Letter from Chandra to his father, 16 October 1929.
39. Balakrishnan (1945), p. 74.
40. Letter from Chandra to his father, 4 January 1930.
41. Quoted in Wali (1991), p. 65.
42. Quoted in Wali (1991), p. 69.
43. Wali (1991), pp. 69–71.
44. Venkataraman et al. (1983), p. 17.
45. Wali (1991), pp. 172–7.
46. Lalitha's two older sisters became doctors, and her younger sister did a master's degree in Sanskrit. Lalitha's was in physics.
47. Chandrasekhar (1977), p. 10.
48. Letter from Chandra to Balakrishnan, 31 July 1930.
49. Letter from Raman to Chandra, 28 July 1930.

CHAPTER 3

1. Eddington (1929b), p. 33.
2. Chandrasekhar (1977), p. 27.
3. Interview by the author with David Dewhirst, 16 September 2002.
4. Chandrasekhar (n.d.5).
5. Letter from Williams to Struve, 13 June 1958.
6. Quoted in Douglas (1956), p. 121.
7. Chandrasekhar (1977), p. 36.
8. Cowling (1966), p. 128.
9. Douglas (1956), p. 2.
10. Chandrasekhar (1987a), p. 95.
11. See Warwick (2003) for a discussion of Eddington's training at Cambridge.

12. Bollobás, in Littlewood (1953), p. 12.
13. Bollobás, in Littlewood (1953), p. 18.
14. Kanigel (1991), p. 143. Kanigel offers an insightful discussion on the personal life of Oxbridge dons during the 1920s and 1930s.
15. See, for example, Littlewood (1953), p. 120.
16. Quoted in Kanigel (1991), p. 139.
17. Quoted in Kanigel (1991), p. 139.
18. Quoted in Kanigel (1991), p. 139.
19. Payne-Gaposchkin (1968), pp. 2–3.
20. Chandrasekhar (1980a), p. 16.
21. Douglas (1956), p. 30.
22. Douglas (1956), p. 7.
23. Quoted in *The Blue* (the journal of Christ's Hospital school), Vol. 75, September 1958, p. 179.
24. From Eddington's cycling diaries, held at Trinity College, Cambridge.
25. Douglas (1956), p. 34.
26. Kanigel (1991), p. 144.
27. Quoted in Douglas (1956), p. 110.
28. At the time, astrophysicists assumed that stars, including the Sun, had much the same chemical make-up as the Earth and thus consisted primarily of oxygen, iron, potassium, calcium, sodium, aluminium, magnesium and silicon. They did not include hydrogen as there is very little in the Earth. It would turn out that healthy stars are actually made up primarily of hydrogen.
29. Even before Heisenberg's discovery of the new atomic physics – quantum mechanics – in 1925, physicists were questioning the accuracy of Bohr's theory. They realised that atoms do not behave like minuscule solar systems and that such visualisations were actually holding back progress. One of the discoveries of quantum theory was that electrons can be waves and so are spread out rather than taking a well-defined orbit around a nucleus. Nevertheless, Bohr's theory was invaluable as a first step and is still useful today as a way of understanding the chemical properties of atoms.
30. Electrons are so minuscule that astrophysicists assumed that they could safely ignore their dimensions and the force with which they repel one another, an acceptable simplification as the positive charge of the nuclei had a neutralising effect.
31. Instead of expressing the perfect gas law as the product of the pressure and volume of a gas in proportion to its temperature, astrophysicists usually express the pressure of a gas in terms of its density, temperature and chemical composition. This is the equation of state of the perfect gas.

    Scientists also studied 'imperfect gases', extending the theory of the perfect gas to incorporate the size of the particles that make up a gas and the forces between them.
32. How bright a star appears to us on the Earth depends on how far away it is. Stars emit a certain amount of light energy per second – their luminosity. The brightness of a star as observed from the Earth, referred

to as the star's apparent brightness, is calculated as its luminosity divided by its distance squared. To measure the distance, astronomers observe how the star's position changes in relation to more distant stars over a period of six months, the time the Earth takes to complete half an orbit around the Sun. This tiny change in position is called stellar parallax. If the star is so far away that its stellar parallax is too small to measure, its distance can found by measuring its brightness, spectral lines and temperature.

33. See Eddington (1919). Modelling a Cepheid as if it were a perfect gas enabled Eddington to calculate that its period times the square root of its density was a constant. This fitted the observation that the star's brightness increased most rapidly after it had passed through its minimum radius. The reason was that its density and temperature were greatest at that point, because its gas particles were squeezed together most tightly.

34. The term 'light' is used for the tiny part of the electromagnetic spectrum which we can see. 'Radiation' is the more general name for the entire spectrum. Wavelength is the distance between the peak of one wave and the next peak, while frequency is the number of peaks that pass a given point per unit time; thus the longer the wavelength of the light, the shorter its frequency. The wavelength of a light wave times its frequency equals the speed of light.

35. To measure a star's surface temperature, astronomers first assume that it radiates as if it was a 'blackbody', a hypothetical object that is both a perfect emitter and a perfect absorber of radiation. The radiation spectrum – or wavelengths – for a black body at different temperatures can be drawn on a graph that compares brightness, or intensity, with wavelength. The curve rises smoothly from zero (at zero wavelength and zero intensity) to a peak, and then tails off to zero at higher wavelengths. The peak is usually not in the middle of the curve but skewed towards low or high wavelengths, depending on the black body's temperature; there are differently shaped curves for different temperatures. Measurements of light intensity from various stars indicate that they do indeed emit light much as black bodies would. This permits astronomers to use the mathematics of black body radiation, developed by the German physicist Max Planck in 1900.

36. Russell (1914), p. 282.

37. Russell (1914), p. 282.

38. Quoted in DeVorkin (2000), p. 126.

39. Some years later, this binary system was renamed 40 Eridani A and 40 Eridani B. The 'o' is the Greek letter omicron.

40. Russell (1941), p. 6.

41. Adams (1914, 1915).

42. Measurement of orbital parameters, such as how long it takes the two stars to make a complete revolution around a common centre (the orbital period), the distance of the binary system from the Earth (found by measurement of stellar parallax) and how much the shape of the

orbit differs from a circle (the orbital eccentricity), gives enough infor- mation to calculate the mass of each star. Each star's radius can be calculated from its surface temperature and luminosity. Average density can be calculated from the ratio of the star's mass to its volume; taking the star to be a sphere, its volume is four times pi times its radius cubed, divided by three.

43. Astronomers measure mass in grams, distance in centimetres and time in seconds. They use the mass, not the weight, of an object because mass remains the same wherever the object happens to be in the universe; its weight, however, varies depending on gravity. On the Moon, for exam- ple, an object weighs a sixth of its weight on Earth, because the Moon's gravity is so much weaker.

44. Eddington (1926), p. 171.

45. The calculation is essentially as follows: the amount of energy generated by a collection of gas particles which has been reduced from a ball with an infinitely large radius to the present radius of the Sun is divided by the average amount of energy emitted by the Sun per second.

46. Eddington (1917b). At this time the only known 'elementary particles' were electrons and protons. No one speculated seriously on their origin.

47. Until the discovery of the neutron in 1932, physicists assumed that the nucleus was made up of protons and electrons. They assumed that the helium nucleus was made up of four protons and two electrons, which would give a net positive charge to balance the negative charge of the two orbiting electrons. Since the electron's mass is minuscule compared with the proton's, it was ignored in these calculations.

48. Aston (1938), p. 106.

49. Eddington (1920), p. 19.

50. Eddington (1926), p. 301.

51. Eddington (1926), p. 17.

52. Eddington (1926), p. 19.

53. Astrophysicists did not speculate seriously at this point in time on the origins of heavier elements. They just assumed they were part of the interstellar dust from which stars were formed.

54. Eddington used data on the brighter of the two stars making up the binary system Capella – Capella A, more than 10^{14} miles away in the constellation Auriga. According to Eddington, the observational data for Capella were 'unusually complete', resulting in extremely precise estimates on its mass, surface temperature, radius and brightness.

55. Eddington (1924a), p. 786.

56. Eddington (1926), pp. 146, 243, and Chandra (1939), p. 278.

57. Eddington (1926), p. 244.

58. Letter from Russell to Payne, 14 January 1925, quoted in DeVorkin (2000), p. 240.

59. Payne (1925), p. 248.

60. Payne-Gaposchkin (1968), p. 17. After four years at Cambridge, Payne was given a certificate which essentially said that if she were a man, she would have been awarded a B.A. Impressed with a lecture given by

Shapley at the 1923 meeting of the British Astronomical Association, Payne approached him and asked if she could continue her research at Harvard. There she was awarded a Ph.D. from Radcliffe (the women's college at Harvard). Payne was the first person to receive a Ph.D. for research conducted at the Harvard Observatory and the first woman to become a full professor at Harvard. Otto Struve, the powerful Director of the Yerkes Observatory, declared her Ph.D. 'the most brilliant thesis ever written in astronomy' (Struve and Zebergs 1962, pp. 220–1).

61. For details of this controversy see DeVorkin (2000), pp. 205–20.
62. Strömgren (1932), pp. 118–22. See also Strömgren (1976), pp. 20–1.
63. Eddington (1932).
64. Quoted in Douglas (1956), p. 40.
65. Quoted in Douglas (1956), p. 40. General relativity predicts that light near the Sun will be deflected by 1.75 seconds while Newton's theory of gravity predicts half that. Difficulties in making the measurements, in particular the very brief time available during an eclipse, led to errors in Eddington's data and his best result was 1.98 plus or minus 0.16 seconds. Nevertheless, his results clearly supported Einstein's theory over Newton's. Measurements which astronomers have made since 1919 on over 380 stars support Einstein's prediction even more strongly. See Weinberg (1972), pp. 188–94, and Misner et al. (1970), pp. 1104–5.
66. *New York Times*, 9 November 1919.
67. Quoted in Douglas (1956), p. 43.
68. Chandrasekhar (n.d.5).
69. Eddington (1936b), p. 390.
70. Assuming that the gas of electrons inside a white dwarf is a perfect gas, Eddington estimated the internal temperature to be about a billion degrees kelvin. See Eddington (1926), p. 174.
71. Letter from Eddington to Adams, 13 January 1924.
72. Letters from Eddington to Adams, 13 January 1924 and 22 March 1924.
73. Eddington (1926), p. 6.
74. Plummer (1948), p. 117.
75. Eddington (1926), p. 172.
76. Eddington (1926), p. 173.

CHAPTER 4

1. Milne (1945a), p. 62.
2. Milne (1923) and (1945b), p. 87.
3. Letter from Chandra to his father, 10 October 1930.
4. Chandra (1980b), p. 4.
5. See Weston-Smith (1990) for details of the development that follows. Meg Weston-Smith is Milne's daughter.
6. Quoted in Weston-Smith (1990), p. 242.
7. When the English scientist H. G .J. Moseley was killed at Gallipoli in 1915, at the age of twenty-eight, it caused uproar in the British scientific

community. The government realised that it should make better use of its scientists in an era in which scientific input was becoming more and more relevant to warfare. Moseley almost certainly would have won a Nobel prize for his work in atomic physics.

8. Bohr achieved this by combining the Newtonian physics for planetary motions with results from quantum physics in a way that guaranteed the stability of atoms. In 1913 other scientists were trying to explain why most atoms were stable. Bohr realised the futility of this approach and instead adopted the daring strategy of taking atomic stability as given. From this he concluded that electrons can occupy only certain orbits, in contrast to the Earth, which can wobble in its journey around the Sun.

9. Letter to Bohr from George Hevesy, 23 September 1913, quoted in Pais (1986), p. 208.

10. In numerical terms, electrons can have half a unit of spin.

11. Fowler's theory of white dwarf stars is based on an equation that relates the pressure of a gas of electrons to its density. The perfect gas equation, in contrast, relates the gas's pressure to its temperature as well as its density. The absence of temperature in Fowler's theory is an unexpected result of quantum physics. When a white dwarf has cooled off completely its temperature is effectively zero. If it were a perfect gas then its gas pressure would also be zero and it would collapse to nothing, as Eddington had predicted. But the Pauli exclusion principle averts this undesirable result by bringing in an outward pressure – degeneracy pressure – which operates even when the star is at zero temperature and prevents it from being crushed as it cools off.

   In other words, even though astronomers measured white dwarfs as being very hot and very dim, physicists could analyse them as if they were cold and dark. When a white dwarf becomes cold and dark, of course, we can no longer see it.

12. Fowler (1926), p. 115.

13. Letter from P. G. Gostline to McCrea, undated, but can be taken as having been written shortly after Milne's death on 21 September 1950.

14. McCrea (1951), pp. 162–3.

15. Chandra (1979b), p. 3, and McCrea (1951), pp. 163–4.

16. Eddington (1930a), p. 809.

17. McCrea (1951), p. 164.

18. Interview with Meg Weston-Smith 18 September 2002.

19. Interview with Meg Weston-Smith, 18 September 2002.

20. McCrea (1951), p. 165.

21. Eddington (1929a).

22. Letter from Milne to Geoffrey Milne, 23 March 1930.

23. See, for example, Milne (1930a), pp. 332–4.

24. Eddington (1930b).

25. Letter from Milne to Herbert Dingle, 1 August 1939.

26. Eddington (1930c), p. 210, letter to the editor dated 19 June 1930.

27. Milne (1930b), p. 4.

28. Jeans (1931), p. 37.

29. Letter from Heisenberg to Pauli, 31 July 1928, in Pauli (1979), p. 467. See also Miller (1994), pp. 30–1. In 1931 Dirac brilliantly cut through this apparent inconsistency. He showed that these weird particles were actually antimatter. They were positrons, antiparticles of electrons, with the same mass as electrons and identical in every respect except that electrons have a negative charge while positrons have a positive charge. Although positrons and protons have the same charge, they are not the same. Among their many differences is that the proton is ten thousand times as massive. When an electron and a positron touch they annihilate each other, producing pure energy in the form of light. That same year, positrons were discovered in cosmic rays by Carl Anderson, working at the California Institute of Technology (Caltech). Eddington had to rephrase his suggestion that stars shine as a result of the annihilation of electrons and protons. In fact, it was as a result of the annihilation of electrons and positrons.

30. Eddington (1958a), pp. xi–xiii.

31. Eddington never specified what the 'E' in E-numbers stood for. Perhaps it stood for 'Eddington'!

32. Eddington (1959), pp. 230–1.

33. Sommerfeld (1923), p. vii. This is an extract from the 'Preface to the First German Edition', of 1919.

34. Eddington (1959), p. 237.

35. Russell (1945), p. 135.

36. Quoted in Kilmister (1994), p. 117.

37. Eddington calculated the rate of recession of galaxies as 467 miles per second for each 19 million trillion miles (a megaparsec) a galaxy is distant from the Earth. Astronomers had measured 340 miles per second per megaparsec. The issue of galaxy recession is taken up in Chapter 13.

38. Eddington assumed that the universe is a degenerate gas, like that inside a metal. He used quantum statistics to calculate the total number of electrons if all levels were filled up to the highest possible level of energy. This he assumed to be the energy of an electron as deduced from the mass–energy relation $E = mc^2$, where m is the electron's mass. However, he fudged the final result in order to obtain the number of electrons he wanted.

39. Quoted in Chandrasekhar (1987a), p. 136. An apt page number!

40. For an example of Pauli's mysticism, see Pauli (1955).

41. Conversation with the late Victor F. Weisskopf, 10 April 1986. Other versions of this story are in Pais (2000), pp. 250–1, and Enz (2002), pp. 533–4.

42. Letter from Pauli to Oskar Klein, a colleague, 18 February 1929, in Pauli (1979), p. 491.

CHAPTER 5

1. Quoted in Ramaseshan (1996), p. 104.
2. Letter from Raman to Chandra, 28 July 1930.

3. Chandrasekhar (n.d.2), p. 2.
4. Chandrasekhar (1977), p. 13.
5. Raman (1930).
6. Chandrasekhar (n.d.2), p. 2.
7. Chandrasekhar (1931c), p. 81.
8. Chandra calculated the limit to be 0.91 times the mass of the Sun. By 1934 he had made a more precise calculation and discovered that the maximum mass depended on the white dwarf's chemical constitution. The exact figure, however, was less important than the fact that there was a maximum mass beyond which a white dwarf would shrink to nothing. The figure currently used as Chandra's maximum mass – 1.4 times the mass of the Sun – first appeared in 1939 in a paper by the Russian physicist George Gamow – see Gamow (1939b). By then the physicist Hans Bethe had discovered the nuclear processes that generate stellar energy, greatly enhancing our knowledge of the constitution of a star.

   Strictly speaking, degeneracy pressure depends on a star's temperature. Fowler and Chandra both restricted themselves to the case of a white dwarf at zero temperature, where the star has cooled off and become rock-hard, and thus where there is maximum degeneracy. This was a reasonable approximation and it kept their equations simple. If all white dwarfs were actually at zero temperature they would be invisible. Nevertheless, for certain temperatures and densities astrophysicists can model white dwarfs as if they were a gas of electrons at zero temperature.
9. Chandrasekhar (1980b), p. 3.
10. Balakrishnan (1983b), p. 3.
11. Chandrasekhar (1977), p. 9.
12. Letter from Chandra to his father, 4 September 1930.
13. Chandrasekhar (n.d.2), p. 3.
14. Stoner (1930). In 1929, Wilhelm Anderson, a thirty-nine-year-old physicist at the University of Tartu in Estonia, informed Stoner that he had omitted to take special relativity into account in his earlier papers – see Anderson (1929). But Stoner quickly found that his inclusion of special relativity was incorrect. Stoner's subsequent correct result for relativistic degeneracy (Stoner 1930) – the same as Chandra's – is sometimes referred to as the Stoner–Anderson equation.
15. Stoner (1930), p. 953. Eddington had been in correspondence with Stoner regarding new data on the mass of white dwarfs. He recommended for publication two of Stoner's subsequent papers on relativistic degeneracy (Stoner 1932a,b).
16. Chandrasekhar (1977), p. 16.
17. Letter from Chandra to his father, 22 November 1930.
18. Chandra often gleefully tells the story (Chandrasekhar n.d.2, p. 3; 1977, pp. 16–18). The referee thought that Chandra's equation of state for a relativistically degenerate gas was incorrect. In the paper as originally submitted Chandra did not derive this result, but he did so in his reply to the referee, who then withdrew the criticism. Chandra felt that he

had to speak out on a point of principle: namely, that while his paper might not be worth publishing because the referee felt it was not interesting, the issue the referee raised was the 'supposed invalidity' of an equation, which was not the case. Chandra never forgot this episode. When he became editor of the *Astrophysical Journal* in 1952, he looked up the correspondence files to find out who the referee was. It turned out to have been the physicist Carl Eckart, who was then at the University of Chicago.

19. Chandrasekhar (1977), pp. 14, 15.
20. Letter from Chandra to his father, 16 September 1930.
21. Letter from Chandra to his father, 10 October 1930.
22. Letter from Chandra to his father, 19 November 1930.
23. Letter from Chandra to Balakrishnan, 16 October 1930.
24. Letter from Chandra to his father, 30 September 1931.
25. Letter from Chandra to his father, 19 November 1930.
26. Letter from Chandra to his father, 26 November 1930.
27. The Dirac equation, which stunningly represented the electron in terms of both relativity theory, which deals with large-scale phenomena, and quantum theory, which deals with atoms.
28. Letter from Chandra to his father, 16 January 1931.
29. Letter from Chandra to Balakrishnan, 21 January 1931.
30. Chandrasekhar (1979b).
31. Letter from Milne to Chandra, 2 November 1930. These papers are Chandrasekhar (1931a,b).
32. Chandrasekhar (1977), p. 49.
33. Milne (1931), p. 34.
34. Eddington (1931a), pp. 34, 35, 36.
35. Jeans (1931), pp. 36, 37, 38. A differential equation is a mathematical way of representing how one quantity behaves when other quantities change. For example, Newton's physics contains a differential equation that shows how an object's position in space varies over time as the object is falling to the ground. In stellar modelling the differential equations are much more complicated. A star's internal pressure depends on its volume and on the mix and distribution of chemical elements, among other variables. There are several different equations that have to be solved, and the quantities they contain are often represented in each of them in complicated ways.
36. Smart (1931), p. 38.
37. Hardy (1931), p. 40.
38. Cowling (1978), p. 20.
39. Letter from Chandra to George Kenat, 28 June 1983.
40. Merrill (1938), p. 84.
41. Letter from Chandra to his father, 10 February 1931. Chandra made a point of misspelling 'occasionally' to irritate his punctilious father. As late as March 1937, when Chandra was at Yerkes Observatory and Otto Struve was director there, Babuji wrote in exasperation, 'I am quite pleased over the life you are leading, both Social and

University at William's Bay. Well, you ought to spell the word "occasion" properly I must say, when you are talking of entertaining Dr. and Mrs. Struve and family' (Letter from Chandra's father, March 1937).

42. Letter from Milne to Chandra, 16 January 1931.
43. Letter from Milne to Chandra, 16 January 1931.
44. Letter from Milne to Chandra, 16 February 1931.
45. Eddington (1931c).
46. Eddington (1931c), p. 446.
47. Chandra would later prove that such temperatures and pressures were indeed possible inside real stars.
48. Eddington (1931b), p. 98.
49. Letter from Eddington to Stoner, 28 February 1932.
50. Letter from Chandra to his father, 30 September 1931.
51. Letter from Chandra to his father, 22 May 1931.
52. Letter from Chandra to Balakrishnan, 12 May 1931.
53. Letter from Chandra to Balakrishnan, 6 June 1931.
54. Letter from Chandra to Balakrishnan, 23 March 1936.
55. Letter from Milne to Chandra, 17 June 1931.
56. Letter from Milne to Dingle, 5 June 1931.
57. Letter from Milne to Dingle, 8 September 1931.
58. Letter from Milne to Dingle, 15 September 1931.
59. Letter from Milne to Chandra, 26 June 1931.
60. Letter from Chandra to his father, 25 June 1931. Chowla returned to the Government College in Lahore and produced a remarkably rich body of research, with several theorems named after him. When the country was partitioned into India and Pakistan, at midnight on 14–15 August 1947, Chowla and his family fled to Delhi. From there they moved to the United States, where Chowla continued his distinguished career. He died on 10 December 1995.
61. Letter from Einstein to Born, 4 December 1926. This is also connected to Heisenberg's uncertainty principle, which asserts that an electron's position and velocity cannot both be measured at the same time with absolute accuracy. In Newtonian physics, however, accurate knowledge of these two quantities is essential to predict an electron's position with precision as it moves through space and time.

Born always thought he was denied the kudos owed him for the discovery of quantum mechanics. He also believed that his rather late Nobel prize (in 1954) was the result of Byzantine plots by major scientists, such as Einstein, who disputed his discovery of the deep meaning of the wave function, combined with the deterioration in his relations with his former assistants, who had bypassed him in fame and professional standing. Being Jewish, Born was forced to leave Germany in 1933 and had some problems obtaining a position. He spent two years at Cambridge as Stokes Lecturer (Chandra had little contact with him there), then six months in Bangalore at the Indian Institute of Science, working with Raman. In 1935, Edmund T. Whittaker obtained for

Born the Tait Professorship in Edinburgh, from which he retired in 1953. Whittaker also obtained a full pension for Born who, along with others who had fled, had lost his possessions and money. Born's co-recipient of the Nobel prize was Walther Bothe. This was not at all to Born's liking, since Bothe had been an active worker in the German atomic bomb project. See, for example, Miller (2002), discussion and references.

62. Letter from Chandra to his father, 8 July 1931.
63. Letter from Chandra to his father, 22 January 1936.
64. Letter from Chandra to his father, 22 September 1931.
65. Letter from Chandra to his father, 14 October 1931.
66. Letter from Chandra's father to Chandra, 17 October 1931.
67. Singh (1963), p. 41.
68. Letter from Chandra to his father, 28 October 1931.
69. Letter from Milne to Chandra, 16 October 1931.
70. Letter from Chandra to his father, 12 November 1931.
71. Letter from Chandra to Ralph Kenat, 28 June 1983.
72. Letter from Chandra to his father, 12 January 1932.
73. Letter from Chandra to his father, 22 January 1932.
74. Letter from Chandra to his father, 26 January 1932.
75. Letter from Chandra to his father, 27 March 1932.
76. Heisenberg had used group theory for just this purpose in 1926 – see Heisenberg (1926). According to quantum mechanics, there are many ways for two or more electrons to combine, depending on their spin and orientation in space. Group theory throws up many combinations, and Pauli's exclusion principle picks out the correct ones to use to explore properties such as how electrons in an atom create spectral lines.
77. Letter from Chandra to his father, 20 April 1932.
78. Letter from Chandra to his father, 16 August 1932.
79. Quoted in Pais (1986), pp. 416–17.
80. Conversation with Victor Weisskopf, 10 April 1986. Sommerfeld's institute was in Munich, and Born's in Göttingen. Munich, Göttingen and Copenhagen were considered the three major centres of atomic physics research.
81. Chandrasekhar (1977), p. 39.
82. Chandrasekhar (1977), p. 20.
83. Quoted in Venkataraman et al. (1983), p. 16.
84. Letter from Chandra to Kenat, 28 June 1983. The published paper is Chandrasekhar (1932).
85. Chandrasekhar (1980a), p. 3.
86. Chandrasekhar (1932), p. 327.
87. Chandrasekhar (1932), p. 326.
88. Letter from Chandra to Kenat, 28 June 1983.
89. Letter from Chandra to his father, 14 December 1932.
90. Letter from Dirac to Bohr, 6 May 1932.
91. Letter from Chandra to his father, 21 October 1932.
92. Letter from Bohr to Dirac, 14 November 1932.

93. Letter from Chandra to his father, 11 November 1932. Chandra succeeded in publishing only one paper on a generalised quantum statistics, in 1929, when he convinced himself that he was on the trail of something new and exciting in theoretical physics. Now his dream had gone up in smoke. Perhaps this was why he did not include the 1929 *Physical Review* paper in a supposedly complete list of published papers in Volume 6 of his *Selected Papers*, so unpleasant were the memories of that episode.

94. Letter from Chandra to Balakrishnan, 2 February 1933.

95. Letter from Chowla to Chandra, 8 September 1935.

96. Letter from Chandra to his father, 17 March 1933.

97. Letter from Chandra's father to Chandra, 4 June 1933.

98. Letter from Chandra to his father, 8 June 1933.

99. Chandrasekhar (1980b), p. 5. Chandra published his Ph.D. research as a series of four papers packed with detailed calculations – Chandrasekhar (1933a).

100. Chandrasekhar (1980a), p. 5.

101. Chandra quoted Raman's words. Letter from Chandra to his father, 19 September 1933.

102. Letter from Chandra to his father, 19 September 1933.

103. Chandrasekhar (1933b), pp. 217–18.

104. Chandrasekhar (1977), p. 21.

105. Letter from Chandra to his father, 12 October 1933.

106. Letter from Milne to Chandra, 20 December 1933.

107. Letter from Chandra's father to Chandra, 23 October 1933.

108. Letter from Chandra to his father, 5 January 1934.

109. Letter from Chandra to his father, 24 June 1934.

110. Chandrasekhar (1934a), p. 98.

111. Wali (1991), pp. 116–17.

112. Letter from Chandra to Balakrishnan, 31 August 1934.

113. This is called polytrope theory. Its most developed form dates from the early twentieth century. Polytrope theory provides methods for calculating a star's interior temperature and pressure from knowledge of these quantities on its surface, assuming that the star is made of a particular sort of gas, such as a perfect gas. Eddington made extensive use of polytrope theory in his research on the Cepheids, in the course of which he systematised the theory to facilitate its use in stellar modelling. The results are in *The Internal Constitution of the Stars*.

114. Letter from Chandra to his father, 14 September 1934.

115. Letter from Chandra's father to Chandra, 24 September 1934.

116. Letter from Chandra to his father, 5 October 1934. There is no known extant correspondence between Chandra and Lalitha for this period.

117. Letter from Chandra's father to Chandra, 19 October 1934.

118. This is Chandrasekhar (1934b). The three major papers he published in 1935 are (1935b,c,e)

119. That is, as in Chandrasekhar (1934a).

120. Chandrasekhar (1934b), pp. 374–5.

121. Chandrasekhar (1934b), p. 377.

122. Chandra deduced the figure for the maximum mass of a white dwarf star to be 5.728 times the mass of the Sun divided by the mean molecular weight squared. The graph he showed at the RAS in January 1935 is generated from the mathematics of his theory and is the definitive demonstration of how a white dwarf star can expire and what might happen if it exceeds Chandra's maximum mass. The vertical axis shows the radius of a white dwarf (R_l) in units of the number l_l, equal to 7.7 × 10^8 centimetres divided by the average molecular weight of the gas of electrons that makes up the star. The horizontal axis shows the mass (M) of the white dwarf in units of the maximum mass for a stable white dwarf (M_3), that is 5.728 divided by the average molecular weight squared. Chandra labelled the vertical and horizontal axes in this very esoteric way for two reasons: to calibrate them using numbers with no units and to show that these ratios emerged naturally from the mathematics of his theory. To draw his curve, he set the average molecular weight at 1. The solid curve represents Chandra's theory and shows very clearly that as the mass of a white dwarf (M) approaches the maximum mass (M_3), the radius (R) becomes zero and the star shrinks to nothing. The broken line (----) shows the application of Chandra's theory when the electrons in a star are moving at speeds considerably less than the speed of light; that is Fowler's theory. We can see that Fowler's inexact theory and Chandra's exact one diverge at a very early point. Electrons move at speeds appreciably slower than that of light only in white dwarf stars of very low mass. The circles show stars at different stages in the diagram. The shaded regions show how quickly the incompressible core in Chandra's theory develops. The broken line continues for ever; thus Fowler's theory does not predict a maximum mass. The dotted line (.....) shows the transition from the core in Fowler's theory to the one in Chandra's. The transition from B to A was required in Milne's theory, which postulated a nest of core within core in order to prevent complete implosion.

123. Letter from Milne to Chandra, 27 September 1934.
124. Letter from Chandra to Kenat, 28 June 1983.
125. Letter from Chandra to his father, 16 November 1934.
126. Letter from Chandra to Balakrishnan, 4 January 1934.

CHAPTER 6

1. Eddington (1935a), p. 38.
2. Eddington (1935a), pp. 38–9.
3. Eddington (1935a), p. 38.
4. Chandrasekhar (1977), p. 31.
5. Letter from Chandra to Rosenfeld, 12 January 1935.
6. Letter from Rosenfeld to Chandra, 14 January 1935.
7. Letter from Rosenfeld to Chandra, 19 January 1935.
8. Eddington (1935b). This paper was followed by Eddington (1935c,d).
9. Eddington (1935a), p. 38. For example, Eddington used an equation for

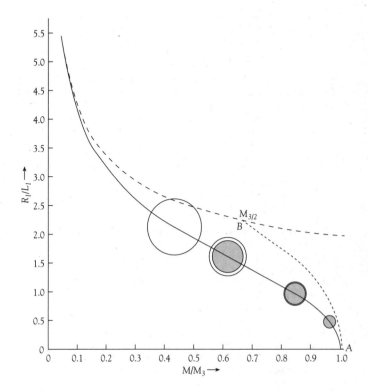

(Courtesy of the Royal Astronomical Society)

the electron's energy made up of relativistic and non-relativistic parts.
10. Eddington (1935b), p. 194.
11. Chandrasekhar (1980b), p. 6.
12. Letter from Chandra to Rosenfeld, 3 February 1935.
13. Letter from Rosenfeld to Chandra, 6 February 1935.
14. Fowler had also dealt with an idealised quantum gas of electrons.
    Eddington went on to derive an alternative equation for Fowler's
    theory, writing the gas pressure in terms of the number of electrons
    in the star and including interactions among all particles in the gas.
    He claimed that this showed that Fowler's theory was in agreement
    with the 'requirements of relativity theory', while Chandra's could not
    produce a result of such 'simplicity' (Eddington 1935b, p. 206).
15. Letter from Rosenfeld to Chandra, 6 February 1935.

16. Letter from Chandra to Rosenfeld, 3 February 1935; Chandrasekhar and Møller (1935d), p. 673.
17. Letter from Rosenfeld to Chandra, 29 January 1935.
18. Letter from McCrea to Chandra, 20 February 1935.
19. Interview by the author with Vandervoort, 5 November 2002.
20. See Chandrasekhar (1980b), p. 6; Wali (1991), p. 145; and Fowler (1936), p. 652.
21. Milne (1935a), p. 52.
22. Letter from Chandra to his father, 9 February 1935.
23. Letter from Chandra's father to Chandra, 18 February 1935.
24. Letter from Chandra to his father, 15 February 1935.
25. Letter from Milne to Chandra, 26 February 1935.
26. See, for example, Chandra (1979b) – a very personal memoir.
27. Letter from Chandra to his father, 14 March 1935.
28. Letter from Chandra to his father, 20 March 1935.
29. Letter from Chandra's father to Chandra, 9 March 1935.
30. Letter from Chandra to his father, 22 March 1935.
31. Letter from Chandra's father to Chandra, 6 May 1935.
32. Letter from Chandra to Rosenfeld, 2 July 1935.
33. Milne (1935c), p. 174.
34. Eddington (1935e), p. 176.
35. Letter from Chandra to his father, 31 May 1935.
36. Letter from Chandra to his father, 7 June 1935.
37. Quoted in Frisch (1979), p. 174.
38. Chandrasekhar (1977), p. 92.
39. Letter from Milne to Chandra, 1 June 1935.
40. Chandrasekhar (1935e), p. 676.
41. Letter from Chandra to Rosenfeld, 24 September 1935.
42. Letter from Chandra to his father, 21 June 1935.
43. Letter from Chandra to his father, 5 July 1935.
44. Chandrasekhar and Møller (1935d).
45. Chandrasekhar and Møller (1935d), p. 676.
46. Eddington (1935c), p. 20.
47. Derivation of the relativistic degeneracy formula rests on a fundamental result of quantum mechanics in which one assumes the electrons to be in a particular space. In this situation their momenta take on certain definite values, in other words, they can be quantised. Eddington wrongly took this to violate the uncertainty principle, according to which position and momentum cannot both be taken as arbitrarily precise.

Eddington's assertion also had to do with the basic assumption that the shape of the space in which electrons are confined is arbitrary, for the sake of calculational ease. It can be a sphere or a cube, whichever is convenient; in the end all references to the exact shape disappear. But Eddington disagreed, as he implied in 1935 (Eddington 1935d, p. 258).

So Chandra persuaded the German physicist Rudolf Peierls to investigate 'whether the pressure–density relation of such a gas enclosed in a certain volume would be independent of the shape of the volume' (Peierls 1936, p. 780). Using relativistic quantum mechanics, Peierls proved this in great detail. He even covered the situation in which, according to relativistic quantum mechanics, forcing an electron to remain enclosed introduces the possibility of creating electron–positron pairs, with the positron escaping and the electron remaining inside. Peierls showed that this deviation from independence of shape was completely negligible.

48. Letter from Chandra to Rosenfeld, 2 July 1935.
49. Letter from Rosenfeld to Chandra, 5 July 1935.
50. According to Einstein's general theory of relativity, the laws of physics are the same for all observers, wherever they are in the universe and however they are moving. Milne contended that it was not true because the distribution of matter is not the same everywhere in the universe. It is concentrated in galaxies, so only observers at the nuclei of galaxies would have the same law of gravitation. Milne's aim was to deduce a new law of gravity which he claimed explained the shapes of galaxies. He did not succeed.
51. Letter from Milne to Geoffrey Milne, 10 August 1932.
52. Einstein (1949), p. 684.
53. Chandrasekhar (1979b), p. 12.
54. Letter from Chandra to Rosenfeld, 26 March 1935.
55. Letter from Chandra to his father, 15 July 1935.
56. McCrea (1935), p. 259.
57. McCrea (1935), p. 259.
58. Chandrasekhar (1977), p. 29.

CHAPTER 7

1. Letter from Chandra to his father, 29 September 1935.
2. Letter from Chandra to his father, 11 September 1935.
3. Letter from Chandra to his father, 12 December 1935.
4. Letter from Chandra to his father, 18 December 1935.
5. Letter from Chandra to his father, 18 December 1935.
6. Letter from Chandra to Balakrishnan, 20 December 1935.
7. Letter from Chandra to his father, 31 December 1935.
8. Letter from Chandra to Balakrishnan, 7 January 1936.
9. Letter from Chandra to his father, 22 January 1936.
10. Quoted from Struve's letter in a letter from Chandra to his father, 27 February 1936.
11. Letter from Chandra to his father, 27 February 1936.
12. Letter from Chandra's father to Chandra, 20 March 1936.
13. Letter from Chandra to his father, 1 March 1936.
14. Letter from Chandra to his father, 23 March 1936. The Oxford University archives list no one with that name on their faculty at that time. Taking into account Chandra's poor English spelling and the different ways in which Indian names were spelled in the West, this is probably a reference to Sarvepalli Radhakrishnan (1888–1975), Spalding Professor of Eastern

Religions and Ethics at Oxford University from 1936 to 1939. Like
Chandra, Radhakrishnan came from Madras. He was a brilliant philoso-
pher-statesman and was President of India from 1962 to 1967.

15. Letter from Chandra to his father, 5 June 1936.
16. Letter from Chandra's father to Chandra, 14 October 1935.
17. Wali (1991), p. 168.
18. Hardy gave a series of lectures on Ramanujan's life and work. Earlier
    that year he had contacted Chandra to discuss his thoughts. He also
    asked whether, on Chandra's next trip to India, he would look for a good
    photograph of Ramanujan for the published version of his lectures
    (Hardy 1940). While he was in India, Chandra sought out Ramanujan's
    widow and found her living in a small apartment in a narrow back alley
    in Triplicane. At first she could not find any photographs of her hus-
    band, but in his passport was a small photograph that in the end Hardy
    used as the frontispiece for his book.
19. Quoted in Douglas (1956), p. 105.
20. Eddington (1936b), p. 390.
21. Eddington (1936b), p. 390–1.
22. Eddington (1936b), p. 391.
23. Kanigel (1991), p. 373.
24. Eddington (1936a), p. 235.
25. Letter from Rosenfeld to Chandra, 11 November 1936.
26. Conversation with Donald E. Osterbrock, 17 June 2004.
27. Letter from Chandra to his father, 12 September 1937.
28. '≈' means 'approximately equals'. Letter from Raman to Chandra,
    28 February 1937.
29. Letter from Trimble to Chandra, 26 May 1953.
30. Eddington (1943).
31. Eddington (1939a), p. 595. In this paper, presented at the 9 June meeting
    of the RAS, Eddington used Fowler's result to conclude that Sirius B has a
    high hydrogen content. He admitted that this ran counter to the usual
    notion of a white dwarf as a burnt-out star. His argument was that a star goes
    through a quasi-white-dwarf stage at the beginning of its life. In this state
    a star is almost all hydrogen. When the star contracts, hydrogen begins to
    burn, eventually forming carbon and nitrogen. At this point the star
    expands, enters the main sequence, and evolves in the usual way. Thus stars
    are white dwarfs twice in their lifespan (a very unlikely scenario).

    Eddington concluded by stating forcefully that the Stoner–Anderson
    formula 'generally used in the theory of white dwarfs is fallacious, and
    that Fowler's original formula was probably correct' (1939a, p. 605). On
    the anomalously high hydrogen content of Sirius B, he speculated – as
    he had in his printed paper – that this would seem reasonable if Sirius B
    had split off from the outer layers of Sirius A.
32. Letter from Chandra to Trimble, 24 June 1953.
33. Letter from Shaler to Chandra, 12 December 1938.
34. Letter from Eddington to Chandra, 22 January 1939.
35. Chandrasekhar (1935e), p. 690. Today astronomers still do not com-

pletely understand the evolution of Wolf–Rayet stars. The current thinking is that they gradually lose their envelopes of atmospheric gases, leaving exposed an interior of nuclear-processed material.

36. Letter from Eddington to Chandra, 22 January 1939.
37. Letter from Eddington to Chandra, 22 January 1939.
38. Letter from Shaler to Chandra, 4 February 1939.
39. Letter from Shaler to Chandra, 11 February 1939.
40. Letter from Chandra to Balakrishnan, 17 June 1939.
41. Chandrasekhar (1980a), pp. 10–11.
42. Chandrasekhar (1977), p. 35.
43. Kuiper (1941), p. 5. Van Maanen's Star was discovered in 1917 by the Dutch astronomer Adriaan Van Maanen (Van Maanen 1917).
44. For white dwarfs that are not part of a binary system, such as Van Maanen's Star, astrophysicists calculate the mass from the measured gravitational redshift and the radius from the stellar parallax, apparent brightness and temperature.
45. According to Fowler's theory, a white dwarf's mass times its radius cubed makes a constant, giving the result that the smaller the radius, the larger the mass. In other words, there can be tiny white dwarfs with a huge mass much bigger than Chandra's limiting mass.
46. Chandrasekhar (1941), p. 42.
47. Chandra's recollection, quoted in Wali (1991), pp. 136–7.
48. Eddington (1941), pp. 66–7.
49. See Chapter 5, note 14.
50. Kuiper (1941), p. 69.
51. Chandrasekhar (1941), p. 64.
52. Eddington (1941), p. 66. Here Eddington showed great prescience. Physicists were already debating the possibility of a neutron star, and, some thirty years later, neutron stars were actually discovered.
53. Atkinson (1939), p. 284.
54. Chandrasekhar (1977), p. 34.
55. Chandrasekhar (1977), p. 35.
56. Letter from Chandra to his father, 30 July 1939.
57. Eddington (1936a), p. 235.
58. Letter from Chandra's father to Chandra, 15 August 1939. Babuji then wrote a longer letter to Chandra on the subject, dated 1 January 1940.
59. Letter from Chandra to his father, 14 February 1940.
60. Peierls, a brilliant young physicist, had already written a paper in response to Eddington's criticism of Chandra's application of quantum mechanics to an enclosed gas of electrons. See Peierls (1936) and note 47 of Chapter 6.
61. Dirac et al. (1942), p. 193.
62. Eddington (1942), pp. 208–9.
63. Salpeter (1961). Peierls had begun to address this (Peierls 1936). He was Salpeter's thesis supervisor in Birmingham in 1946–8, and interested Salpeter in this problem. Interview with Salpeter, 31 July 2002, as well as Salpeter (1999), p. 28.
64. Letter from Struve to Raman, 24 May 1941.

65. This correspondence has been collected in Ramaseshan (1996).
66. Ramaseshan (1996), p. 105.
67. Quoted in Wali (1991), p. 253.
68. Ramaseshan (1997), p. 105.
69. Letter from Chandra to Raman, 29 November 1948.
70. Letter from Milne to Chandra, 15 March 1944.

CHAPTER 8

1. Milne (1947), p. 584.
2. Milne (1947), p. 586.
3. Chandrasekhar (1980a), p. 8.
4. Chandrasekhar (1977), p. 36.
5. Chandrasekhar (1977), pp. 30, 36.
6. Chandrasekhar (1979a), p. 7.
7. Letter from Milne to Chandra, 13 October 1939.
8. Letter from Milne to Chandra, 2 September 1944.
9. Letter from Milne to Chandra, 2 September 1944.
10. Letter from Milne to Chandra, 14 November 1936.
11. Chandrasekhar (1979b), p. 9.
12. Letter from Milne to Chandra, 24 February 1935.
13. Chandrasekhar (1979b), p. 8.
14. Letter from Chandra to Cowling, 18 February 1977.
15. Letter from Cowling to Chandra, 2 February 1977.
16. Chandrasekhar (1979a), pp. 5–6.
17. Chandrasekhar (1979a), p. 7.
18. Chandrasekhar (1979a), pp. 7–8.
19. Plaskett (1951), p. 172.
20. Chandrasekhar (1977), p. 37.
21. Quoted in Douglas (1956), p. 170.
22. Quoted in Douglas (1956), p. 152. See Eddington (1946) for his fullest development of the fundamental theory.
23. Chandrasekhar (1987a), p. 139.
24. Eddington (1958b), p. 126.
25. Quoted in Douglas (1956), p. 101.
26. Dirac (1937), p. 323. See also Kragh (1990), Chapter 11.
27. Bohr's remark was reported by Gamow in a letter of 1 September 1967 to the editor of Science, Philip H. Abelson; quoted in Kragh (1990), p. 233.
28. Chandrasekhar (1977), p. 51.
29. Chandrasekhar (1937), p. 757.
30. Chandrasekhar (1937), p. 757.
31. Chandrasekhar (1977), p. 50.
32. Chandrasekhar (1983a).
33. Chandrasekhar (1988), p. vii.
34. The Lorentz transformation, named after the brilliant Dutch physicist Hendrik Lorentz, is a mathematical means for observers in laboratories

in relative motion to compare measurements such as lengths and times. Each observer would, for example, measure the same velocity of light. The Lorentz transformation is at the heart of the special theory of relativity and so, too, of Chandra's theory of white dwarf stars.

35. Letter from Eddington to Walter Adams, 20 April 1941.
36. Letter from Eddington to Adams, 29 January 1941.
37. Quoted in Chandrasekhar (1987a), p. 140.
38. Douglas (1956), p. 183.
39. Letter from Winifred Eddington to Shapley, 27 November 1944.
40. Russell (1945), p. 133.
41. Chandrasekhar (1977), p. 49.
42. Compton (1904), p. 247; see also Kanigel (1991), pp. 99–102.
43. Slater (1936), p. 47.
44. Chandrasekhar (1977), p. 54.
45. Quoted in Wali (1991), p. 114.
46. Chandrasekhar (1977), p. 54.
47. Interview by the author with Michael Turner, 30 October 2002.
48. Littlewood (1953), p. 136.
49. Littlewood (1953), p. 136.
50. Letter from Chandra to his father, 24 April 1936.
51. Wali (1991), p. 237. Chandra's Uncle Raman suffered similar humiliating experiences during a visit to America in the 1920s. In Boston he was turned away from several hotels until in the end a taxi driver took him to one run by a Japanese couple in the suburbs (Wali 1991, pp. 236–7).
52. Quoted in Douglas (1956), p. 155.

CHAPTER 9

1. Milne (1932), pp. 346–7.
2. Zwicky studied with Peter Debye and Paul Scherrer, two leaders in the field. Debye was awarded the 1936 Nobel Prize for Chemistry.
3. Letter from Chandra to his father, 10 October 1930.
4. Osterbrock (1997), pp. 58–61.
5. Osterbrock (1996), p. 310.
6. Greenstein (1982), p. 26.
7. Osterbrock (2001).
8. A supernova was also discovered in 1572, by the Danish astronomer Tycho Brahe, and another in 1604, by his former assistant, Johannes Kepler. All these events were, of course, seen by the naked eye without the aid of a telescope, which was not yet invented. The first supernova to be observed with a telescope was in the Andromeda Galaxy in August 1885.
9. Since light takes time to travel across space, we observe the universe in the distant past.
10. See, for example, Struve and Zebergs (1962), p. 340.
11. Baade and Zwicky (1934).
12. Bethe (1993), p. 29.
13. Osterbrock (1996), pp. 58–64.

14. Zwicky (1939), p. 726.
15. Chandrasekhar (1935c), p. 258.
16. Chandrasekhar (1941), p. 47.
17. Quoted in Danin (1989), p. 81.
18. Pellam (1989), p. 198.
19. Landau (1932), p. 458.
20. Landau (1932), p. 458.
21. Landau (1932), p. 459.
22. See also Thorne (1994), pp. 184–7, and correspondence in Khalatnikov (1989), pp. 308–19.
23. Landau went on to unravel the puzzle of superconductivity, for which he won a Nobel prize in 1962. Kapitza received his in 1978.
24. Bethe (1966), p. 49.
25. Letter from Gamow to Chandra, 7 January 1938.
26. Letter from Gamow to Struve, 17 September 1938; quoted in DeVorkin (2000), p. 254.
27. Rosenfeld (1966), p. 483.
28. This was before the discovery of the neutron, so the alpha particle was assumed to be made up of four protons and two electrons to give it a double positive charge. When it is the nucleus of a helium atom, its double positive charge neutralises the double negative charge of the two atomic electrons. Rutherford obtained alpha particles from a gas of radon atoms, although he did not understand the process which caused this emission of particles.

    This was not explained until after the discovery of the neutron, in the 1930s, when physicists reached a consistent understanding of the nucleus. They reasoned that certain nuclei are unstable or 'radioactive' depending on the numbers of protons and neutrons. To make themselves more stable the nuclei emit alpha particles, electrons, positrons, neutrons, protons or gamma rays.
29. See also Stuewer (1986).
30. Frisch (1979), p. 72. See also Rhodes (1986), pp. 370–1, and Khriplovich (1992).
31. Quoted in Khriplovich (1992), p. 30.
32. Bethe and Critchfield (1938).
33. Chandrasekhar (1977), p. 77.
34. Bethe (1966), p. 4.
35. Bethe (1966), p. 4.
36. Bethe (1966), p. 50.
37. Gamow (1968), p. 68.
38. First, Bethe had to prove that the proton–proton reaction could proceed fast enough to produce helium and thus start the process of fusion. This was the main conclusion of the Bethe–Critchfield paper. They established a chain of nuclear reactions beginning with two protons fusing together and momentarily forming a 'diproton'. One of the protons is transformed into a neutron, a positron and a neutrino through one sort of beta decay (the other sort is mentioned below). The positron and neutrino fly around the interior of the star, but the other proton stays

with the neutron to form a deuteron, which then fuses with another proton and neutron to form the nucleus of a helium atom. This is how stars with masses comparable to the Sun's produce energy and shine.

Pauli proposed the neutrino in 1930 in response to problems thrown up by the process that occurs inside a nucleus in which a neutron is transformed into a proton and electron – i.e. beta decay. When it was first discovered, beta decay seemed to violate the law of conservation of energy in that the energy content of the nucleus beforehand was greater than the sum of the nucleus's energy afterwards and the energy of the escaping electron. (Electrons do not exist in the nucleus; the electron is created and ejected at the instant the neutron decays.) Pauli made the audacious suggestion that the simplest way to balance these two energies was to assume that a hitherto unknown particle with no mass or charge must exist, and that what actually happens in beta decay is that a neutron becomes a proton, an electron *and* a neutrino. The neutrino was actually discovered more than two decades later, in 1956.

39. As Bethe described it, he examined systematically 'reactions between protons and other nuclei, going up the periodic system' (1972, p. 222) that were possible at the high temperatures inside stars much brighter than the Sun. He discovered a self-sustaining chain of reactions in which hydrogen is burnt to produce helium over millions of years, a timescale in agreement with the lifetime of main sequence stars of this type. The chain begins with hydrogen interacting with carbon, which makes up about 1% of a young star. Helium is the end product of a sequence of reactions which produce nitrogen and oxygen. The sequence is, in fact, a cycle. Carbon is reproduced and acts as a catalyst – keeping the reaction going and using up protons. This carbon–nitrogen–oxygen (CNO) cycle can continue for a million years. Bethe's theory was consistent with the lifetime of stars such as Sirius A, around twice the mass of the Sun, and the hot young Y Cygni, in the constellation Cygnus, some twenty times the mass of the Sun and 600 times more luminous (more recent measurements have raised the figure to 10,000 times more luminous than the Sun).

40. Eddington (1926), p. 306; Chandrasekhar (1977), p. 39.

41. Bethe (1939). His paper was not published until March 1939, though it had been received at the *Physical Review* on 7 September 1938. It had been entered for the A. Cressy Morrison Prize of the New York Academy of Sciences, which required that papers be unpublished. Bethe won.

42. Chandrasekhar, Gamow and Tuve (1938).

43. Bethe (1939), p. 456.

44. Rhodes (1986), p. 444.

45. Oppenheimer (1963), p. 18.

46. See note 29 of Chapter 4.

47. After Bethe departed, Oppenheimer and his group began looking at this problem. Oppenheimer insisted on checking with experimentalists the various nuclear reactions his group had in mind for how energy is produced in stars more massive than the Sun. According to Robert Serber,

someone gave Oppenheimer incorrect data, and this led the group away from a carbon cycle. 'Bethe didn't have that data' and therefore assumed that this set of reactions would work. See Serber (1992), p. xxvi.

48. Herken (2002), p. 290.
49. Serber (1992), p. xxxii.
50. Oppenheimer and Serber (1938).
51. Oppenheimer and Volkoff (1939).
52. Zwicky (1939), p. 743.
53. Chandrasekhar (1977), p. 92.
54. Chandrasekhar (1977), p. 79.
55. Interview by the author with Donald E. Osterbrock, 24 March 2003. He was a student of Chandra's from 1949 to 1952.
56. Letter from Einstein to Schwarzschild, 9 January 1916.
57. Letter from Eddington to his mother, 5 August 1913.
58. Eddington (1926), p. 6.
59. Oppenheimer and Snyder (1939), p. 456.
60. Einstein (1939), p. 936.
61. Alvarez (1987), p. 78.
62. Shock waves are undulations that travel through a medium faster than sound. Their familiar signature is the sonic boom produced by jet planes breaking the sound barrier. Radiative transfer is the study of how heat and radiation moves through material, like the radiation escaping from a star.
63. Letter from Bethe to Chandra, 26 April 1944.
64. Chandrasekhar (1977), p. 98.
65. Letter from Bethe to Chandra, 27 September 1944.
66. Chandrasekhar (1977), p. 98.
67. Chandrasekhar (1977), p. 99.

CHAPTER 10

1. Chandrasekhar (1977), p. 34.
2. Chandrasekhar (1977), p. 78.
3. Chandrasekhar (1975a), p. 3.
4. Chandrasekhar (1943).
5. Gamow (1939).
6. Chandrasekhar and Schönberg (1942).
7. Chandra's astrophysical research entailed much numerical computing, carried out on electric computing machines. In his early days at Yerkes, his graduate students assisted him on particularly lengthy calculations. He also hired full-time computers, as the people who used these machines were called, whose salaries were paid out of observatory funds. Donna Elbert, a native of Williams Bay, was Chandra's best-known computer. She worked for him for over thirty years, starting in 1949 at Yerkes, and followed him to the Chicago campus. See Elbert (1997), and Osterbrock (1999), pp. 212–13.
8. See Sandage and Schwarzschild (1952), and Osterbrock (2001).
9. Astrophysicists first predicted that white dwarfs ended their lives by

becoming diamond stars in the 1960s, but it was only in February 2004 that astronomers found one. In the constellation Centaurus, 300 trillion miles away, the white dwarf star catalogued as BPM 37093 has a diamond core of some 10 billion trillion trillion carats. BPM 37093 is estimated to be 1.1 times the mass of the Sun and 90% crystallised. It has been nicknamed Lucy, after the Beatles song.

10. Chandrasekhar (1977), pp. 106, 101.
11. Chandrasekhar (n.d.3), p. 4.
12. Telegdi (1997), p. 207.
13. Vandervoort (n.d.), p. 12.
14. Vandervoort (n.d.), p. 18.
15. In daily life we distinguish between left and right – our heart is on our left side. But it was always assumed that the laws of physics could not; this is 'conservation of parity'. Violation of this law would mean, for example, that an experiment and its mirror image might yield different results. Lee and Yang predicted in 1956 that on the atomic level, in reactions such as beta decay, nature could in fact distinguish between left and right and parity conservation could be violated. Later that year their theory was verified. Two sets of experiments were set up which were mirror images of each other, in which nuclei of cobalt underwent beta decay. The meters in the experiments registered different numbers of electrons emitted by the cobalt nuclei, thus providing definite proof of an asymmetry between left and right which would have to be allowed for in quantum physics. Lee and Yang's work was so important that they won the Nobel Prize for Physics the following year.
16. Osterbrock (1999), pp. 205–6, and personal communications.
17. Chandrasekhar (1977), p. 151.
18. Comment by the physicist Valentine Telegdi, a colleague of Chandra's and Fermi at the University of Chicago, quoted in Wali (1991), p. 19. The papers they wrote were Chandrasekhar and Fermi (1953a,b).
19. As recalled by Hutchins some years later and quoted in Wali (1991), p. 236.
20. Letter from Chandra to Hutchins, 20 December 1950.
21. Letter from Hutchins to Chandra, 28 December 1950.
22. Although Fermi invited Chandra to transfer to the physics department in 1952, Chandra may simply have delayed making the switch final until 1964. Conversation with Donald E. Osterbrock.
23. Urey won the 1934 Nobel Prize for Chemistry for his discovery of deuterium, or heavy water, essential for the production of nuclear weapons. Libby became a Nobel laureate in chemistry in 1960 for discovering radiocarbon dating, and Goeppert-Mayer in physics, in 1963, for research on nuclear structure.
24. Interview with Roger Hildebrand by the author, 31 October 2002.
25. After a certain amount of time, nuclei of radioactive elements emit elementary particles and radiation and transform themselves into other elements, thus achieving stability. The time that half a sample of radioactive nuclei takes to decay is known as the 'half life' of the nucleus. The half life of any nucleus is a property of the nucleus itself and does not depend on

how many nuclei are packed together or on the temperature of the sample. The half life of uranium is 4 billion years. As this is the age of the Earth, only half the original amount of uranium on Earth has decayed so far.

26. Fermi had assumed that he had discovered an element that contained one more proton than uranium and with a half life of 13 minutes. He took this to be element 93, one beyond uranium, which is designated element 92 in the periodic table because it has 92 protons. In his data analysis Fermi had failed to compare what he thought was a new element with all known elements, and stopped at lead, which was assumed to be the final step in the disintegration of a radioactive element. What he had actually done was to split the uranium nucleus into two large fragments, each radioactive and lighter than lead, namely barium and krypton.

27. Any sample of naturally occurring uranium contains 99.3% of ^{238}U and 0.7% of its isotope ^{235}U. Bohr had discovered in 1939 that ^{235}U can be split more easily, as the neutrons required to split it travel more slowly than those required for ^{238}U.

28. Quoted in Rhodes (1986), p. 417.

29. Quoted in Rhodes (1995), p. 303.

30. Letter from Wheeler to the head of the Princeton University physics department, 8 January 1951. Quoted in Galison and Bernstein (1989), p. 320.

31. ^{239}Pu, produced in reactors by colliding fast neutrons with ^{238}U, undergoes fission as efficiently as ^{235}U, and turned out to be easier to produce. ^{235}U was manufactured mainly by separating it from the heavier isotope ^{238}U by gaseous diffusion techniques implemented on a massive scale at Oak Ridge, Tennessee. ^{239}Pu was produced in nuclear reactors at Hanford, Washington, and then chemically separated at the University of Chicago's metallurgical laboratory.

32. Fuch's Communist leanings led to his fleeing Germany in 1933. In England he impressed several important scientists, including Max Born, who was then in Edinburgh. Eventually Fuchs was recruited into the British atomic bomb project and then the Manhattan Project at Los Alamos. As a reward for his research, in 1943 Fuchs was granted British citizenship. All the while he was passing key information to the Soviets. After the war he continued to supply them with details of hydrogen bomb development. Fuchs's espionage activities were suspected upon decipherment of the decoding of Soviet documents known as the Verona cables in 1948, and in 1950 he was arrested. At the time he was head of the theoretical physics division of the Atomic Energy Establishment at Harwell in England. Fuchs was released after serving nine years of a fourteen-year sentence and returned to East Germany, where he resumed his physics career. In 1953, in the United States, Julius and Ethel Rosenberg were executed for committing espionage at a level that paled in comparison.

33. Quoted in Herken (2002), p. 219.

34. Ulam (1976), p. 60.

35. Ulam (1976), p. 61.
36. Ulam's wife Françoise recalled his flash of inspiration: see Rhodes (1995), p. 463.
37. USAEC (1954), p. 251.
38. Quoted in Holloway (1994).
39. Holloway (1994).
40. Sakharov returned to physics research in 1965 and published several important papers on particle physics and cosmology. At the end of the 1960s he decided to take a bold stand on violations of human rights in the Soviet Union, for which he was awarded the Nobel Peace Prize in 1975. He died in Moscow on 14 December 1989.
41. Sakharov (1990), p. 94.
42. See Khariton et al. (1993), Khariton and Smirnov (1998). American scientists claimed that it was not a true hydrogen bomb, but a conventional atomic bomb boosted by thermonuclear fuel, set off by chemical explosives. The Americans took its low yield as proof that the Russians had not yet discovered the Teller–Ulam design. Soviet scientists who took part in the project have recently claimed the credit for exploding the first 'real' hydrogen bomb, because theirs was the first with the potential to be weaponised.
43. Sakharov (1990), p. 102.
44. Wheeler and Ford (1998), p. 208.
45. Khariton and Smirnov (1998).
46. Russian scientists claim that it remains unsurpassed in terms of yield, deriving 97% of its energy from thermonuclear reactions. In contrast, for example, more than 75% of Mike's yield originated in the fission of its atomic-bomb trigger, followed by another round of fission reactions as fast neutrons from the combined trigger and fusion explosion in turn fissioned the uranium tamper (made of ^{238}U) encasing the bomb.
47. Interview by the author with Blinnikov, 27 November 2002.
48. Bethe (1993). 'Gadget' was the bomb's nickname at Los Alamos.
49. Christy (1994).
50. Wheeler et al. (1958), p. 137.
51. The computer simulation being described is for a star 18 times the mass of the Sun – see Herant et al. (1997). See Woosley et al. (2002) for the milestones in the life of a star twenty-five times the mass of the Sun. See also Harrison et al. (1965).
52. The mass of the iron core at this point is actually less than the Chandrasekhar limit, so it really should not begin to collapse yet. This puzzled astrophysicists for several years. By the late 1980s they realised that the Chandrasekhar limit has to handled with care. Chandra had deduced it for the 'pristine' case of relativistic electrons which are not interacting with one another – as Chandra states in his early papers, for an 'ideal' white dwarf.

    Various corrections had been introduced since the early 1960s. Salpeter, for example, calculated corrections to take account of electron interactions (Salpeter 1961, Hamada and Salpeter 1961). One of Eddington's criticisms had been that Chandra ignored the interactions between electrons.

Among the factors that have to be taken into account to calculate a 'realistic' Chandrasekhar limit for cores inside stars are the following. While the surface of a white dwarf is not under pressure, this is not the case for the iron core. This alone produces a Chandrasekhar maximum mass below the expected core mass, which turns out to be about 1.3 times the mass of the Sun. The electrical interactions between the tightly packed iron nuclei, the heat content of the core and the effect of general relativity also have to be taken into account. All of this alters the 'traditional' Chandrasekhar maximum mass from 1.4 times the mass of the Sun for white dwarfs to as low as 1.15 times the mass of the Sun for the degenerate iron core. See Woosley et al. (2002), esp. pp. 1045–7.

53. Wheeler et al. (1958), p. 137.

CHAPTER 11

1. Wheeler et al. (1958), p. 147.
2. Quotations from Stirling Colgate in this chapter are from interviews held by the author in Los Alamos, 29–30 March 2003, and email communications.
3. No one knows exactly what would happen if a hydrogen bomb were exploded in space. Almost certainly gamma rays, radioactivity and many high-energy particles would burst out – from lighter ones, such as electrons and protons, to heavy nuclei created by the fission–fusion–fission process. In this process, the first fission ignites the fusion fuel. The second is caused by additional fast neutrons that blow off the outer layer (tamper) of ^{238}U, there to reflect neutrons back into the original fission.
4. The neutrino is a neutral particle with no mass. It was postulated in 1930 by Pauli and subsequently discovered, to great excitement, in 1956, a year before the B^{2}FH paper was published. See Chapter 4, note 29.
5. Physicists assume that neutrons and protons are made up of quarks, fundamental building blocks that cannot be broken down any further. But quarks have never been observed as free, isolated objects. To account for this, and to create a theory of how elementary particles such as neutrons and protons interact to form a nucleus, physicists postulated that quarks were always 'confined' within the particles they constitute. Observing them as free objects would be a great advance in understanding how to formulate a theory of elementary particles. The stars may present that opportunity.
6. Email from White to the author, 20 September 2003.
7. Fowler and Hoyle proposed that core bounce occurs after alpha particles (helium nuclei) have been cut to pieces by high-energy radiation, thus increasing the density and number of particles in the core and preventing further compression. But they did not do any computer simulations. They were amazed when Colgate and White found that the star does not stop collapsing at that point. The explosion that Fowler and Hoyle had predicted turned out to be an implosion. See Fowler and Hoyle (1964).
8. Email from White to the author, 6 June 2003.

9. Interview with Colgate, 29–30 March 2003.
10. Thorne (1994), p. 239.
11. Colgate and White (1966), p. 643. Thanks to Richard White for pointing out Chandra's addition. Despite the fact that they had adapted bomb codes for their research, there were no problems with the censor, who studied every paper published by Livermore personnel. By 1965 it was clear to everyone that compressible hydrodynamics, which was taken into account in their equation of state, was an important aspect of nuclear weapons.
12. Like planets, stars spin. The Sun completes a rotation every 25 days, white dwarfs from once every few hours to once every few days.
13. Astrophysicists suspect that the radio signal from a pulsar comes from charged particles near its north and south magnetic poles, accelerated by the rapidly rotating star's intense magnetic field. The line connecting the north and south poles is tilted away from the line around which the star rotates, and as a result the pulsar's beam sweeps across the sky. When the Earth is in the path of one of the beams, the star appears to pulse on and off.
14. In 1974 Hewish received the Nobel Prize for Physics for discovering pulsars. There was some controversy over why Bell did not share the prize. Martin Ryle, another Cambridge radio astronomer, was co-winner.

CHAPTER 12

1. The Schwarzschild radius is the distance from the centre of the black hole to a surface that came to be called the 'event horizon', the horizon where the event occurs, as it were. Stars do not disappear when they fall into a black hole, but continue to exist in the same way that the Moon still exists even when it is over the horizon.
2. Thorne (1994), pp. 248–51.
3. Eddington (1924b). For details see Misner et al. (1970), pp. 828–32.
4. Penrose (1969).
5. It is difficult to measure the distances of the different galaxies from Earth, with the result that there may be a 10% margin of error in the Hubble constant. Moreover, since the universe may not always have been expanding at the same rate, it is possible that the Hubble constant was different in the past. Having established the Hubble constant, we can very easily work out the age of the universe to be 13 to 16 billion years (taking into account the 10% margin of error).

In the 1950s there were two conflicting theories of the origin of the universe. According to the Steady-State theory, the universe is the same today as it always was. The Big Bang theory attempted to incorporate Hubble's discovery and predicted a cosmic microwave background temperature of about 3 degrees kelvin. (An advocate of the steady state, Fred Hoyle, coined the term 'Big Bang' in a somewhat derogatory manner on a BBC radio programme in 1950, to distinguish it from his own theory.) The most vigorous proponent of the Big Bang was Gamow. See Singh

(2004) for an enlightening discussion of the history of the Big Bang theory.

In 1965 Arno Penzias and Robert Wilson discovered radiation with a temperature of 3 degrees kelvin that seems to permeate the universe and is believed to be the 'echo' of the Big Bang. Amazingly, this was exactly the temperature that the Big Bang theory predicted. Beginning as an incredibly hot, small, dense mass, as the universe expanded it cooled to the 3 degrees kelvin it is today. Penzias and Wilson's work was the impetus for modern research into the origin of the universe and is considered to be the turning point in modern cosmology. See Weinberg (1977) for an account of the discovery of the 3 degree background radiation and its immediate effect on astrophysics.

In 2003 the Wilkinson Microwave Anisotropy Probe satellite made the most precise measurements to date of the universe's temperature and thus of its age: in line with previous estimates, it turned out to be 13.7 billion years, with a margin of error of a few hundred million years.

6. Radar is a form of radio waves emitted from a radio transmitter–receiver, like a two-way telescope, and received back if they are reflected by objects.

7. One of these was the radio wave source first detected by Jansky back in 1932. As happened to Chandra, no one would believe his discovery.

8. The term 'optical' is used because the counterparts are 'seen' with a telescope and can be photographed, although they cannot be seen with the naked eye.

9. Baade and Minkowski (1954).

10. In the 1980s infrared telescopes showed that what Baade and Minkowski had interpreted as two galaxies in collision is in fact an optical illusion caused by the absorption of light by the dust band running across the face of the galaxy containing Cygnus A.

11. The numbers are the designations in the *Third Cambridge Catalogue* of radio wave sources, assembled by a group working under the radio astronomer Martin Ryle.

12. Conversation with Gary Steigman, 9 July 2003.

13. Schmidt (1963).

14. Schmidt (1963).

15. Greenstein and Matthews (1963).

16. Schmidt (1963).

17. Greenstein and Schmidt (1964).

18. Colgate (1967). See also comments in Burbridge and Burbridge (1967), p. 193. Richard White recalls that this paper was so complicated that the original referee at Livermore requested his help. White in turn requested advice from experts at Livermore in the various fields on which Colgate drew. The collective recommendation of the 'board' of referees was that, although parts needed clarification, the paper was worthy of publication. No one ever managed to read it through completely.

19. Hoyle and Fowler (1963), p. 170. See Burbridge and Burbridge (1967), Chapter 17, for other proposals for energy generation.

20. Fowler (1984), p. 141.

21. Fowler (1966).
22. Chandrasekhar (1977), pp. 149–50.
23. Chandrasekhar (1977), p. 79.
24. Chandrasekhar (1977), p. 154.
25. Chandrasekhar (1964a,b).
26. Chandrasekhar (1977), p. 153.
27. Interview by the author with Richard White, 8 July 2003.
28. Interview with Richard White, 8 July 2003. At this stage Halton C. Arp, a thirty-seven-year-old observational astronomer at the Hale Observatory, put himself forward as the leading advocate of gravitational redshift as part of the explanation for quasars. If this were the case, the quasar would not need to be so massive as to make it unstable, but would be extremely dense. This would result in a sufficiently high gravitational redshift to remove the need to suppose that quasars were moving at very high speeds or were very far away from our Galaxy. Greenstein and Schmidt had already advanced strong arguments against this position (Greenstein and Schmidt 1964). But Arp soldiered on, and even produced photographs showing quasars apparently in nearby galaxies. Hubble's law, he argued, could not apply to quasars because their redshifts suggested they were farther away. There had to be some as yet undiscovered law that would place them closer to Earth and prove that they were not as luminous as they appeared. His arguments were so persuasive that even astronomers like the Burbidges were prepared to hedge their bets and declare that they could 'see no conclusive answer'. Cecilia Payne, then America's premier woman astronomer, wrote in her 1970 astronomy textbook that scientists might have to alter their view of galaxy recession. See Burbidge and Burbidge (1967), p. 220, and Payne-Gaposchkin and Haramundanis (1970), p. 562.

    Chandra was less charitable. He remembered how, as managing editor of the *Astrophysical Journal*, he 'used to reject Arp's papers outright, several of them'. He agreed to publish only Arp's data and photographs, but with no theoretical interpretation. Eventually astronomers realised that most of Arp's data could be explained away. In the two-dimensional photographs, quasars that happened to be in the background of low-redshift galaxies could look as if they were part of the galaxy in the foreground. See Chandrasekhar (1977), pp. 133, 157.
29. See Thorne (1994), pp. 256–7, and Wheeler and Ford (1998), pp. 296–7, for this story.
30. Kerr (1963).
31. According to Euclidean geometry, the sum of the three angles inside a triangle is always 180°. Such a triangle can be drawn on a flat surface, no matter how it is oriented in relation to the Earth. In 1838 the astronomer-mathematician Bessel – who predicted that Sirius A has a companion – had the idea of applying this law of Euclidean geometry to the triangle employed in measuring the stellar parallax of the star 61 Cygni. Two sides of the triangle were the distances from the star to

the Earth. The sightings were taken six months apart so that the Earth was at opposite sides of its orbit around the Sun. The third side was the diameter of the Earth's orbit. Bessel found that the sum of the angles was still 180°, with little margin of error. From this he concluded that the physical nature of space was Euclidean. Thus we say that three-dimensional space is Euclidean and flat. If Bessel had obtained a result greater than 180° he would have inferred that space is curved like the surface of a sphere.

One of the extraordinary things about general relativity is that Einstein discovered that he had to write its equations in a geometry in which the interior angles of a triangle did not have to add up to 180° – a 'non-Euclidean' geometry. This turned out to be the proper way to represent the fabric of spacetime, and it led to the prediction of physical phenomena caused by the extreme curvature of space around massive bodies such as stars. One was the deflection of starlight passing by the Sun, which Eddington observed in 1919. See Miller (1972, 2000).

32. He coined the phrase in about 1975. See Thorne (1994), p. 275.

33. Chandrasekhar (1975b), p. 54.

34. Electric and magnetic disturbances travel through space in the form of waves at 186,000 miles per second, which we observe as light waves.

35. In 1978, Russell Hulse and Joseph Taylor obtained indirect evidence for gravitational waves by observing the binary pulsar PSR 1913+16, made up of two neutron stars. General relativity has to be used in the analysis of such a system. Hulse and Taylor found that the rate of decrease of the system's orbital period was consistent with the emission of gravitational radiation. For discoveries such as their exploration of this highly interesting binary system, they shared the 1993 Nobel Prize for Physics. More direct tests for gravitational waves are in Chapter 14.

36. Chandrasekhar (1983b), 'Prologue'.

37. Chandrasekhar (1972), p. 173.

38. Supergiants such as HDE 226868 and Betelgeuse are rare stars a thousand times larger than the Sun and a hundred thousand times as luminous; if the Sun were replaced by a supergiant, the Earth's orbit would be inside the star. HDE 226868 appears bluish.

39. Scott et al. (2004), pp. 317–18.

40. NASA (1972), Chapter 17, pp. 13–17.

41. Chandrasekhar (1978), esp. pp. 405, 415–17.

42. Chandrasekhar (1977), p. 77.

43. The jets are the means by which energy is extracted from the region around a supermassive black hole and transferred to the radio lobes. How does the black hole produce these jets? One possibility is that the crowding of gas particles in the inner portion of the accretion disk leads to such high pressures that matter has to be expelled to prevent an explosion. Particles seek the path of least resistance, so, rather than ploughing back through the accretion disk, excess matter shoots off at right angles to it.

CHAPTER 13

1. Quoted in Wali (1991), p. 300.
2. Quoted in Wali (1991), p. 301.
3. Chandrasekhar, 20 January 1976, Box 77 (Addenda), Joseph Regenstein Library.
4. These were the Reissner–Nordström equations, representing a Schwarzschild black hole with electrical charge. Chandra was also working on the Kerr–Newman equations, which are analogues of the Reissner–Nordström equations but for a Kerr black hole. See Chandrasekhar (1977), p. 166.
5. Chandrasekhar, 16 November 1976, Box 77 (Addenda).
6. Chandrasekhar (1977), p. 173.
7. Chandrasekhar, 18, 19 October 1980, Box 77 (Addenda).
8. Chandrasekhar (1977), p. 56.
9. Interview by the author with Lalitha Chandrasekhar, 1 November 2002.
10. Chandrasekhar, 5 June 1982, Box 77 (Addenda).
11. Chandrasekhar, 28 November 1982, Box 77 (Addenda).
12. Chandrasekhar (1983b), p. 529.
13. Quoted in Rees (1997), p. 184.
14. Quoted in Rees (1997), p. 185.
15. Chandrasekhar (1977), p. 155.
16. Johansson (1993), p. 136.
17. Chandrasekhar (1983b), p. 140.
18. Wali (1991), pp. 294–8.
19. Interviews by the author with Gary Steigman, 1 August 2002, and Michael Turner, 30 October 2002.
20. Vandervoort (n.d.), p. 39.
21. In his acceptance speech at Yerkes, Fowler lamented that Fred Hoyle had not shared the prize, especially since he had had such critical input in their earlier work, in, for example, the B^{2}FH paper. Most probably Hoyle was excluded for two reasons. He continued to support the steady-state universe in the face of overwhelming evidence for the Big Bang; and he complained vociferously to the Nobel committee because it had overlooked Jocelyn Bell's part in the discovery of pulsars.
22. Quoted in Wali (1991), p. 295.
23. Interview by the author with Valeria Ferrari, 8 December 2002.
24. Xanthopoulos (1991), p. xvi.
25. Hawking and Israel (1987).
26. Quoted in Dalitz (1997), p. 152.
27. Swerdlow (1997).
28. Chandrasekhar, 4 May 1991, Box 77 (Addenda).
29. Chandrasekhar (n.d.3), p. 2.
30. Chandrasekhar (1995), 'Prologue'.
31. Interview by the author with Chandra, 31 October 1993.
32. Westfall (1996), p. 702. See, for example, Gleick (2003) an illuminating biography of Newton.

33. See, for example, Chandrasekhar (1987c), p. ix, and Chandrasekhar (1983a), p. 140.
34. Hardy (1941), p. 91.
35. Letter from Chandra to Balakrishnan, 17 June 1939.
36. Chandrasekhar (n.d.4), p. 1.
37. Chandrasekhar (1977), p. 161.
38. Chandrasekhar (1992), p. 135.
39. Chandrasekhar (1979c), p. 69, and Chandrasekhar (1992), p. 135.
40. Chandrasekhar (1992), p. 138, and Chandrasekhar (1979c), p. 70. See Francis Bacon, *Essays on Beauty*, Essay 43 (published 1925), and Heisenberg (1971), p. 68.
41. Chandrasekhar (1992), p. 135.
42. According to quantum theory there is no such thing as empty space. Space is seething with matter and antimatter, appearing and disappearing, constantly being created and annihilated. Imagine that somewhere near the event horizon, the black hole's intense gravitational field spawns a particle and its antiparticle. During their fleeting existence, one falls through the event horizon into the black hole. The other cannot now combine with its partner and be annihilated, and wanders off into space. According to Einstein's equation $E = mc^2$, the particle that escapes removes energy from the black hole. The same thing happens when two quanta (particles of light) are created and one escapes. Hawking hypothesises that black holes will – given time – evaporate by energy 'leakage' as particles escape, leaving their partners behind. The escaping particles constitute Hawking radiation.

The evaporation time is so long – more than 10^{67} years as compared to the age of the universe, 13 billion (just over 10^9) years – that black holes formed from collapsed stars simply do not have the vast time required to evaporate in this way. Strangely enough, it turns out that lightweight black holes with masses a billion billion (10^{18}) times less than the Sun have an evaporation time comparable to the age of the universe. These 'mini' black holes could have formed at the time of the big bang, when the universe was an ultradense, incredibly hot, seething soup in which quantum fluctuations occurred with abandon. Some of them should now be on the verge of evaporating in a blast of Hawking radiation, made up of particles and gamma rays, that ought to be picked up by orbiting observatories. So far, however, nothing like this has been observed.
43. Chandrasekhar (1977), p. 161.
44. Chandrasekhar (1992), p. 138. See Heisenberg (1974), Chapter 13.
45. Khan and Penrose (1971).
46. Interview with Valeria Ferrari, 8 December 2002.
47. Chandrasekhar (1992), p. 143.
48. ε is in fact a quantity that plays a role in the mathematics of general relativity. It is called the Ernst function.
49. Chandrasekhar (1992), p. 138.
50. Interview by the author with Valeria Ferrari, 16 July 2003.
51. Chandrasekhar (1983), p. 637.

52. Interview by the author with Lalitha Chandrasekhar, 1 November 2002, and Lalitha Chandrasekhar (n.d.).

CHAPTER 14

1. The orbiting observatory had been in planning since 1977 and was originally known as the Advanced X-ray Astrophysics Facility. A new name was chosen in a competition won by Jaltila van der Veen of the Physics Department at the University of California, Santa Barbara, and Tyrel Johns, a high-school student from Laclede, Idaho. Their entries paid tribute to Chandra's discovery of the maximum mass for white dwarfs.

The Chandra X-ray Observatory was placed in a highly elliptical orbit: at its closest point it is 6000 miles above the Earth, at its furthest 86,400 miles, a third of the way to the Moon. It takes 64 hours to circle the Earth once. The elongated orbit carries Chandra's sensitive X-ray detectors well beyond interference from the Earth's radiation belts, permitting it to make 55 hours of continuous observations in each orbit. The observatory itself is 45 feet long and 64 feet wide with its solar panels deployed, and weighs over 4 tons. It has a minimum lifetime of five years. At the heart is an X-ray telescope made up of exquisitely machined and polished mirrors weighing about a ton. The mirrors focus X-rays into a high-resolution camera with the accuracy you would need to hit a hole in one from Los Angeles to San Francisco.

The Hubble Space Telescope and the Chandra X-ray Observatory often complement one another. While Hubble operates in the near infrared, visible and ultraviolet ranges, Chandra is for the X-ray range. To obtain a picture of the full spectrum of radiation, Hubble and Chandra combine their data with observations made with the Very Large Array, twenty-seven linked radio telescopes in the desert near Socorro, New Mexico. Hubble, the first of NASA's so-called Great Observatories, was placed in orbit by the space shuttle *Discovery* in 1990. Next up was the Compton Gamma Ray Observatory, placed in orbit by *Atlantis* in 1991. After nine productive years the mission ended. The Spitzer Space Telescope (formerly the Space Infrared Telescope Facility) was launched on 25 August 2003.

The Chandra X-ray Observatory also often works in conjunction with XMM-Newton, built by the European Space Agency (ESA) and placed in orbit on 10 December 1999. The proximity of the two launch dates was a historical accident. Delays in the launch of Chandra were due to the *Challenger* disaster in 1986, initial problems with Hubble's mirrors, and vagaries of United States funding of NASA programmes. XMM stands for X-ray Multi-Mirror Mission, referring to its combination of three X-ray telescopes. Aware of NASA's plans, ESA designed XMM-Newton to complement Chandra. XMM-Newton's detecting apparatus was set up to collect the maximum number of X-rays and make more detailed measurements of faint X-ray sources than Chandra can.

2. Lalitha Chandrasekhar (n.d.).

3. An active galaxy is a galaxy emitting huge amounts of radiation and radio waves; the active galactic nucleus is the energy source at its heart, generating vast amounts of energy, more than can be explained simply as the product of nuclear reactions. Quasars are one sort of active galaxy. Another are galaxies that are less luminous than quasars but show signs of intense and violent activity at their centre; they also emit radio waves. These are known as Seyfert galaxies because they were first identified by Carl Keener Seyfert at the Mount Wilson Observatory in 1943.

4. Thorne (1994, Chapter 10; 1998) and conversations with Keith Mason, 23 February 2004.

5. Rees (1998), p. 99.

6. Schödel et al. (2002).

7. Rosat (Röntgen Satellite – named after Konrad Röntgen, the discoverer of X-rays in 1895) was launched on 1 June 1990 and ended its mission on 12 February 1999.

8. Giddings and Thomas (2002), Dimopoulos and Landsberg (2001).

9. For more on Hawking radiation see Chapter 13, note 42. In an extraordinary about face, Hawking recently retracted his breathtaking claim that the very existence of black holes violates the most fundamental theory of matter, quantum theory. According to quantum theory, information can become inaccessible but can never be lost. Any process must be traceable back to its origins; it should always be possible to 'run the film backwards'. But black holes as predicted by general relativity – Kerr black holes – are featureless, except for their mass and spin. All the information is lost about whatever falls in – be it elephants, cars or stars – because nothing can escape from a black hole. But although this information is inaccessible, at least it must remain inside in some form or other. The problem arose when Hawking brought quantum theory into the picture and proved that black holes can evaporate in the form of Hawking radiation, which is random and featureless and so offers no clue as to what is inside. He predicted that eventually all black holes will evaporate along with their information. But if they destroy information in this way, then that subverts quantum theory. This was Hawking's conclusion thirty years ago.

   Hawking now claims that an event horizon never actually forms and so there are no real black holes. Instead, after emitting a certain amount of Hawking radiation, black holes will open up and release the information that fell in. Details have yet to be sorted out and alternatives eliminated.

10. Current theories on the cutting edge of astrophysics and physics, such as string theory, assume that there are seven dimensions in addition to the four we experience, making eleven in all. The new equations that emerge from formulating physics in more than four dimensions are expected to be those at the basis of electromagnetic theory and the electroweak theory, among others. Theories based on eleven dimensions are spoken of as grand unified theories that can deal with quantum and

gravitational phenomena in a way that avoids singularities. See, for example, Greene (2000).

11. This singularity could arise from the collapse of spacetime due to the bunching up of high-energy X-rays from matter sucked in at speeds approaching that of light.

12. Light takes 26,000 years to reach Sagittarius A* from Earth, but only objects with no mass can travel that fast. A spaceship has to accelerate gently over a very long period of time to reach a speed close to that of light so that astronauts can survive. Astronauts' training to go the Moon involved whirling them in giant centrifuges fast enough to put them under a crushing gravitational force of twelve to fourteen times their body weight, about the maximum humans can endure. For further discussion of black holes, wormholes and time travel see Thorne (1994), Chapter 14, 'Wormholes and time machines', and Gott (2002). Or just watch the film *Back to the Future*!

13. This slowing down of time has been verified in experiments. Elementary particles moving at speeds close to that of light – such as in cosmic rays – decay slower than those at rest in the laboratory.

14. For amusing and insightful discussions of the conundrums involved in backwards time travel – such as killing your mother before you were born – see Thorne (1994) and Gott (2002). Again I recommend the film *Back to the Future*.

15. Chandrasekhar, 9 September 1978, Box 77 (Addenda).

16. Interview by the author with Lalitha Chandrasekhar, 1 November 2002, and Lalitha Chandrasekhar (n.d.).

17. From Chandrasekhar (1980a), p. 14. Lalitha added pp. 14–15 and endnote 6 on pp. 17–18. They were driving from Chicago through New York State, towards Binghampton.

18. Chandrasekhar (1980a), p. 17.

19. McCrea (1996), p. 123.

20. McCrea (1996), p. 123.

21. Cronin (1998).

22. McCrea (1998). McCrea had a distinguished career in a series of universities and had the reputation of being a diligent ambassador for science.

23. From interviews by the author with James Cronin, 4 November 2002, and Michael Turner, 30 October 2002.

24. Interview by the author with Cronin, 4 November 2002. The biography of Schrödinger that Chandra had read was Moore's (1991).

25. Lalitha Chandrasekhar (1996).

26. Chandrasekhar (1977), p. 29.

27. Letter from Eddington to Chandra, 4 July 1938.

28. Letter from Eddington to Chandra, 2 September 1943.

29. Lalitha Chandrasekhar (1996), p. 276. See also Chandrasekhar (n.d.3).

30. Chandrasekhar (1977), p. 156.

31. Lalitha Chandrasekhar (1996), p. 276.

32. Conversation with Gary Steigman, 1 August 2002.

33. See Osterbrock (1999) for a fairly complete list of Chandra's Ph.D. students, postdocs and collaborators.
34. This is Candlestickmaker (1972). See also Osterbrock (1999), pp. 228–9.
35. Vandervoort (n.d.), p. 33.

APPENDIX A

1. Kuiper (1941), p. 33.
2. For ease of presentation, Chandra based the graph for his theory on a mean molecular weight of 1, which happens to represent a white dwarf made up of 100% hydrogen. But Kuiper found that the mean molecular weight as calculated according to Chandra's theory from the measured values for the radius and mass of Sirius B led to a mean molecular weight of about 1.43 (giving a hydrogen content of 40%). This value for the mean molecular weight gives a maximum mass for Sirius B of 2.86 times the mass of the Sun. See Kuiper (1941), p. 32.
3. Kuiper (1941), p. 33.
4. Greenstein et al. (1971). For further discussion of Adams' problems in taking measurements of Sirius B, see Greenstein (1985).
5. Gamow (1939b), p. 722. Astrophysicists can derive an equation for the mean molecular weight in such a way that it depends only on the star's hydrogen content, by taking an average of the amounts of chemical elements in the star. The result depends on the degree to which atoms are stripped of their electrons. If they are assumed to be totally stripped – as is pretty much the case in the hot interior of a white dwarf – then the mean molecular weight is 2 divided by the sum of 1 plus the percentage of hydrogen. Gamow's paper was not mentioned at the Paris meeting, nor cited by Kuiper in the proceedings, published two years later.
6. Marshak (1940), Lee (1950).
7. The quotations are from Greenstein (1985, p. 280; 1974, p. 35).

APPENDIX B

1. Email from White to the author, 30 May 2003.
2. Interview by the author, 25 March 2003.
3. Woosley et al. (2002), p. 1015.
4. In the years after the Colgate–White paper, much work was done in refining neutrino mechanisms and other matters such as the rotation and magnetic field of a star to see how a supernova could be generated (Bethe 1990, Woosley et al. 2002). Among the major figures in this research were David Arnett, Bethe, Jim Wilson, T. A. Weaver and Woosley.
5. The neutrino detectors were in the northern hemisphere, and SN 1987A was visible only from the southern. But this is irrelevant when it comes to detecting neutrinos, which pass through the Earth as if it were not there.

6. To be more precise, the number of neutrinos emitted was one followed by fifty-eight zeros, or 10^{58}.

7. 'New supernova models take on third dimension', interview in News and Public Affairs – News Releases, Los Alamos National Laboratory, 4 June 2002:
   www.lanl.gov/worldview/news/releases/archive/02-061.shtml

8. In a degenerate gas of electrons the pressure and volume do not depend on its temperature. As a result, when a white dwarf's mass exceeds the Chandrasekhar limit, there is no safety valve allowing the pressure to decrease as the temperature increases. The star's core cannot expand and cool off. This is not the case in a perfect gas, where an increase in the gas's temperature brings about changes in its volume and pressure. When a white dwarf accretes enough mass from its companion to exceed the Chandrasekhar limit, the build-up in temperature leads to a series of uncontrolled reactions in the carbon and oxygen core which eventually blows the star to pieces.

9. Type II supernova explosions take place in massively heavy stars. The light that the explosion emits contains spectral lines of hydrogen. White dwarfs have only a little hydrogen on their surface, so these lines are missing from the spectra of Type Ia supernovae. Types Ib and Ic supernovae are the explosions of heavier stars stripped of their hydrogen layer by stellar winds or by having it drawn off by a companion star; the light from these explosions has spectral lines characteristic of helium.

BIBLIOGRAPHY

Chandra's letters and archival material are in the Chandrasekhar Archives, Joseph Regenstein Library, University of Chicago. Milne's correspondence and papers are on deposit at the Bodleian Library, Oxford University. Eddington's letters to Walter Adams are from the Huntington Library, San Marino, California. Eddington's cycling diaries and other correspondence are on deposit at Trinity College, Cambridge University. Stoner's correspondence is at the Stoner Archives, Leeds University.

ABBREVIATIONS

AIP: Niels Bohr Library, American Institute of Physics, College Park, MD, USA

OHP: Oral History Project, California Institute of Technology Archives

Adams, Walter Sydney (1914), 'An A-type star of very low luminosity', *Publications of the Astronomical Society of the Pacific*, **26**, 198.

—— (1915), 'The spectrum of the companion of Sirius', *Publications of the Astronomical Society of the Pacific*, **27**, 236–7.

Alvarez, Luis (1987), *Alvarez: Adventures of a Physicist* (New York: Basic Books).

Anderson, Wilhelm (1929), 'Über die Grenzdichte der Materie und der Energie', *Zeitschrift für Physik*, **56**, 851–6.

Aston, Francis William (1938), 'Forty years of atomic theory', in Joseph Needham and Walter Pagel (eds.), *Background to Modern Science* (London: Macmillan).

Atkinson, Robert d'Escourt (1931), 'Atomic synthesis and stellar energy, I', *Astrophysical Journal*, **73**, 250–95.

—— (1939), 'A model symposium', *The Observatory*, **62**, 281–4.

Ayyar, C. Subramanyan (n.d.), 'Details of the biography of C. Subramanya Ayyar, B.A.' Chandrasekhar Archives/Box 6/Folder 5.

Baade, Walter and Rudolph Minkowski (1954), 'Identification of the radio sources in Cassiopeia, Cygnus A, and Puppis A', *Astrophysical Journal*, **119**, 206–14.

—— and Fritz Zwicky (1934), 'Supernovae and cosmic rays', *Physical Review*, **45**, 138.

Balakrishnan (1945), 'Subrahmanyan Chandrasekhar', *Trevini*, **17**, 73–85.

—— (1983a), 'My brother Chandra', *Aside*, December, pp. 18–29.

—— (1983b), 'Chandrasekhar re-reminisced', typescript manuscript, Chandrasekhar Archives/Box 1/Folder 10.

Bethe, Hans (1939), 'Energy production in stars', *Physical Review*, **55**, 434–56.

—— (1966), interview by Charles Weiner and Jagdesh Mehra, 27 October 1966, AIP.

—— (1972), interview by Charles Weiner, 8–9 May 1972, AIP.

—— (1990), 'Supernova mechanisms', *Reviews of Modern Physics*, **62**, 801–66.

—— (1993), interview by Judith Goodstein, 28 February 1993, OHP.

—— and Charles L. Critchfield (1938), 'The creation of deuterons by proton combination', *Physical Review*, **54**, 248–54.

Burbidge, Geoffrey and Margaret Burbidge (1967), *Quasi-Stellar Objects* (New York: Freeman & Company).

Candlestickmaker, S. (1972), 'On the imperturbability of elevator operators: LVII', *Quarterly Journal of the Royal Astronomical Society*, **13**, 63–6. [The real author was J. Sykes.]

Chadwick, James (1961), 'Frederick John Marrian Stratton', *Biographical Memoirs of Fellows of the Royal Society*, **7**, 281–93.

Chandrasekhar, Lalitha (1996), 'Our song'. Talk given after the Symposium Banquet, 14 December, 1996. Videotape on deposit at the Chandrasekhar Archives. [Page numbers in the notes refer to a published version in Robert M. Wald (ed.), *Black Holes and Relativistic Stars* (Chicago: University of Chicago Press), pp. 273–8.]

—— (n.d. but probably 1999), Unpublished speech and other notes on the occasion of the TRW hospitality reception for winners of the competition to name NASA's orbiting X-ray observatory.

Chandrasekhar, Subrahmanyan (1928), 'The thermodynamics of the Compton effect with reference to the interior of stars', *Indian Journal of Physics*, **3**, 241.

—— (1929a), 'A generalized form of the new statistics', *Physical Review*, **34**, 1204–11.

—— (1929b), 'The Compton scattering and the new statistics', *Proceedings of the Royal Society* A, **125**, 231–7.

—— (1930), 'The ionization-formula and the new statistics', *Philosophical Magazine*, **9**, 292–9.

—— (1931a), 'The highly collapsed configurations of a stellar mass', *Monthly Notices of the Royal Astronomical Society*, **91**, 456–66.

—— (1931b), 'The dissociation formula according to relativistic mechanics', *The Observatory*, **91**, 446–55.

—— (1931c), 'The maximum mass of ideal white dwarfs', *Astrophysical Journal*, **74**, 81–2.

—— (1932), 'Some remarks on the state of matter in the interior of stars', *Zeitschrift für Astrophysik*, **5**, 321–7.

—— (1933a), 'The equilibrium of distorted polytropes', *Monthly Notices of the Royal Astronomical Society*, **93**, 390–405, 449–61, 462–71, 539–74.

—— (1933b), 'The equilibrium of distorted polytropes', *The Observatory*, **56**, 215–19.

—— (1934a), 'The physical state of matter in the interior of stars', *The Observatory*, **57**, 93–9.

—— (1934b), 'Stellar configurations with degenerate cores', *The Observatory*, **57**, 373–7.

—— (1935a), 'Stellar configurations in degenerate cores', *The Observatory*, **58**, 37.

—— (1935b), 'The highly collapsed configurations of a stellar mass (second paper)', *Monthly Notices of the Royal Astronomical Society*, **95**, 207–25.

—— (1935c), 'Stellar configurations with degenerate cores', *Monthly Notices of the Royal Astronomical Society*, **95**, 226–60.

—— and Christian Møller (1935d), 'Relativistic degeneracy', *Monthly Notices of the Royal Astronomical Society*, **95**, 673–6.

—— (1935e), 'Stellar configurations with degenerate cores (second paper)', *Monthly Notices of the Royal Astronomical Society*, **95**, 676–93.

—— (1937), 'The Cosmological Constants', *Nature*, **139**, 757–8.

—— (1938), with George Gamow and Merle A. Tuve, 'The Problem of Stellar Energy', *Nature*, **141**, 982.

—— (1939), *An Introduction to the Study of Stellar Structure* (Chicago: University of Chicago Press).

—— (1941), 'The white dwarfs and their importance for theories of stellar evolution', in Lundmark et al., pp. 41–8; discussion on pp. 49–50; discussion of Eddington's lecture on pp. 64–9.

—— and Mario Schönberg (1942), 'On the evolution of the main-sequence stars', *Astrophysical Journal*, **96**, 161–72.

—— (1943) *The Principles of Stellar Dynamics* (Chicago: University of Chicago Press).

—— (1950), *Radiative Transfer* (Oxford: Oxford University Press).

—— and Enrico Fermi (1953a), 'Magnetic fields in spiral arms', *Astrophysical Journal*, **118**, 113–15.

—— and Enrico Fermi (1953b), 'Problems of gravitational stability in the presence of a magnetic field', *Astrophysical Journal*, **118**, 116–41.

—— (1961), *Hydrodynamic and Hydromagnetic Stability*: (Oxford: Clarendon Press).

—— (1964a), 'Dynamical instability of gaseous masses approaching the Schwarzschild limit in general relativity', *Physical Review Letters*, **12**, 114–16.

—— (1964b), 'The dynamical instability of gaseous masses approaching the Schwarzschild limit in general relativity', *Astrophysical Journal*, **140**, 417–33.

—— (1972), 'The increasing rôle of general relativity in astronomy', *The Observatory*, **92**, 160–74. [This is Chandra's Halley Lecture for 1972, delivered in Oxford, 2 May 1972.]

—— (1975a), 'How I came periodically to change the area of my active interest after writing a book'; in Chandra Archives, Addenda Box 70/Folder 6, dated 8 January 1975.

—— (1975b), 'Shakespeare, Newton, and Beethoven, or patterns of creativity.' The Nora and Edward Ryerson Lecture, delivered at the University of Chicago, 22 April 1975. Reprinted in Chandrasekhar (1987c), pp. 29–58. [Page numbers in the notes refer to Chandra's book.]

—— (1977), interview by Spencer Weart, 17 May 1977, AIP.

—— (1978), 'The Kerr metric and stationary axisymmetric gravitational fields', *Proceedings of the Royal Society* A, **358**, 405–20.

—— (1979a), 'I. Edward Arthur Milne', dictated to Lalitha on 22 January 1979, edited and transcribed by her; in Chandra Archives, Addenda Box 77/Folder 5 – 'Recollections of Milne, Eddington and Fowler'.

—— (1979b), 'Edward Arthur Milne – recollections and reflections', manuscript in the AIP Niels Bohr Library and in the Chandrasekhar Archives, University of Chicago. [Written at the request of Milne's grand-daughter, Miranda Weston-Smith. I have assigned the date 1979 because this is a manuscript version of Chandra (1987b), presented on 6 December 1979 at Oxford University.]

—— (1979c), 'Beauty and the quest for beauty in science'. Lecture delivered at the Fermi Accelerator Laboratory, Batavia, Illinois, at the symposium in honour of Robert R. Wilson, 27 April 1979. Reprinted in Chandrasekhar (1987c), pp. 59–73. [Page numbers in the notes refer to Chandra's book.]

—— (1980a), 'II. Sir Arthur Stanley Eddington, dictated to Lalitha on 3 July 1980, edited and transcribed by her'; in Chandra Archives, Addenda Box 77/Folder 5 – 'Recollections of Milne, Eddington and Fowler'.

—— (1980b), 'III. Ralph Howard Fowler', dictated to Lalitha on 4 September 1980, edited and transcribed by her; in Chandra Archives, Addenda Box 77/Folder 5 – 'Recollections of Milne, Eddington and Fowler'.

—— (1980c), interview by D. J. R. Bruckner, 6 August 1980; in Chandra Archives, Box 2/Folder 10.

—— (1983a), 'On stars, their evolution and their stability', Nobel Lecture delivered on 8 December 1983, in Tore Frängsmyr (ed.), *Nobel Lectures in Physics: 1981–1990* (Singapore: World Scientific, 1993), pp. 142–63; 'Subrahmanyan Chandrasekhar', an autobiographical statement, pp. 139–41.

—— (1983b), *The Mathematical Theory of Black Holes* (Oxford: Clarendon Press).

—— (1987a), 'Eddington: The most distinguished astrophysicist of his time', in Chandrasekhar (1987c), pp. 92–143. [Chandra's Centenary Lectures in Memory of Arthur Stanley Eddington, delivered at Cambridge University, 19 and 21 October 1982.]

—— (1987b), 'Edward Arthur Milne: His part in the development of modern astrophysics', in Chandrasekhar (1987c), pp. 74–91. [This is Chandra's

Milne Lecture, delivered at Oxford University, 6 December 1979.]

—— (1987c), 'Truth and Beauty: Aesthetics and Motivations in Science (Chicago: University of Chicago Press).

—— (1988), 'Foreword' to the reissue of Arthur Stanley Eddington, The Internal Constitution of the Stars (Cambridge: Cambridge University Press, 1st edn, 1926; 2nd edn, 1930), pp. vii–xii.

—— (1992), 'The series paintings of Claude Monet and the landscape of general relativity', lecture delivered at the Inter-University Centre for Astronomy and Astrophysics, Pune, India, 28 December 1992. Reprinted in S. Chandrasekhar, Selected Papers, Volume 7 (Chicago: University of Chicago Press, 1997), pp. 135–67. [Page numbers in the notes refer to the reprint volume.]

—— (1995), Newton's Principia for the Common Reader (Oxford: Oxford University Press).

—— (n.d.1), 'Science in pre- and post-independent India'; in Chandra Archives, Addenda/Box 23/Folder 1.

—— (n.d.2), Fragment beginning 'I shall strictly adhere . . .'; in Chandra Archives, Addenda Box 70/Folder 6.

—— (n.d.3), 'How I came to writing my book on the Principia'; in Chandra Archives, Addenda Box 70/Folder 7.

—— (n.d.4), 'Historical notes on some astrophysical problems'; in Chandra Archives, Addenda Box 70/Folder 6.

—— (n.d.5), Handwritten reminiscence; in Chandra Archives, Box 2/Folder 1.

Christy, Robert F. (1994), interview by Sandra Lippincott, 21 June 1994, OHP.

Colgate, Stirling and Montgomery H. Johnson (1960), 'Hydrodynamic origin of cosmic rays', Physical Review Letters, 5, 235–8.

—— and Richard White (1966), 'The hydrodynamic behavior of supernovae explosions', Astrophysical Journal, 143, 626–81.

—— (1967), 'Stellar coalescence and the multiple supernova interpretation of quasi-stellar sources', Astrophysical Journal, 150, 163–92.

Compton, Herbert (1904), Indian Life in Town and Country (New York: G. P. Putnam's).

Cowling, T. G. (1966), 'The development of the theory of stellar structure', Quarterly Journal of the Royal Astronomical Society, 7, 121–37.

—— (1978), interview by David DeVorkin, 22 March 1978, AIP.

Cronin, James W. (1998), 'Subrahmanyan Chandrasekhar', The Observatory, 118, 24.

Daltiz, Richard H. (1997), 'Some recollections of S. Chandrasekhar', in Wali (1997), pp. 142–55.

Danin, D. S. (1989), 'The passionate sobriety of youth', in I. M. Khalatnikov (ed.), Landau: The Physicist and the Man (Oxford: Pergamon Press), pp. 78–83.

DeVorkin, David H. (2000), Henry Norris Russell: Dean of American Astronomers (Princeton: Princeton University Press).

Dimopoulos, Savas and Greg Landsberg (2001), 'Black holes at the Large

Hadron Collider', *Physical Review Letters*, **87**, 161602–4.

Dirac, P. A. M. (1937), 'The cosmological constants', *Nature*, **139**, 323–4.

——, R. Peierls and M. H. L. Pryce (1942), 'On Lorentz invariance in the quantum theory', *Proceedings of the Cambridge Philosophical Society*, **38**, 193–200.

Douglas, Allie Vibert (1956), *The Life of Arthur Stanley Eddington* (London: Thomas Nelson & Sons Ltd).

Eddington, Arthur Stanley (1917b), 'The radiation of the stars', *Nature*, **99**, 445.

—— (1918), *Report on the Relativity Theory of Gravitation* (London: Physical Society of London).

—— (1919), 'Globular clusters, Cepheid variables, and radiation', *Nature*, **103**, 25–7.

—— (1920), 'The internal constitution of the stars', *Nature*, **106**, 14–20.

—— (1923), *Mathematical Theory of Relativity* (Cambridge: Cambridge University Press).

—— (1924a), 'The relation between the masses and luminosities of the stars', *Nature*, **113**, 786–8.

—— (1924b), 'A comparison of Whitehead's and Einstein's formulas', *Nature*, **113**, 192.

—— (1926), *The Internal Constitution of the Stars* (Cambridge: Cambridge University Press, 1st edn, 1926; 2nd edn, 1930; reissued 1988 with a 'Foreword' by Chandrasekhar).

—— (1927), *Stars and Atoms* (New Haven, CT: Yale University Press).

—— (1929a), 'The masses, luminosities and effective temperatures of the stars', *The Observatory*, **52**, 349.

—— (1929b), *Seen and Unseen World* (London: Allen & Unwin).

—— (1930a), 'The effect of stellar boundary conditions: A reply', *Monthly Notices of the Royal Astronomical Society*, **90**, 808–9.

—— (1930b), 'The problem of stellar luminosity', *Nature*, **125**, 489.

—— (1930c), 'The connection of mass with luminosity of stars', *The Observatory*, **53**, 208–11.

—— (1931a), 'The analysis of stellar structure', *The Observatory*, **54**, 34–6.

—— (1931b), 'Upper limits to the central temperature and density of a star', *The Observatory*, **54**, 97–9.

—— (1931c), 'Upper limits to the central temperature and density of a star', *Monthly Notices of the Royal Astronomical Society*, **91**, 444–6.

—— (1932), 'The hydrogen content of stars', *Monthly Notices of the Royal Astronomical Society*, **92**, 471–81.

—— (1935a), 'Relativistic degeneracy', *The Observatory*, **58**, 37–9.

—— (1935b), 'On "relativistic degeneracy"', *Monthly Notices of the Royal Astronomical Society*, **95**, 194–206.

—— (1935c), 'Note on "relativistic degeneracy"', *Monthly Notices of the Royal Astronomical Society*, **96**, 20–1.

—— (1935d), 'The pressure of a degenerate electron gas and related problems', *Proceedings of the Royal Society* A, **152**, 253–72.

—— (1935e), 'The external radii of stars of the generalised standard model',

The Observatory, **58**, 176.

—— (1936a), *Relativity Theory of Protons and Electrons* (Cambridge: Cambridge University Press).

—— (1936b), 'Constitution of the stars', *The Scientific Monthly*, November 1936, pp. 385–95.

—— (1939a), 'The hydrogen content of white dwarf stars in relation to stellar evolution', *Monthly Notices of the Royal Astronomical Society*, **99**, 595–606.

—— (1939b), 'The hydrogen content of white dwarf stars in relation to stellar evolution', *The Observatory*, **62**, 171–4.

—— (1942), 'On Lorentz invariance in the quantum theory, II', *Proceedings of the Cambridge Philosophical Society*, **38**, 201–9.

—— (1943), Review of Chandra's *The Principles of Stellar Dynamics*, *Nature*, **151**, 91.

—— (1946), *Fundamental Theory* (Cambridge: Cambridge University Press). [Edited and seen through the press by Edmund T. Whittaker, a long-time colleague of Eddington's and Professor of Mathematics at Edinburgh University. Whittaker chose the book's title.]

—— (1958a), *The Nature of the Physical World* (1st edn, Cambridge: Cambridge University Press, 1928: republished, Ann Arbor: University of Michigan Press). [Based on Eddington's Gifford Lectures, delivered January to March 1927 at University of Edinburgh.]

—— (1958b), *The Expanding Universe* (1st edn, 1933, Cambridge: Cambridge University Press; republished, 1958: Ann Arbor: University of Michigan Press). [From a series of lectures presented in 1932 at the International Astronomical Union, Cambridge, MA.]

—— (1958c), *The Philosophy of Physical Science* (1st edn, 1939, Cambridge: Cambridge University Press; republished 1958, Ann Arbor: University of Michigan Press). [Based on the Tanner Lectures, Trinity College, Cambridge, Easter Term 1938.]

—— (1959), *New Pathways in Science* (1st edn, 1935, Cambridge: Cambridge University Press).

Einstein, Albert (1923), 'Fundamental ideas and problems of the theory of relativity', in *Nobel Lectures: Physics 1901–1921* (Amsterdam: Elsevier, 1967), pp. 482–90.

—— (1939), 'On a stationary system with spherical symmetry consisting of many gravitating masses', *Annals of Mathematics*, **40**, 922–36.

—— (1949), 'Reply to criticisms', in P. A. Schilpp (ed.), *Albert Einstein: Philosopher-Scientist* (New York: Open Court), p. 684.

Elbert, Donna (1997), 'On working with Chandra', in Wali (1997), pp. 41–5.

Enz, Charles P. (2002), *No Time to be Brief: A Scientific Autobiography of Wolfgang Pauli* (Oxford: Oxford University Press).

Fowler, Ralph H. (1926), 'On dense matter', *Monthly Notices of the Royal Astronomical Society*, **87**, 114–22.

—— (1936), *Statistical Mechanics*, 2nd revised edn (1st edn, Cambridge: Cambridge University Press, 1929).

Fowler, William A. and Fred Hoyle (1964), 'Neutrino processes and pair for-
    mation in massive stars and supernovae', *Astrophysical Journal
    Supplement*, **9**, 201–318.
—— (1966), 'The stability of supermassive stars', *Astrophysical Journal*, **144**,
    180–200.
—— (1984), interview by John Greenberg, 31 May 1984, OHP.
Frisch, Otto (1979), *What Little I Remember* (Cambridge: Cambridge
    University Press).
Galison, Peter and Barton Bernstein (1989), 'In any light: Scientists and the
    decision to build the superbomb, 1942–1954', *Historical Studies in the
    Physical Sciences*, **19**, Part 2, 267–349.
Gamow, George (1939), 'Physical possibilities of stellar evolution', *Physical
    Review*, **55**, 718–25.
—— (1968), interview by Charles Weiner, 25 April 1968, AIP.
Gandhi, Mohandas K. (1983), *Autobiography: The Story of My Experiments
    with Truth* (1st edn, London: Public Affairs Press, 1948; reprinted, New
    York: Dover Publications).
Giddings, Steven and Scott Thomas (2002), 'High energy colliders as black
    hole factories: The end of short distance physics', *Physical Review* D, **65**,
    056010-1–056010-12.
Gleick, James (2003), *Isaac Newton* (New York: HarperCollins).
Gott, J. Richard (2002), *Time Travel in Einstein's Universe: The Physical
    Possibilities of Travel Through Time* (London: Phoenix).
Greene, Brian (2000), *The Elegant Universe: Superstrings, Hidden Dimensions,
    and the Quest for the Ultimate Theory* (London: Vintage).
Greenstein, Jesse and Thomas A. Matthews (1963), 'Red-shift of the unusual
    radio source: 3C 48', *Nature*, **197**, 1041–2.
—— and Maarten Schmidt (1964), 'The quasi-stellar radio sources 3C 48
    and 3C 273', *Astrophysical Journal*, **140**, 1–34.
——, J. B. Oke and Harry Shipman (1971), 'Effective temperature, radius,
    and gravitational redshift of Sirius B', *Astrophysical Journal*, **169**, 563–6.
—— (1974), interview by Paul Wright, 31 July 1974, AIP.
—— (1982), interview by Rachel Prud'homme, 16 March 1982, OHP.
—— (1985), 'On the redshift of Sirius B', *Quarterly Journal of the Royal
    Astronomical Society*, **26**, 279–88.
Hamada, T. and Edwin E. Salpeter (1961), 'Models for zero-temperature
    stars', *Astrophysical Journal*, **134**, 683–98.
Hardy, G. H. (1931), 'The analysis of stellar structure', *The Observatory*, **54**,
    40.
—— (1940), *Ramanujan* (Cambridge: Cambridge University Press).
—— (1941), *A Mathematician's Apology* (Cambridge: Cambridge University
    Press).
Harrison, B. Kent, Kip S. Thorn, Masamo Wakano and John A. Wheeler
    (1965), *Gravitational Theory and Gravitational Collapse* (Chicago:
    University of Chicago Press).
Hawking, Stephen and Werner Israel (eds.) (1987), *300 Years of Gravitation*
    (Cambridge: Cambridge University Press).

Heisenberg, Werner (1926), 'Mehrkörperprobleme und Resonanz in der Quantenmechanik. II', Zeitschrift für Physik, **41**, 239–67.

—— (1962), interview by Thomas S. Kuhn, 30 November 1962, AIP.

—— (1971), Physics and Beyond: Encounters and Conversations, translated by P. Heath (New York: Harper).

—— (1974), Across the Frontiers, translated by P. Heath (New York: Harper Torchbooks).

Herant, Marc, Stirling Colgate and Chris Fryer (1997), 'Neutrinos and supernovae', Los Alamos Science, No. 25, 164–79.

Herken, Gregg (2002), Brotherhood of the Bomb: The Tangled Lives and Loyalties of Robert Oppenheimer, Ernest Lawrence, and Edward Teller (New York: Henry Holt).

Holloway, David (1994), 'How the bomb saved Soviet physics', Bulletin of the Atomic Scientists, **50**(6) <http://www.thebulletin.org/issues/1994/nd94/nd94Holloway.html>

Hoyle, Fred and William A. Fowler (1960), 'Nucleosynthesis in supernovae', Astrophysical Journal, **132**, 565–90.

—— (1963), 'On the nature of strong radio sources', Monthly Notices of the Royal Astronomical Society, **125**, 169–76.

Jeans, James (1931), 'The analysis of stellar structure', The Observatory, **54**, 36–8.

Johansson, Sven (1993), 'The Nobel Prize for Physics', in Tore Frängsmyr (ed.), Nobel Lectures in Physics: 1981–1990 (Singapore: World Scientific, 1993), pp. 135–7.

Kanigel, Robert (1991), The Man Who Knew Infinity: A Life of the Genius Ramanujan (New York: Charles Scribner's Sons).

Kerr, Roy (1963), 'Gravitational field of a spinning mass as an example of algebraically special metrics', Physical Review Letters, **11**, 237–8.

Khalatnikov, I. M. (1989), 'Landau, Bohr, and Kapitza: Letters 1936–41', in I. M. Khalatnikov (ed.), translated by J. B. Sykes, Landau: The Physicist and the Man (Oxford: Pergamon Press), pp. 308–19.

Khan, K. A. and Roger Penrose (1971), 'Scattering of two impulsive gravitational plane waves', Nature, **229**, 185–6.

Khariton, Yuli, Kitor Adamskii and Yuri Smirnov (1998), 'The way it was', Bulletin of the Atomic Scientists, **54**(6) <http://www.bullatomsci.org/issues/1996/nd96/nd96khariton.html>

—— and Yuri Smirnov (1993), 'The Khariton version', Bulletin of the Atomic Scientists, **49**(3) <http://www.thebulletin.org/issues/1993/may93/may93Khariton.html>

Khriplovich, Iosif B. (1992), 'The eventful life of Fritz Houtermans', Physics Today, **45**, 29–37.

Kilmister, Clive W. (1994), Eddington's Search for a Fundamental Theory: A Key to the Universe (Cambridge: Cambridge University Press).

Kragh, Helge (1990), Dirac: A Scientific Biography (Cambridge: Cambridge University Press).

Kuiper, Gerard P. (1941), 'White dwarfs: Discovery, observations, surface conditions', in Lundmark et al., pp. 3–39, and discussion session, pp. 64–9.

Landau, Lev Davidovich (1932), 'On the theory of stars', *Physikalische Zeitschrift der Sowjetunion*, 1, 285–8.

—— (1938), 'Origin of stellar energy', *Nature*, 141, 334–5.

Lang, Kenneth and Owen Gingerich (1979), *A Source Book in Astronomy and Astrophysics, 1900–1975* (Cambridge, MA: Harvard University Press).

Lee, T. D. (1950), 'Hydrogen content and energy-productive mechanism of white dwarfs', *Astrophysical Journal*, 111, 625–40.

Littlewood, J. E. (1953), *Littlewood's Miscellany*, edited and with a foreword by Béla Bollobás (Cambridge: Cambridge University Press).

Lundmark, Knut, Serge Gaposchkin, Bengt Edlén, C. P. Gaposchkin, F. J. M. Stratton and C. S. Beals (eds.) (1941), *Novae and White Dwarfs*, Colloque International d'Astrophysique, 17–23 July 1939, Paris, under the presidency of Henry Norris Russell (Paris: Hermann).

McCrea, William H. (1935), 'The International Astronomical Union meeting in Paris 1935', *The Observatory*, 58, 256–65.

—— (1951), 'Edward Arthur Milne: Obituary notice', *Monthly Notices of the Royal Astronomical Society*, 111, 161–70.

—— (1996), 'Subrahmanyan Chandrasekhar', *The Observatory*, 116, 121–4.

—— (1998), 'Sir William McCrea replies', *The Observatory*, 118, 24.

McVittie, George C. (1978), interview by David DeVorkin, 21 March 1978, AIP.

Marshak, Robert Eugene (1940), 'The internal temperature of white dwarf stars', *Astrophysical Journal*, 92, 321–46.

Merrill, Paul W. (1938), *The Nature of Variable Stars* (New York: Macmillan).

Miller, Arthur I. (1972), 'The myth of Gauss' experiment on the Euclidean nature of physical space', *ISIS*, 63, 345–8.

—— (1994), *Early Quantum Electrodynamics: A Source Book* (Cambridge: Cambridge University Press).

—— (1998), *Albert Einstein's Special Theory of Relativity: Emergence (1905) and Early Interpretation (1905–1911)* (Reading, MA: Addison-Wesley, 1981; reprinted, London: Springer-Verlag).

—— (2000), *Insights of Genius: Imagery and Creativity in Science and Art* (Cambridge, MA: MIT Press; 1st edn, 1996, New York: Springer-Verlag).

—— (2002), 'Erotica, aesthetics and Schrödinger's wave equation', in Graham Farmeloe (ed.), *It Must be Beautiful: Great Equations in Modern Science* (London: Granta), pp. 80–101.

Milne, Edward Arthur (1923), 'Those in authority: R. H. Fowler M.A. (Trinity), Senior Lecturer', *The Granta*, 32, 469.

—— (1930a), 'The analysis of stellar structure', *The Observatory*, 53, 330–4.

—— (1930b), 'The analysis of stellar structure', *Monthly Notices of the Royal Astronomical Society*, 91, 4–55.

—— (1931), 'The analysis of stellar structure', *The Observatory*, 54, 34.

—— (1932), 'The theory of stellar structure II (Energy-generation)', *Zeitschrift für Astrophysik*, 5, 337–47.

—— (1935a), 'The configuration of stellar masses', *The Observatory*, 58, 52.

—— (1935b), 'The external radii of stars of the generalised standard model', *The Observatory*, **58**, 174–6.

—— (1945a), 'R. H. Fowler', *Obituary Notices of Fellows of The Royal Society*, **5**, 61–78.

—— (1945b), 'Ralph H. Fowler', *Monthly Notices of the Royal Astronomical Society*, **105**, 80–7.

—— (1947), 'Sir James Jeans', *Obituary Notices of Fellows of the Royal Society*, **5**, 573–89.

Misner, Charles, Kip S. Thorne and John A. Wheeler (1970), *Gravitation* (New York: Freeman).

Moore, Walter (1991), *Schrödinger: Life and Thought* (Cambridge: Cambridge University Press).

NASA (1972), *Apollo 15 Preliminary Science Report*, Special Publication SP-289 (Washington, DC: Science and Technical Information Office).

Oppenheimer, J. Robert and Robert Serber (1938), 'On the stability of stellar neutron cores', *Physical Review*, **54**, 504.

—— and Hartland Snyder (1939), 'On continued gravitational contraction', *Physical Review*, **56**, 455–9.

—— and Georg M. Volkoff (1939), 'On massive neutron cores', *Physical Review*, **55**, 374–81.

—— (1963), interview by Thomas S. Kuhn, 18 and 20 November 1963, AIP.

Osterbrock, Donald E. (1996), 'Walter Baade, observational astrophysicist, (2): Mount Wilson 1931–1937', *Journal of the History of Astronomy*, **27**, 301–48.

—— (1997) *Yerkes Observatory, 1892–1950: The Birth, Near Death, and Resurrection of a Scientific Research Institution* (Chicago: University of Chicago Press).

—— (1999) 'Chandra and his students at Yerkes Observatory', in G. Srinivasan (ed.), *From White Dwarfs to Black Holes: The Legacy of S. Chandrasekhar* (Chicago: University of Chicago Press), pp. 199–237.

—— (2001), 'Who really coined the word supernova/Who first predicted neutron stars?', *Bulletin of the American Astronomical Society*, **33**, 1330–1.

Pais, Abraham (1986), *Inward Bound: On Matter and Forces in the Physical World* (Oxford: Oxford University Press).

—— (2000), *The Genius of Science: A Portrait Gallery of Twentieth-Century Physicists* (Oxford: Oxford University Press).

Pauli, Wolfgang (1955), 'The influence of archetypal ideas on the scientific theories of Kepler', in C. G. Jung and W. Pauli, *The Interpretation of Nature and the Psyche* (New York, Pantheon), pp. 219–79.

—— (1979), *Wissenschaftlicher Briefwechsel mit Bohr, Einstein, Heisenberg, u.a. I: 1919–1929* (A. V. Hermann, K. v. Meyenn and V. F. Weiskopf, eds.), (Berlin: Springer-Verlag).

Payne, C. (1925), 'The relative abundances of the elements', in *Stellar Atmospheres*, Harvard Observatory Monograph No. 1 (Cambridge, MA: Harvard University Press). Excerpt in Lang and Gingerich (1979), pp. 245–8.

Payne-Gaposchkin, C. (1968), interview by Owen Gingerich, 5 March 1968, AIP.

—— and Katherine Haramundanis (1970), *Introduction to Astronomy* (Englewood Cliffs, NJ: Prentice-Hall).

Peierls, Rudolf (1936), 'Note on the derivation of the equation of state for a degenerate relativistic gas', *Monthly Notices of the Royal Astronomical Society*, **96**, 780–4.

Pellam, J. R. (1989), 'Lev Davidovich Landau', in I. M. Khalatnikov, *Landau: The Physicist and the Man* (New York: Pergamon Press), pp. 198–204.

Penrose, Roger (1965), 'Gravitational collapse and space-time singularities', *Physical Review Letters*, **14**, 57–9.

—— (1969), 'Gravitational collapse: The role of general relativity', *Revista del Nuovo Cimento*, Numero speciale, **1**, 252–76.

Plaskett, Henry H. (1951), 'Edward Arthur Milne', *Monthly Notices of the Royal Astronomical Society*, **111**, 170–2.

Plummer, H. C. (1948), 'Arthur Stanley Eddington', *Obituary Notices of the Royal Society*, **5**, 113–25.

Raman, C. V. (1930), 'The molecular scattering of light', Nobel Lecture delivered 11 December 1930, Stockholm, Sweden. In *Nobel Lectures in Physics, 1922–1941* (New York: Elsevier, 1965), pp. 267–75.

Ramaseshan, S. (1996), 'S. Chandrasekhar and C. V. Raman – some letters', *Current Science*, **70**, 104–7.

—— (1997), 'Chandrasekhar – some reminiscences', in Wali (1997), pp. 101–12.

Rees, Martin (1997), 'Reminiscing about Chandra's research', in Wali (1997), pp. 183–6.

—— (1998), 'Astrophysical evidence for black holes', in R. W. Wald (ed.), *Black Holes and Relativistic Stars* (Chicago: University of Chicago Press), pp. 79–101.

Rhodes, Richard (1986), *The Making of the Atomic Bomb* (New York: Simon & Schuster).

—— (1995), *Dark Sun: The Making of the Hydrogen Bomb* (New York: Simon & Schuster).

Rosenfeld, Léon (1966), 'Nuclear physics, past and future', in M. Nève de Mévergnies, P. Van Assche and J. Vervier (eds.), *Nuclear Structure Study with Neutrons* (Amsterdam: North Holland).

—— (1967), 'Niels Bohr in the thirties', in S. Rozenthal (ed.), *Niels Bohr: His Life and Work as Seen by His Friends and Colleagues* (New York: Wiley), pp. 114–36.

Rüger, Alexander (1988), 'Atomism from cosmology: Erwin Schrödinger's work on wave mechanics and space-time structure', *Historical Studies in the Physical Sciences*, **18**, 377–401.

Russell, Henry Norris (1913), 'Giant and dwarf stars', *The Observatory*, **36**, 325.

—— (1914), 'Relations between spectra and other characteristics of stars', *Popular Astronomy*, **22**, 275–94.

—— (1941), 'Address', in Lundmark et al., pp. 1–6.

—— (1945), 'Arthur Stanley Eddington, 1882–1944', *Astrophysical Journal*, **101**, 133–5.

Saha, Meghnad (1920), 'Ionisation in the solar chromosphere', *Philosophical Magazine*, **40**, 479–88.

Sakharov, Andrei (1990), *Memoirs* (New York: Knopf).

Salpeter, Edwin E. (1961), 'Energy and pressure of a zero-temperature plasma', *Astrophysical Journal*, **134**, 669–82.

—— (1999), 'Neutron stars before 1967 and my debt to Chandra', in G. Srinivasan (ed.), *From White Dwarfs to Black Holes: The Legacy of S. Chandrasekhar* (Chicago: University of Chicago Press), pp. 27–9.

Sandage, Allan R. and Martin Schwarzschild (1952), 'Inhomogeneous stellar models II: Models with exhausted cores in gravitational contraction', *Astrophysical Journal*, **116**, 463–76.

Schmidt, Maarten (1963), 'A star-like object with large red-shift', *Nature*, **197**, 1040.

Schödel, R. et al. (2002), 'A star in a 15.2-year orbit around the supermassive black hole at the centre of the Milky Way', *Nature*, **419**, 694–6.

Scott, David R., Alexei Leonov and Christine Toomey (2004), *Two Sides of the Moon: Our Story of the Cold War Space Race* (New York: Simon & Schuster).

Serber, Robert (1992), *The Los Alamos Primer* (Berkeley, CA: University of California Press). [Based on a set of five lectures given by Serber in the first two weeks of April 1943 as an indoctrination course in connection with the starting of the Los Alamos Project. 'The Los Alamos primer' is the first five chapters which are Serber's lectures, annotated by him. There is an introduction by Richard Rhodes, pp. ix–xxi, and a preface by Serber, pp. xxii–xxxiii.]

Singh, Amar Kuman (1963), *Indian Students in Britain: A Survey of Their Adjustment and Attitudes* (London: Asia Publishing House).

Singh, Simon (2004), *Big Bang: The Most Important Scientific Discovery of All Time and Why You Need to Know About It* (London: Fourth Estate).

Slater, Gilbert (1936), *Southern India: Its Political & Economic Problems* (London: George Allen & Unwin).

Smart, W. G. M. (1931), 'The analysis of stellar structure', *The Observatory*, **54**, 38.

Sommerfeld, Arnold (1923), *Atomic Structure and Spectral Lines*, translated by Henry L. Brose from the 3rd German edition (London: Methuen).

—— (1928a), 'Zur Elektronentheorie der Metalle auf Grund der Fermischen Statistik', *Zeitschrift für Physik*, **47**, 1–32.

—— (1928b), 'Zur Elektronentheorie der Metalle auf Grund der Fermischen Statistik, insbesondre über den Volta-Effekt', *Zeitschrift für Physik*, **47**, 38–60.

Stoner, Edmund C. (1929), 'The limiting density in white dwarf stars', *Philosophical Magazine*, 7, 63–70.

—— (1930), 'The equilibrium of dense stars', *Philosophical Magazine*, **9**, 944–63.

—— (1932a), 'The minimum pressure of a degenerate gas', *Monthly Notices of the Royal Astronomical Society*, **92**, 651–61.

—— (1932b), 'Upper limits for densities and temperatures in stars', *Monthly Notices of the Royal Astronomical Society*, **92**, 662–76.

Strömgren, Bengt (1932), 'The opacity of stellar matter and the hydrogen content of the stars', *Zeitschrift für Astrophysik*, **4**, 118–52.

—— (1976), interview by Lillian Hoddeson and Gordon Baym, 6 and 13 May 1976, AIP.

Struve, Otto and Velta Zebergs (1962), *Astronomy of the Twentieth Century* (New York: Macmillan).

Stuewer, Roger H. (1986), 'Gamow's theory of alpha decay', in Edna Ullman-Margalit (ed.), *The Kaleidoscope of Science* (Englewood Cliffs, NJ: Humanities Press), pp. 147–86.

Swerdlow, Noel M. (1997), 'Chandrasekhar's research on Newton's *Principia*', in Wali (1997), pp. 201–5.

Telegdi, Valentin (1997), 'Recollections about Chandra', in Wali (1997), pp. 206–9.

Teukolsky, Saul A. (1997), 'Chandra at Caltech', in Wali (1997), pp. 76–9.

Thorne, Kip S. (1994), *Black Holes and Time Warps: Einstein's Outrageous Legacy* (New York: Norton).

—— (1998), 'Probing black holes and relativistic stars with gravitational waves', in R. M. Wald (ed.), *Black Holes and Relativistic Stars* (Chicago: University of Chicago Press), pp. 41–78.

Ulam, Stanislaw (1976), *Adventures of a Mathematician* (New York: Scribner's).

USAEC (United States Atomic Energy Commission) (1954), *In the Matter of J. Robert Oppenheimer* (Cambridge, MA: MIT Press).

Vandervoort, Peter O. (n.d.), 'S. Chandrasekhar: Incidental lessons'. Unpublished manuscript in the possession of Professor Vandervoort at the University of Chicago.

Van Maanen, Adriaan (1917), 'Two faint stars with large proper motion', *Publications of the Astronomical Society of the Pacific*, **29**, 258–9.

Venkataraman, Janaki, Sridevi Rao and Nirupama Nityanandan (1983), 'Our Nobel family', *Aside*, December, pp. 12–18.

Wali, Kameshwar C. (1991), *Chandra: A Biography of S. Chandrasekhar* (Chicago: University of Chicago Press).

—— (ed.) (1997), *S. Chandrasekhar: The Man Behind the Legend* (London: Imperial College Press).

Warwick, Andrew (2003), *Masters of Theory: Cambridge and the Rise of Mathematical Physics* (Chicago: University of Chicago Press).

Weinberg, Steven (1972), *Gravitation and Cosmology: Principles and Applications of the General Theory of Relativity* (New York: John Wiley & Sons).

—— (1977), *The First Three Minutes: A Modern View of the Origin of the Universe* (New York: Basic Books).

Westfall, Robert S. (1996), 'Technical Newton', *Isis*, **87**, 701–6.

Weston-Smith, Meg (1990), 'E. A. Milne and the creation of air defence: Some letters from an Unprincipled Brigand, 1916–1919', *Notes and Records of the Royal Society*, **44**, 241–55.

Wheeler, John A., Kent Harrison and J. A. Wakano (1958), 'Matter-energy at high density: End point of thermonuclear evolution', in *Solvay – Onzième Conseil de Physique – La Structure et l'évolution de l'univers* (Brussels: R. Stoops), pp. 125–41; see also 'Discussion of Wheeler's report', pp. 147–8.

—— and Ken Ford (1998), *Geons, Black Holes & Quantum Foam: A Life in Physics* (New York: Norton).

Wolpert, Stanley (1997), *A New History of India*, 5th edn (Oxford: Oxford University Press; 1st edn, 1977).

Woosley, Stan F., A. Heger and T. A. Weaver (2002), 'The evolution and explosion of massive stars', *Reviews of Modern Physics*, **74**, 1015–71.

Xanthopoulos, Basilis C. (1991), 'Foreword', in S. Chandrasekhar, *Selected Papers, Volume 6: The Mathematical Theory of Black Holes and of Colliding Plane Waves* (Chicago: University of Chicago Press).

Zwicky, Fritz (1939), 'On the theory and observation of highly collapsed stars', *Physical Review*, **55**, 726–43.

INDEX